CALCIUM OXALATE

in

BIOLOGICAL SYSTEMS

Edited by
Saeed R. Khan

CRC Press
Taylor & Francis Group
Boca Raton London New York

CRC Press is an imprint of the
Taylor & Francis Group, an **informa** business

T0174836

CRC Press
Taylor & Francis Group
6000 Broken Sound Parkway NW, Suite 300
Boca Raton, FL 33487-2742

© 1995 by Taylor & Francis Group, LLC
CRC Press is an imprint of Taylor & Francis Group, an Informa business

First issued in paperback 2019

No claim to original U.S. Government works

ISBN-13: 978-0-367-44886-8 (pbk)
ISBN-13: 978-0-8493-7673-3 (hbk)

**Visit the Taylor & Francis Web site at
http://www.taylorandfrancis.com**

**and the CRC Press Web site at
http://www.crcpress.com**

Library of Congress Card Number 95-36917

Library of Congress Cataloging-in-Publication Data

Calcium oxalate in biological systems/ [edited by] Saheed Khan.
     p. cm.
   Includes bibliographical references and index.
   ISBN 0-8493-7673-4 (permanent paper)
   1. Calcium oxalate--Physiological aspects. 2. Calcium oxalate--Pathophysiology.
 3. Kidneys--Calculi--Etiology. I. Khan, Saeedur R.
   [DNLM: 1. Calcium Oxalate. QU 98 C144 1995]
   QP535.C2C2639   1995
   574.19'214--dc20
   DNLM/DLC
   for Library of Congress   95-36917
                                                     CIP

PREFACE

Calcium oxalate is quite common in nature and is found in almost all types of living beings, micro-organisms, fungi, plants and animals including humans. In plants where a majority of the families of seed plants contain calcium oxalate deposits, this sparingly soluble salt plays diverse roles such as storing excess calcium, forming exoskeleton or making plants less palatable to foraging animals. In man and other mammals, oxalate is endogenously produced as well as obtained from the food. Since it can not be metabolized, oxalate is excreted in the urine. Urinary overexcretion of oxalate may result in crystal deposition in the kidneys, formation of kidney stones and eventually in renal failure. Grazing on calcium oxalate containing plants such as halogeton, greasewood and soursob can cause nephrolithiasis and renal failure and decimate large herds of cattle. In humans, more than 75% of the urinary stones contain calcium oxalate as their main constituent. The cost of managing human urinary stone disease in United States alone is estimated to be more than 2.4 billion dollars per year. Thus calcium oxalate is of major biological and economic importance.

The past ten to fifteen years have seen a tremendous increase in research in this area and a substantial amount of information is currently available about oxalic acid and calcium oxalate. But workers involved in oxalate research are dispersed in highly diverse fields such as botany, biochemistry, cell and molecular biology, mycology, microbiology, physical biochemistry, urology, nephrology and human and animal pathology. Thus the information about various aspects of oxalate metabolism and synthesis and crystallization of calcium oxalate is scattered in the literature specific to many disciplines. This book, for the first time, brings current information about calcium oxalate in one volume. Subjects selected for inclusion in this volume include *in vitro* crystallization, biosynthesis of calcium oxalate in plants and fungi, oxalate degrading bacteria, molecular genetics of primary hyperoxaluria type 1, oxalate transport across cellular membranes, oxalate measurement in body fluids, modulators of calcium oxalate crystallization, *in vivo* models of calcium oxalate kidney stone formation, and calcium oxalate crystal/epithelial cell interaction. All the subjects have been comprehensively discussed by internationally recognized leaders in the fields.

I would like to thank all the authors for taking the time from their busy schedules to write critical articles with up-to-date information. I would also like to thank them for timely submission of their manuscripts. My purpose in carrying out this project of bringing currently available information about calcium oxalate in one book was to foster systematic research in this field. I hope this volume will contribute to that end.

Saeed R. Khan
Gainesville, Florida

THE EDITOR

Saeed R. Khan, Ph.D., is an Associate Professor of Pathology at the University of Florida, Gainesville, Florida. He received B.Sc. from Agra University, India; M.Sc. from Peshawar University, Pakistan and Ph.D. from University of Florida, Florida. His postdoctoral training was carried out at the University of Adelaide, Australia. Prior to joining the faculty at University of Florida he served at the Islamabad University, Pakistan and King Abdulaziz University, Jeddah, Saudi Arabia.

Dr. Khan's professional interests include biomineralization, calcification, and nephrolithiasis. He is particularly interested in the pathogenesis of calcium oxalate urinary stones and has developed animal models to investigate the mechanisms involved in stone formation.

Dr. Khan is a member of the Microscopy Society of America, the American Society for Investigative Pathology, the American Urological Association, the American Society of Nephrology, the Histochemical Society, the American Society for Bone and Mineral Research, and New York Academy of Sciences and is a past president of Florida Society for Electron Microscopy. He is a member of the editorial board of Scanning Microscopy International and Italian Journal of Mineral and Electrolyte Metabolism.

Dr. Khan was the Vice-Chairman of the first Gordon Conference on Calcium Oxalate, held in 1986 and has organized a symposium on Stones and Crystals annually since 1984. He has delivered invited and plenary lectures and taken part in round table discussions on mechanisms involved in calcium oxalate nephrolithiasis during many international gatherings in USA, Europe and Asia. His research has been funded through grants from the United States National Institutes of Health. He is the author of over 100 articles and book chapters.

LIST OF CONTRIBUTORS

James H. Adair, Ph.D.
Department of Materials
Science and Engineering
University of Florida
Gainesville, Florida 32611

Milton J. Allison, Ph.D.
National Animal Disease
Agricultural Research Service
U.S. Department of Agriculture
Ames, Iowa 50010

Howard J. Arnott, Ph.D.
Department of Biology
The University of Texas
at Arlington
Arlington, Texas 76091

Bruno Baggio, M.D., D.Sc.
Universitá di Padova
Istituto di Medicina Interna
Via Giustiniani 2
Padova, Italy

Nancy A. Cornick, M.S.
National Animal Disease Center
Agricultural Research Service
U.S. Department of Agriculture
Ames, Iowa 50010

Steven L. Daniel, Ph.D.
Lehrstuhl für Ökologische
Mikrobiologie
BITÖK, Universitat Bayreuth
Bayreuth, Germany

Christopher J. Danpure, Ph.D.
MRC Protein Translocation
Group, Department of Biology
University College London
London WC1E 6BT, UK

Vincent R. Franceschi, Ph.D.
Washington State University
Electron Microscopy Center
Pullman, Washington 99164

Robert W. Freel, Ph.D.
Department of Medicine/
Nephrology Division
University of California
at Irvine
Irvine, California 92717

Giovanni Gambaro, M.D., Ph.D.
Universitá di Padova
Istituto di Medicina Interna
Via Giustiniani 2
Padova, Italy

Phulwinder K. Grover, Ph.D.
Department of Surgery
Flinders Medical Centre
Bedford Park, South Australia
Australia

Craig F. Habeger, B.S.
Department of Materials
Science and Engineering
University of Florida
Gainesville, Florida 32611

Raymond L. Hackett, M.D.
Department of Pathology
College of Medicine
University of Florida
Gainesville, Florida 32610

Marguerite Hatch, Ph.D.
Department of Medicine/
Nephrology Division
University of California
at Irvine
Irvine, California 92717

Harry T. Horner, Ph.D.
Department of Botany and
Bessey Microscopy Facility
Iowa State University
Ames, Iowa 50011

John P. Kavanagh, Ph.D.
Department of Urology
University Hospital of South
Manchester
Manchester, M20 8LR, UK

Saeed R. Khan, Ph.D.
Department of Pathology
College of Medicine
University of Florida
Gainesville, Florida 32610

Dirk J. Kok, Ph.D.
Department of Endocrinology
and Metabolic Diseases
University Hospital
2300 RC Leiden
The Netherlands

Richard V. Linhart
Department of Materials
Science and Engineering
University of Florida
Gainesville, Florida 32610

Frank Loewus, Ph.D.
Institute of Biological
Chemistry
Washington State University
Pullman, Washington 99164

Martino Marangella, M.D.
Nephrology Division and Renal
Stone Laboratory, Ospedale
Mauriziano Umberto 1
Largo Turati, Torino, Italy

Dawn S. Milliner, M.D.
Division of Nephrology
Mayo Clinic
Rochester, Minnesota 55902

Michele Petrarulo, M.Sc.
Nephrology Division and Renal
Stone Laboratory, Ospedale
Mauriziano Umberto 1
Largo Turati, Torino, Italy

Gillian Rumsby, Ph.D.
Department of Chemical
Pathology
University College London
Hospitals
London W1P 6DB, UK

Rosemary L. Ryall, Ph.D.
Department of Surgery
Flinders Medical Centre
Bedford Park, South Australia
Australia

Paula N. Shevock, B.S.
Department of Pathology and
Laboratory Medicine
College of Medicine
University of Florida
Gainesville, Florida 32610

Alan M.F. Stapleton, M.D.,
Ph.D.
Department of Surgery
Flinders Medical Centre
Bedford Park, South Australia

Bruce L. Wagner, M.S.
Department of Botany and
Bessey Microscopy Facility
Iowa State University
Ames, Iowa 50011

TABLE OF CONTENTS

Chapter 1
Calcium Oxalate Crystallization *in vitro* 1
John P. Kavanagh

Chapter 2
Inhibitors of Calcium Oxalate Crystallization 23
Dirk J. Kok

Chapter 3
Mechanisms of Calcium Oxalate Aggregation in the Biophysical
Environment 37
James H. Adair, Richard V. Linhart, Craig F. Habeger and Saheed R. Khan

Chapter 4
Calcium Oxalate Formation in Higher Plants 53
Harry T. Horner and Bruce L. Wagner

Chapter 5
Calcium Oxalate in Fungi 73
Howard J. Arnott

Chapter 6
Oxalate Biosynthesis and Function in Plants and Fungi 113
Vincent R. Franceschi and Frank A. Loewus

Chapter 7
Oxalate Degrading Bacteria 131
Milton J. Allison, Steven L. Daniel and Nancy A. Cornick

Chapter 8
Epidemiology of Calcium Oxalate Urolithiasis in Man 169
Dawn S. Milliner

Chapter 9
Enzymology and Molecular Genetics of Primary Hyperoxaluria Type 1.
Consequences for Clinical Management 189
Christopher J. Danpure and Gillian Rumsby

Chapter 10
Cellular Abnormalities of Oxalate Transport in Nephrolithiasis 207
Bruno Baggio and Giovanni Gambaro

Chapter 11
Oxalate Transport Across Intestinal and Renal Epithelia **217**
Marguerite Hatch and Robert W. Freel

Chapter 12
Oxalate Measurement in Biological Fluids **239**
Martino Marangella and Michele Petrarulo

Chapter 13
Urinary Macromolecules in Calcium Oxalate Stone and Crystal
Matrix: Good, Bad, or Indifferent **265**
Rosemary L. Ryall and Alan M. F. Stapleton

Chapter 14
Lipid Matrix of Urinary Calcium Oxalate Crystals and Stones **291**
Saheed R. Khan

Chapter 15
Urate and Calcium Oxalate Stones: A New Look at an Old
Controversy **305**
Phulwinder K. Grover and Rosemary L. Ryall

Chapter 16
The Role of Crystal-Cell Attachment and Retention in Stone Disease **323**
Raymond L. Hackett and Paula N. Shevock

Chapter 17
Animal Model of Calcium Oxalate Nephrolithiasis **343**
Saheed R. Khan

To my wife Patricia and children Omar and Ameena Khan

Chapter 1

CALCIUM OXALATE CRYSTALLIZATION *IN VITRO*

John P. Kavanagh, Department of Urology,
University Hospital of South Manchester, Manchester M20 8LR, U.K.

I. INTRODUCTION

Calcium oxalate crystallization in vitro is usually carried out in the context of investigating urolithiasis. Applications range from studying fundamental physical chemistry in simple solutions to developing clinically meaningful tests using urine. Many methods have been used and this has sometimes led to difficulties in interpretation; problems can arise when authors use different nomenclature and units for seemingly similar phenomena, or worse, use the same terms to express different ideas. Such difficulties are almost inevitable; some methods can only characterise crystallization in a broad sense without distinguishing between nucleation, growth or aggregation (also referred to as agglomeration), while others aim to focus on a particular process. As these semantic obstacles are present in the literature and are likely to be repeated in the future, it is important that writers and readers have a clear understanding of the principles involved in the methods used. It is equally important to be clear about which crystallization processes are taking place, how they interact and the affects of the system and measuring technique.

The aim of this chapter is to review many of the methods used for *in vitro* crystallization studies, with the emphasis on how nucleation, growth and aggregation can be measured, either separately or in combination.

Small differences in sample collection and preparation, concentrations of reactants, reaction vessel geometry, stirring mechanisms, quality of seed crystals or surfaces (if any), induction times, fixed time or continuous analysis and the method for monitoring the crystallization process can give rise to inconsistent results. Some methods claim particular physiological relevance, while others seek clarity by using well defined conditions. This might be reflected in the use of fresh or diluted urine as opposed to artificial urine, or it could be in an attempt to model some features of stone formation *in vitro* instead of using a simple crystallization vessel. The conclusions reached by particular experiments are therefore often system dependent.[1,2]

The goal of most of these studies is to advance the understanding, prevention or treatment of urolithiasis. There is a place for experimental extremes and variations, as long as the results are interpreted with due care.

II. THE IMPORTANCE OF SUPERSATURATION

Supersaturation is the driving force for crystallization. Without it crystal growth (enlargement through solute deposition) can not occur. The initiation of crystallization (nucleation) is similarly dependent on supersaturation, which must be sufficiently high so that the energy given up in depositing the crystal nuclei is greater than that required to form its surface. The boundary at which this occurs is known as the metastable limit (ML). Lowering the energy required for surface formation will lower this limit, so surfaces presented by components of the crystallizing medium or environment can act as heterogeneous nucleators. It is widely agreed that in in vivo and in most in vitro experiments, crystallization of calcium oxalate is heterogeneously nucleated. Very high supersaturations, in very clean solutions, would be required for homogeneous nucleation.[3]

An appreciation of metastability is fundamental to the understanding of how crystallization is initiated in various systems. Figure 1 shows a mechanical analogy of metastability. In the model, crystal growth is envisaged as a cylinder moving from left to right across a plane. The height (h) of the cylinder represents the supersaturation and the frictional resistance between it and the plane is inversely related to the propensity for nucleation. When h is less than or equal to 1, the system is undersaturated or at saturation and is stable, no crystallization will take place. As h increases above 1, the system becomes increasingly unstable. The weight will eventually begin to roll (nucleation) and once started it will continue (growth) until it comes back to equilibrium at $h=1$ (saturation=1). The point at which the cylinder starts to roll will depend on the frictional resistance between it and the plane. A low friction would correspond to a high propensity to nucleate or a low ML; a high frictional resistance would correspond to a system with a higher ML.

The degree of supersaturation is therefore a critical factor in determining the rate and extent of solute precipitation (both growth and nucleation). How this supersaturation is developed, controlled and allowed to change during an experiment, will influence the size, number and form of the crystals produced under different conditions. It is also an important factor in determining the aggregation properties of the system[4,5]. In a previous review[6] the changes in supersaturation were used as the basis of a classification of different methods into three broad categories and this will be followed again here. The system may be i) initially supersaturated and the supersaturation allowed to decay; ii) initially undersaturated or saturated and concentrated to bring about supersaturation; iii) controlled by continuous additions to maintain a constant supersaturation (Figure 2).

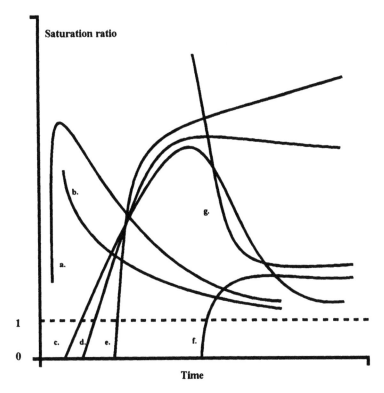

Figure 1.   Some saturation profiles possible from different calcium oxalate crystallization experiments. Batch crystallisers (**a, b**) are characterised by saturation decay. If below the metastable limit (**b**) they may be seeded to initiate crystallization. The supersaturation may be raised relatively slowly by concentration of reactants (**c, d, e**), in which case the outcome will depend on the rate of concentration, nucleation would start earlier if seeded. Continuous crystallization (**f, g**) can generate a stable saturation.

III. CALCIUM OXALATE CRYSTALLIZATION METHODS

A. SUPERSATURATION ALLOWED TO DECAY

This type of experiment is performed as a batch crystallization, without further addition once the crystallization has been initiated. This category of methods is certainly the most widely applied to calcium oxalate and urolithiasis. Ryall[7] has recently critically reviewed their strengths and weaknesses and pointed out that most of our knowledge about urinary crystallization inhibitors and their role in stone formation has come from these systems, while other methods have largely been useful for confirmation. The induction of crystallization (seeded or unseeded) in these methods is of great importance as it determines whether nucleation plays a significant role.

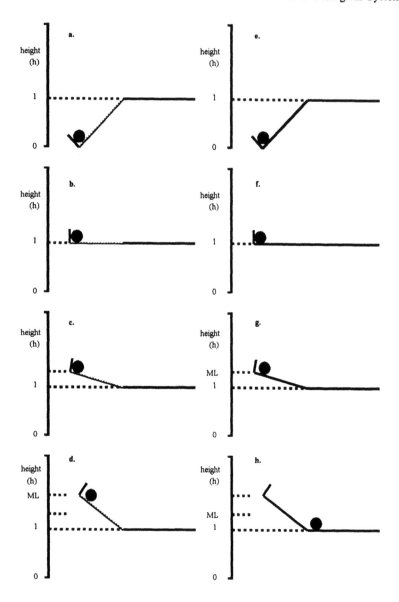

Figure 2. A mechanical analogy of supersaturation and metastability. Crystallization is viewed as the cylinder moving from left to right. The saturation ratio is analogous to the height (*h*), increasing **a** through **d** and **e** through **h**. The system becomes increasingly unstable at *h*>1. The point at which the cylinder begins to roll (onset of nucleation) is dependent on the frictional resistance. **a-d** have a greater friction than **e-f** and therefore a higher metastable limit (*ML*).

1. Seeded Systems
a. Seeded with Calcium Oxalate

The basic approach here is to prepare a metastable solution and introduce seeds of crystalline calcium oxalate. This will allow crystallization to proceed without nucleation. In the mechanical analogy (Figure 1) it is as though the cylinder (between $h>1$ and $h<ML$) is given a nudge to set it rolling. Furthermore, if the deposition of calcium or oxalate, or their reduction in solution is measured, then it is usually assumed that a pure growth effect is being quantified. This may be a little simplistic as concurrent aggregation may influence the simultaneous growth. Changes in surface area associated with aggregation will be reflected in the growth kinetics and there may be local changes in supersaturation which will also affect the overall growth.[8,9]

In Table 1, a number of examples of this system are given. It is not intended to be a comprehensive survey but does show some of the variations that have been applied. Most investigators have used buffered saline as the medium to determine quantitative effects of added inhibitors or diluted urine. Seeded crystallization rates in whole urine have sometimes been studied[10,11] and if allowed to reach equilibrium, empirical determination of urinary saturation is possible.[10,12]

A great advantage of this type of system is in the analysis of the growth kinetics, which are usually found to be second order, suggesting a surface controlled mechanism.[13] This allows the calculation of physico-chemically meaningful constants for growth and inhibition. The methods for collecting and handling data from these experiments can be very diverse and these details can have implications for derivation of constants and interpretation of the results. This has been discussed more fully recently.[6]

The seeded batch crystallizer behaves in a predictable and understandable manner because nucleation is avoided. This has been very useful in comparing, purifying and characterising urinary crystallization growth inhibitors but usually at the expense of neglecting aggregation. It is noteworthy that in their early experiments, Robertson et al. did not attempt to distinguish between growth and aggregation, treating the change in particle size distribution (PSD) as a combined estimate.[14,15] Ryall et al. later went on to show that PSD results can be used to give separate estimates of growth and aggregation.[16-21] Both of these processes can also be evaluated as distinct but interrelated processes using the kinetic analysis of calcium uptake by crystals, developed in Leiden.[8,9,22]

b. Seeded Systems Studied at Saturation

In a saturated suspension of crystals, growth or nucleation can not occur. This can sometimes be useful. Hess et al.[37] have developed a turbidimetric assay for aggregation, studied as a single process.[39,42,43] Other applications for studying calcium oxalate crystals at saturation are for

Table 1

Summary of Some Batch Crystallization Experiments, Seeded with Calcium Oxalate

Reference	Method[a,b]	Cryst'n[c,d]	Time[e,f]	Seed mg/ml[g]	Initial Ca mM	Initial Ox mM	T°C	pH[h]	Application
Robertson et al 1972[14]	PSD, No. >20μm	(G and A)	4 hr	0.02	1.0	0.2	?	6.0	Effects of 5% urine
Robertson et al 1973[15]	PSD, Vol. >20μm	(G and A)	4 hr	0.025	1.0	0.2	37	6.0	Effects of inhibitors
Nancollas & Gardner 1974[23]	Conductivity	G	V	Variable	Variable	Variable	Variable	?	Kinetic analysis
Gill et al 1974[10]	^{14}C-Ox in supernatant	G	2 hr	0.2	Variable	Variable	37	Variable	**Whole urine**
Meyer & Smith 1975[24]	Ca (by AAS) in filtrate	G	V	0.06	0.44	0.44	37	6 to 7	Method development
Meyer & Smith 1975[25]	Ca (by AAS) in filtrate	G	V	0.06	0.43	0.44	37	6.7	Effects of 1.4 to 10% urine and inhibitors
Pak et al 1975[26]	Ca (by AAS) and Ox (chemically) in filtrate	G	V	0.5	0.3	0.3	37	6.0	Effects of diphosphonate
Gill et al 1977[27]	^{14}C-Ox in supernatant	G	2 hr	0.01	Variable	Variable	37	5.8	Effects of urinary macromolecules
Ito & Coe 1977[28]	^{14}C-Ox in filtrate	G	40 min	0.06	1.0	0.2	24	5.7	Effects of 8% urine and inhibitors
Ligabue et al 1979[29]	^{14}C-Ox in supernatant	G	3.5 hr	0.06	1.0	0.2	37	6.0	Effects of 2% urine and inhibitors
Werness et al 1981[30]	Ca (by AAS) in filtrate	G	V	0.04	0.46	0.43	37	6.0	Effects of inhibitors
Curreri et al 1981[31]	Ca (by AAS) in filtrate	G	V	0.06	0.3	0.3	38	6.5	Effects of citrate
Tiselius & Fornander 1981[32]	^{14}C-Ox in filtrate and in supernatant	G	2hr and V	0.04	1.0	0.2	37	6.0	Effects of 2% urine, comparison of methods
Nakagawa et al 1981[33]	^{14}C-Ox in filtrate	G	40 min	0.125	0.833	0.167	37	5.7	Inhibitor purification
Ryall et al 1981[16]	PSD, No. and Vol.	G and A	2 hr	0.02	1.0	0.2	37	6.0	Effect of 2 to 10% urine

a. AAS, atomic absorption spectroscopy. b. PSD, particle size distribution. c. Crystallization processes considered; G, growth, A, aggregation, (G and A), growth and aggregation considered as a combined process. d. Nucleation is assumed not to be present; even if the aim is to measure a pure growth effect, aggregation may occur and influence growth. e. Initial measurements are normally made at a series of various times. f. V, measurements made at a series of various times. g. Other characteristics of the seed are also important (e.g. age, size, homogeneity, etc). h. Various buffers and ionic strengths have been used.

Table 1 (cont.)
Summary of Some Batch Crystallization Experiments, Seeded with Calcium Oxalate

Reference	Method[a,b]	Cryst'n[c,d]	Time[e,f]	Seed mg/ml[g]	Initial Ca mM	Initial Ox mM	T°C	pH[h]	Application
Ryall et al 1981[17]	PSD, No. and Vol.	G and A	V	0.02	1.0	0.2	37	6.0	Effects of 1% urine and inhibitors
Fellström et al 1982[34]	14C-Ox in filtrate	G	1 hr and V	0.136	1.55	0.1	?	6.0	Effects of urine fractions (10%)
Will et al 1983[22] Blomen et al 1983[8] Bijvoet et al 1983[9]	45Ca uptake by (filtered) crystals	G and A	V	Variable	Variable	Variable	37	6.0	Method development. Variations in seed & soln. composition
Ryall et al 1986[20]	PSD, iterative program based on No. and Vol.	G and A	V	0.02	1.0	0.2	37	6.0	Method development
Nakagawa et al 1987[55]	O.D. 214nm (Ox absorbance)	G	V	0.012	1.0	1.0	37	7.2	Dissociation constant of nephrocalcin
Kok et al 1988[36]	45Ca uptake by (filtered) crystals	G and A	V	0.14	0.372	0.372	37	6.0	Effect of inhibitors
Hess et al 1989[37]	O.D. 620nm (turbidity)	A	V	0.8	Saturated (0.14)	Saturated (0.14)	37	5.7 or 7.2	Method development, effects of inhibitors
Baumann et al 1990[11]	Ca-electrode	G	V	1.0	Variable	Variable	37	Variable	Whole urine
Tiselius et al 1990[38]	45Ca in solution	G	30 min	0.01	0.97	Variable	?	58	Dialysed urine (87%)
Tiselius et al 1993[39]	O.D. 690nm (turbidity)	A	5 min	3.0	Saturated	Saturated	37	7.2	Effects of citrate and dialysed urine
Yamaguchi et al 1993[40]	PSD, particles >9μm	(G and A)	4 hr	0.04	1.0	0.2	37	6.0	Effect of stone matrix components
Wolf & Stoller 1994[41]	Ca-electrode	G	V	0.25 Calculi granules	0.3	0.3	25	6 to 7	Inibitors preabsorbed to seeds

a. AAS, atomic absorption spectroscopy. b. PSD, particle size distribution. c. Crystallization processes considered; G, growth, A, aggregation, (G and A), growth and aggregation considered as a combined process. d. Nucleation is assumed not to be present; even if the aim is to measure a pure growth effect, aggregation may occur and influence growth. e. Initial measurements are normally also required. f. V, measurements made at a series of various times. g. Other characteristics of the seed are also important (e.g. age, size, homogeneity, etc). h. Various buffers and ionic strengths have been used.

measuring adsorption to crystals[31,44,45] or for measuring zeta potentials. This used to require laborious microscopic observations[31,46,47] but can now be performed by laser scattering, which is simpler and gives more information.[48,49]

c.   Seeded with Heterogeneous Nucleators

The final example in Table 1[41] used natural calculi as the source of seed crystals instead of the conventional preparations of well characterised calcium oxalate monohydrate (COM) seeds. Many other possible initiators of the crystallization can be introduced (e.g. single large crystals of calcium oxalate dihydrate (COD),[50] but if they are not calcium oxalate then they are probably acting as heterogeneous nucleators. In the mechanical analogy (Figure 1) it is as though oil is applied to the surface of the plane, allowing the cylinder to start rolling at a lower height than would otherwise be the case. Seed crystals e.g. uric acid, calcium phosphate or silica[51] or membrane fragments[52] can be introduced into a metastable solution, or the potential nucleant can be attached to a surface, such as a silicon wafer[53] or sepharose beads.[54] Of course, in these examples the emphasis of the experiments has changed from a study of growth and aggregation to a study of heterogeneous nucleation. Non-particulate additions to a metastable solution may provoke nucleation by a salting out mechanism[55,56] as might be the case for urate promotion of crystallization from urine.[57,58]

2.   Unseeded Batch Crystallizers

In these methods, the supersaturation is (usually quickly) raised above *ML*, initiating nucleation, and the crystallization (nucleation, growth and aggregation) proceeds without any further additions. By confining observations to the very early stages, nucleation processes can be concentrated upon. Estimating *ML*, and investigating means of raising or lowering this supersaturation at which nucleation starts is a valuable role for these methods.

a.   Determination of the Metastable Limit (ML)

In principle, this is simply a matter of progressively raising the supersaturation until crystallization begins to take place, but the *ML* does not represent a well defined boundary.[3] In practice the measured value is dependent on the techniques used. Such factors as the rate of raising the supersaturation and agitation of the solution will be very significant.[56] More obviously, the means of identifying the onset of crystallization and the time of observation will be crucial. Any detection method will have a limiting sensitivity which must be overcome by sufficient nucleation and growth (and possibly aggregation if the detection method is based on particle size). The time required for the limiting sensitivity to be breached is the most important component of the induction time, but, especially at low

Table 2
Some Methods for Determination of Metastable Limits in Urine

Reference	Urine Pre-treatmet	Super-saturation raised by	Detection of critical point	Time	Expression of results
Pak & Holt 1976[12]	Centrifugation & filtration	Ox added to a series of urine aliquots	Visual	3 hr	[a]Formation product, ratio (FPR) = Critical($[Ca] \times [Ox]$) / urine($[Ca] \times [Ox]$)
Briellmann et al 1985[60] & Schnierle et al 1992[61]	None	Pumped infusion of Ox	Detectable change in turbidity	Continuous monitoring	[b]Tolerable oxalate, compared against standard curve for artificial urine, taking into account urine [Ca]
Ryall et al 1985[62] & 1986[63]	Centrifugation & filtration	Ox added to a series of urine aliquots	Particles $>2\mu m$	30 min	[b]Change in [Ox] or critical($[Ca] \times [Ox]$)
Tiselius 1985[64]	Acidify, store frozen, adjust to pH 5.8, dilute to a fixed creatinine concentration or 80% fresh urine	Timed additions of Ox (at 5 min intervals)	100 particles in the range 3.5 to $5\mu m$	2 min	[b]Calcium oxalate crystallization risk (CaOx$_{cr}$) = 1/change in [Ox]
Kohri et al 1991[65]	None	Ca & Ox added to a series of urine aliquots	Visual	5 min	Three point scale of increasing metastability
Rodgers et al 1993[66]	Filtration	Ox added to a series of urine aliquots	Dramatic change in turbidity	30 min	[b]Change in [Ox]

All these methods will to some extent reflect growth and aggregation as well as the onset of nucleation. **a.** [ ] represents activities. **b.** [ ] represents concentrations.

supersaturations, there may also be a latent period before rapid desaturation takes place.[56] Again these factors can be described by the mechanical analogy (Figure 1). The speed at which the plane is raised, any vibrations, the way of deciding if the cylinder has started to roll and the length of time it is observed to reach this decision, will all influence the empirically determined height at which rolling starts. Nevertheless, for any particular set of conditions, consistent results can be obtained and useful comparisons made. Table 2 lists some examples of methods that have been described for application to urine. Consideration of these suggests that, although nucleation must have started in order for the detection criterion to be matched, a variable degree of growth and aggregation may also have taken place and this will be reflected in the reported results. Of course, these or similar methods can also be applied in defined solutions to study the effect of various additives.[27] As another indicator of the nucleating potential of a system it is possible to bring the calcium and oxalate concentrations to some chosen values and measure the induction time for measurable crystallization to start.[59]

b.   Crystallization from above Metastable Limit (ML)

Once some estimate of *ML* has been made, then crystallization can be induced by raising the supersaturation to some predetermined value known to exceed *ML* or the supersaturation can be raised to some fixed value related to *ML* (e.g. *ML* plus 0.3mM oxalate).[62] As in seeded experiments, the progress of the crystallization can be followed by many methods (e.g. particle counting or visual examination, calcium or oxalate depletion, calcium-electrode, or turbidity). The reaction can be monitored continuously, at a series of times or at only one or two time points and there are accordingly many ways to derive quantitative expressions of the results.[6] Interpretation is, however, more difficult than with a seeded method because of the simultaneous nucleation, growth and aggregation.

When the method involves the direct or indirect assessment of calcium or oxalate depletion or deposition then aggregation is usually ignored and the results are taken to be an expression of both nucleation and growth.[69-71] A method which combines measurement of oxalate depletion with particle size discrimination (by filtration) was used to quantify aggregation, as a non-aggregated ratio. Particles > 20μm were assumed to be aggregated.[72]

Counting and sizing the crystals allows particle numbers and volumes to be distinguished. An increase in precipitated volume results from nucleation <u>and</u> growth, while a change in crystal numbers will result from nucleation <u>and</u> aggregation. A further complication is that the relative contributions to the PSD of nucleation, growth and aggregation will change during the course of the reaction. Changes in total volume, after the initial nucleating phase, are usually related to growth and changes in numbers in the later stages of the reaction are loosely equated with aggregation. Often the problems in interpretation are recognised and results are discussed in qualitative terms[62,73,74] rather than trying to define a growth rate or inhibition index, or apply the methods applicable for distinguishing growth and aggregation in seeded systems.[16,18,20,21] In some cases, however, quantitative values have been derived from particle volumes and numbers and ascribed separately to growth and aggregation.[47,75,76] Crystals are usually counted and sized with a Coulter Counter (or equivalent apparatus) and the conclusions drawn are often supported by microscopic examination. This may sometimes be particularly relevant (e,g, if Tamm Horsfall protein is present[77]). Quantitative microscopy has also been applied[78,79] and this can enable single crystals and aggregates to be directly counted and an aggregation coefficient calculated.[79]

The nucleating, growing and aggregating particles in suspension can also be detected by light scattering or absorbance (nephelometry or turbidity). Although this gives less information than a Coulter Counter analysis, it can be performed continuously and is particularly useful for studying the early phase of the crystallization. The results have been interpreted as being due to growth and nucleation,[78,80] nucleation,[66] mainly growth,[59] or aggregation.[81] Agitation or stirring of the suspension will have a great influence on the outcome of these experiments and this has not always been adequately described in the published reports. Most of these studies have quantified the initial linear increase in turbidity[59,66,78] but Ebisuno et al.[81] allowed aggregation to develop over a 10 minute period with stirring and then observed the decrease in turbidity as the suspension settled over the next 10 minutes without stirring. This is similar to the method of Hess et al.[37] but performed in a supersaturated solution which is also nucleating and growing.

An advantage of unseeded batch crystallization is the ease of application to whole urine. The phrase "whole urine" is often a misnomer as filtration or centrifugation are frequently used to remove debris and larger particulates and it is recognised that these procedures will also remove most of the Tamm Horsfall protein.[77] Nevertheless, these methods have been used to try to distinguish the calcium oxalate crystallization properties of urine from stone formers and healthy controls.[60-62,64,68] The usual method for increasing the supersaturation is by oxalate addition, but the effect of these additions on the supersaturation will also be dependent on the urinary calcium, other electrolytes and macromolecules. A recent approach has been to titrate urine with a calcium solution and measure ionised calcium with a specific electrode.[82] This study is concerned with the pre-crystallization phase of increasing supersaturation, but is interesting because of the noted difficulty, with some samples, of identifying a precipitation point.

B. SUPERSATURATION DEVELOPED SLOWLY

The manner and time scale over which the supersaturation is brought about will affect the different crystallization processes in different ways, the PSD, crystal phase and habit may all vary with subtle variations in method.[55,56] There is a seemingly endless variety of methods available to raise the supersaturation, three distinctions can be identified; semi-batch methods, with additional reactants being fed into the system; supersaturation raised by diffusion; concentration by solvent (water) removal.

1. Semi batch crystallization

Some of the methods discussed in the above section might more properly be included here, as they involve continuous[60,61] or stepped[64] increases in oxalate, but they were primarily concerned with the onset of crystallization rather than its progress. Grases et al.[83,84] set up saturated solutions, with separate feeds of calcium and oxalate. A constant pump rate was used giving initial changes of about 0.7 mM/hr oxalate, 7 mM/hr calcium[84] and 4 mM/hr oxalate, 16 mM/hr calcium[83] (perhaps 0.4 mM/hr and 1.6 mM/hr if the initial volume was 500 ml, not 50 ml as stated). No attempt to quantify the crystallization processes was made, the crystals were examined at intervals by microscopy and the aggregates described. On morphological grounds they distinguished between primary and secondary agglomeration, concluding that, in this system primary agglomeration dominates. The secondary aggregation they describe is dependent on contact between crystals in suspension and corresponds to the process usually considered in calcium oxalate crystallization. Primary agglomeration represents an irregular growth of crystals from pre-formed crystals, and probably corresponds to secondary nucleation.[55,56] An osmotic pump has been used to deliver calcium (at 0.04 mM/hr) into an oxalate solution, which generated calcium oxalate trihydrate (COT) crystals adhering to the tip of the delivery port.[85] Another slow delivery system is filter wicks, transporting calcium and oxalate feed solutions into a beaker containing suspended glass fibres on which crystals grow. The calcium oxalate on the fibres can then be quantified as an overall indicator of crystallization.[86-88]

2.   Supersaturation raised by diffusion

Supersaturation development in these experiments is usually slow and the crystallization may be allowed to proceed for days or weeks. Solid calcium chloride has been embedded in a polymeric matrix and allowed to diffuse out into an oxalate solution,[85] generating COM. Controlled delivery of oxalate to the surface of a calcium containing solution can be achieved by allowing diethyl-oxalate to hydrolyse. The calcium solution can be layered on the heavier and immiscible diethyl-oxalate, producing large crystals (up to 2 mm in some circumstances) over 3 to 4 weeks. The type of crystals formed (COD or COT) can be controlled by the choice of calcium concentration.[85,89] The hydrolysis of diethyl-oxalate has been used to generate oxalic acid in the vapour phase which then diffuses into a solution of calcium chloride.[90] The crystallization within these systems is usually characterised descriptively by light or electron microscopy, without trying to quantify particular processes. From such observations of crystal sizes and numbers, Deganello[90] was able to conclude that nephrocalcin in solution inhibited the growth of calcium oxalate but when attached to the surface of the crystallization chamber it promoted nucleation.

Diffusion of one[91] or both[89,92] reactants into a gel made up of agar,[93] gelatin[89] or silica[91] can be used to raise the supersaturation sufficiently for crystallization to take place. Seeding the gel with crystals will both reduce the time required and eliminate nucleation. As the growing seeds are immobilised in the gel, aggregation should not occur and pure growth effects can be studied. The extent of crystallization can be measured by the change in optical density and reactions conveniently performed in 96 well microtitre plates, permitting the easy analysis of many permutations of inhibitors and conditions.[94,95] In this gel crystallization model, oxalate is usually included within the gel (but this can be reversed[96]). If urine is being used as the source of calcium, then the outcome (expressed as a growth rate relative to a control) will be dependent on the urinary calcium concentration as well as other urinary factors such as inhibitors.[95] A flow model for crystallization in gels has been developed[97] in which a supersaturated solution is formed just above the gel (which may include seed crystals) and conducted over the surface. Crystals form in or just on the top of the gel. The relative merits of these two gel systems has recently been reviewed.[98] It is suggested that the presence of the gel and the diffusion processes involved in the crystallization may mimic some aspects of stone formation, where growing calculi might be covered in a mucinous boundary through which to the supersaturated surrounding liquor must pass.[98]

3. Concentration by water removal

Slow evaporation (in the region of 0.04 ml/day) from dilute solutions of calcium and oxalate will eventually generate sufficient supersaturation for crystallization.[99] A more rapid removal of water is thought to recreate some aspects of urine formation within kidney tubules and can be achieved by reverse osmosis[100,101] or evaporation under reduced pressure at 37°C.[102-107] The first applications of this method[102-104] concentrated urine to some fixed osmotic pressure (e.g. 1200 mOsmol/kg) at about 2 ml/min, allowed the sample to stand for 1 hour and then examined the crystals produced. In a similar approach, Rodgers and Wandt[105] noted the difficulty in controlling the rate of evaporation and therefore in making quantitative comparisons. In a recent development, urine is concentrated as before, but crystallization is induced by freezing and thawing the sample.[106,107] In these cases crystals were counted directly by microscopy and an increase in particle density was taken to indicate promotion of nucleation and a decrease in crystal size was interpreted as growth inhibition, but of course any aggregation would also affect the crystal numbers and sizes.

C. STEADY STATE SUPERSATURATION SYSTEMS

1. Constant Composition Crystallization

In this method (reviewed by Khan and Opalko[108]), calcium and oxalate solutions are separately pumped into a reaction vessel (which may be seeded) and the calcium ion activity monitored by a specific electrode. As calcium and oxalate crystallize the electrode triggers the pumps to replenish the solution, thereby maintaining a constant composition, which is more representative of the conditions within the kidney than a supersaturation decay method. The rate of delivery from the pumps gives the rate of consumption of calcium and oxalate and can be used to study reaction kinetics and inhibitor activities.[109-112] A recent application is of interest because it tests three inhibitors by this method alongside a seeded batch system analysed by particle sizes and numbers; reaching essentially the same conclusions about growth inhibition.[48] When seeded, the constant composition method is assumed to be measuring growth, although aggregation may be occurring and having an indirect influence on the outcome. Nucleation may not always be absent and the induction time can be used to investigate heterogeneous nucleation.[113-115] Another study of nucleation involved the use of a few large COM seed crystals.[116] These were found to promote secondary nucleation, giving an exponential increase in calcium and oxalate consumption, rather than the usual linear reaction. Rate constants were calculated from linearised slopes in a log plot.

2. Mixed Suspension Mixed Product Removal (MSMPR)

The application to urolithiasis research of this widely used crystallization technique was pioneered by Finlayson[117] who recognised parallels between it and some aspects of fluid dynamics in the collecting ducts and renal pelvis. In this method, reactants are fed continuously into the crystallization chamber and some take-off mechanism maintains a constant volume. When well mixed, the crystals reach an equilibrium PSD after about 10 volumes have passed through the chamber. At this point a constant supersaturation will have developed, at a much lower level than would be calculated from the input concentrations. The equations governing analysis of the PSD can readily be applied to give both a nucleation rate and a growth rate.[55,56,118] Aggregation phenomena can not easily be accommodated in this method, although it has sometimes been taken into account.[119,120] A drawback of early versions of the MSMPR technique was the difficulty of applying it to urine, which was overcome with the development of a mini-crystallizer and redesign of the feed stocks and flow rates.[121] This version has been used with artificial urine,[122] 92% urine[123] and parallel chambers have been introduced to improve comparability.[6] In a separate development, pooled urine has been used with an in-line particle sizer[124] and in another variation, calculi fragments have been suspended in the crystallizer and their change in weight measured.[125] A recent review summarises the history of calcium oxalate applications of the MSMPR method.[126]

IV. PHYSIOLOGICAL RELEVANCE

A number of the methods described above display features which are claimed to model some aspect of urine development, urine passage through the kidney or stone formation from urine. Whether this gives real advantages or offers only illusory belief of physiological relevance is open to debate. It is clear that no one method will be able to mimic all the important aspects of stone formation. On the other hand, some models may allow particular insights that would not follow from conventional methods. Studies of crystallization and flow of calcium oxalate in tubes[127,128] have suggested that crystals *in vivo* would not flow freely through renal tubules, even if the relative diameters seem adequate, instead they may spend much of their time in close association with the walls of the tubules. A model has been designed specifically to reproduce some of conditions experienced by papillary and calyceal stones during the early stages of their generation.[129] In this system, a supersaturated artificial urine is produced by mixing reagents and allowing this to flow over a silicate sphere which acts as a substrate for the crystallization. A second sphere is partially immersed in the run-off from the first. From microscopic examination of the crystals, it was suggested that secondary aggregation is of minor importance in stone formation, a conclusion at odds with widely held beliefs derived from more orthodox techniques.

The purpose of the techniques and investigations discussed in this chapter is to understand and resolve clinical problems, so it always important to keep in mind the physiological significance of the findings. All *in vitro* methods for calcium oxalate crystallization will have their limitations, but with care in interpretation, they can contribute to the goals of urolithiasis research and clinical practice.

REFERENCES

1. **Baumann, J.M.**, How reliable are the measurements of crystallization conditions in urine?, *Urol. Res.*, 1988, 16, 133.
2. **Rodgers, A.L. and Ball, D.**, Studies of urinary macromolecules: an urgent appeal for a standard reference crystallization model, in *Urolithiasis 2*, edited by Ryall, R., Bais, R., Marshall, V.R., Rofe, A.M., Smith, L.H. and Walker, V.R., Plenum Press, New York and London, 1994, 209.
3. **Finlayson, B.**, Physicochemical aspects of urolithiasis, *Kidney Int.*, 1978, 13, 344.
4. **Hounslow, M.J., McLaughlin, G.J., Olley, J.E., Bramley, A.S. and Ryall, R.L.**, A general aggregation mechanism for calcium oxalate, in *Urolithiasis 2*, Ryall, R., Bais, R., Marshall, V.R., Rofe, A.M., Smith, L.H. and Walker, V.R., Eds., Plenum Press, New York and London, 1994, 173.
5. **Bramley, A.S., Hounslow, M.J., Paterson, W.R. and Ryall, R.L.**, The role of solution composition in calcium oxalate crystal enlargement, in *Urolithiasis: consensus and controversies*, Rao, P.N., Kavanagh, J.P. and Tiselius, H.-G., Eds., Lithotriptor Unit, South Manchester University Hospital, Manchester, U.K., 1995, 294.
6. **Kavanagh, J.P.**, Methods for the study of calcium oxalate crystallisation and their application to urolithiasis research, *Scann. Microsc.*, 1992, 6, 685.
7. **Ryall, R.L.**, Batch crystallizers; long shots and shortcomings, in *Urolithiasis: consensus and controversies*, Rao, P.N., Kavanagh, J.P. and Tiselius, H.-G., Eds., Lithotriptor Unit, South Manchester University Hospital, Manchester, U.K., 1995, 65.
8. **Blomen, L.J.M.J., Will, E.J., Bijvoet, L.M. and van der Linden, H.**, Growth kinetics of calcium monohydrate. II. The variation of seed concentration, *J. Cryst. Growth*, 1983, 64, 306.

9.   **Bijvoet, O.L.M., Blomen, L.J.M.J., Will, E.J. and van der Linden, H.,** Growth kinetics of calcium oxalate monohydrate. III. Variation of solution composition, *J. Cryst. Growth,* 1983, 64, 316.

10. **Gill, W.B., Silvert, M.A. and Roma, M.J.,** Supersaturation levels and crystallization rates of calcium oxalate from urines of normal humans and stone formers determined by a ^{14}C-oxalate technique, *Invest. Urol.,* 1974, 12, 203.

11. **Baumann, J.M., Ackermann, D. and Affolter, B.,** Rapid method of measuring the inhibition of calcium-oxalate monohydrate growth in urine, *Urol. Res.,* 1990, 18, 219.

12. **Pak, C.Y.C. and Holt, K.,** Nucleation and growth of brushite and calcium oxalate in urine of stone-formers, *Metabolism,* 1976, 25, 665.

13. **Nancollas, G.H., Smesko, S.A., Campbell, A.A., Richardson, C.F., Johnsson, M., Iadiccico, R.A., Binette, J.P. and Binette, M.,** Physical chemical studies of calcium oxalate crystallization, *Am. J. Kidney Dis.,* 1991, 17, 392.

14. **Robertson, W.G. and Peacock, M.,** Calcium oxalate crystalluria and inhibitors of crystallization in recurrent renal stone-formers, *Clin. Sci.,* 1972, 43, 499.

15. **Robertson, W.G., Peacock, M. and Nordin, B.E.C.,** Inhibitors of the growth and aggregation of calcium oxalate crystals in vitro, *Clin. Chim. Acta,* 1973, 43, 31.

16. **Ryall, R.L., Bagley. C.J. and Marshall, V.R.,** Independent assessment of the growth and aggregation of calcium oxalate crystals using the Coulter counter, *Invest. Urol.,* 1981, 18, 401.

17. **Ryall, R.L., Harnett, R.M. and Marshall, V.R.,** The effect of urine, pyrophosphate, citrate, magnesium and glycosaminoglycans on the growth and aggregation of calcium oxalate crystals in vitro, *Clin. Chim. Acta,* 1981, 112, 349.

18. **Ryall, R.L., Ryall, R.G. and Marshall, V.R.,** Interpretation of particle growth and aggregation patterns obtained from the Coulter counter: a simple theoretical model, *Invest. Urol.,* 1981, 18, 396.

19. **Ryall, R.L., Harnett, R.M. and Marshall, V.R.,** The effect of monosodium urate on the capacity of urine, chondroitin sulphate and heparin to inhibit calcium oxalate crystal growth and aggregation, *J. Urol.,* 1986, 135, 174.

20. **Ryall, R.G., Ryall, R.L. and Marshall, V.R.,** A computer model for the determination of extents of growth and aggregation of crystals from changes in their distribution, *J. Cryst. Growth,* 1986, 76, 290.

21. **Hounslow, M.J., Ryall, R.L. and Marshall, V.R.,** At last, a non-iterative program to calculate growth and aggregation rates, in *Urolithiasis,* Walker, V.R., Sutton, A.L., Cameron, E.C.B., Pak, C.Y.C. and Robertson, W.G. Eds, Plenum Press, 1989, 147.

22. **Will, E.J., Bijvoet, O.L.M., Blomen, L.J.M.J. and van der Linden, H.,** Growth kinetics of calcium oxalate monohydrate, *J. Cryst. Growth,* 1983, 64, 297.

23. **Nancollas, G.H. and Gardner, G.L.,** Kinetics of crystal growth of calcium oxalate monohydrate, *J. Cryst. Growth,* 1974, 21, 267.

24. **Meyer, J.L. and Smith, L.H.,** Growth of calcium oxalate crystals. I. A model for urinary stone growth, *Invest. Urol.,* 1975, 13, 31.

25. **Meyer, J.L. and Smith, L.H.,** Growth of calcium oxalate crystals. II. Inhibition by natural urinary crystal growth inhibitors, *Invest. Urol.,* 1975, 13, 36.

26. **Pak, C.Y.C., Ohata, M. and Holt, K,** Effect of diphosphate on crystallization of calcium oxalate in vitro, *Kidney Int.,* 1975, 7, 154.

27. **Gill, W.B., Karesh, J.W., Garsin, L. and Roma, M.J.,** Inhibitory effects of urinary macromolecules on the crystallization of calcium oxalate, *Invest. Urol.,* 1977, 15, 95.

28. **Ito, H. and Coe, F.L.,** Acidic peptide and polyribonucleotide crystal growth inhibitors in human urine, *Am. J. Urol.,* 1977, 233, F455.

29. **Ligabue, A., Fini, M. and Robertson, W.G.,** Influence of urine on "in vitro" crystallization rate of calcium oxalate: determination of inhibitory activity by a [^{14}C]oxalate technique, *Clin. Chim. Acta,* 1979, 98, 39.

30. **Werness, P.G., Bergert, J.H. and Lee, K.E.,** Urinary crystal growth: effect of inhibitor mixtures, *Clin. Sci.,* 1981, 61, 487.

31. **Curreri, P.A. Onoda, G. and Finlayson, B.,** A comparative appraisal of citrate on whewellite seed crystals, *J. Cryst. Growth,* 1981, 53, 209.

32. **Tiselius, H.-G. and Fornander, A.M.**, Evaluation of a routine method for determination of calcium oxalate crystal growth inhibition in diluted urine samples, *Clin. Chem.*, 1981, 27, 565.

33. **Nakagawa, Y., Margolis, H.C., Yokoyama, S., Kézdy, F.J., Kaiser, E.T. and Coe, F.L.**, Purification and characterization of a calcium oxalate monohydrate crystal growth inhibitor from human kidney tissue culture medium, *J. Biol. Chem.*, 1981, 256, 3936.

34. **Fellström, B., Backman, U., Danielson, B.G., Holmgren, K., Ljunghall, S. and Wikström, B.**, Inhibitory activity of human urine on calcium oxalate crystal growth: effects of sodium urate and uric acid, *Clin. Sci.*, 1982, 62, 509.

35. **Nakagawa, Y., Ahmed, M., Hall, S.L., Deganello, S. and Coe, F.L.**, Isolation from human calcium oxalate renal stones of nephrocalcin, a glycoprotein inhibitor of calcium oxalate crystal growth. Evidence that nephrocalcin from patients with calcium oxalate nephrolithiasis is deficient in γ-carboxyglutamic acid, *J. Clin. Invest.*, 1987, 79, 1782.

36. **Kok, D.J., Papapoulos, E., Blomen, L.J.M.J. and Bijvoet, O.L.M.**, Modulation of calcium oxalate monohydrate crystallization kinetics in vitro, *Kidney Int.*, 1988, 34, 346.

37. **Hess, B., Nakagawa, Y. and Coe, F.L.**, Inhibition of calcium oxalate monohydrate crystal aggregation by urine proteins, *Am. J. Physiol.*, 1989, 257, F99.

38. **Tiselius, H.-G., Fornander, A.M. and Nilsson, M.A.**, Effects of urinary macromolecules on the crystallization of calcium oxalate, *Urol. Res.*, 1990, 18, 381.

39. **Tiselius, H.-G., Fornander, A.M. and Nilsson, M.A.**, The effects of citrate and urine on calcium oxalate crystal aggregation, *Urol. Res.*, 1993, 21, 363.

40. **Yamaguchi, S., Yoshioka, T., Utsunomiya, M., Koide, T., Osafune, M., Okuyama, A. and Sonoda, T.**, Heparan sulfate in the stone matrix and its inhibitory effect on calcium oxalate crystallization, *Urol. Res.*, 1993, 21, 187.

41. **Wolf, J.S. and Stoller, M.L.**, Inhibition of calculi fragment growth by metal-bisphosphonate complexes demonstrated with a new assay measuring the surface activity of urolithiasis inhibitors, *J. Urol.*, 1994, 152, 1609.

42. **Hess, B.**, The role of Tamm-Horsfall glycoprotein and nephrocalcin in calcium oxalate monohydrate crystallization processes, *Scann. Microsc.*, 1991, 5, 689.

43. **Hess, B., Nakagawa, Y., Parks, J.H. and Coe, F.L.**, Molecular abnormality of Tamm-Horsfall glycoprotein in calcium oxalate nephrolithiasis, *Am. J. Physiol.*, 1991, 260, F569.

44. **Angell, A.H. & Resnick, M.I.**, Surface interaction between glycosaminoglycans and calcium oxalate, *J. Urol.*, 1989, 141, 1255.

45. **Utsunomiya, M., Koide, T., Yoshioka, T., Yamaguchi, S. and Okuyama, A.**, Influence of ionic strength on crystal adsorption and inhibitory activity of macromolecules, *Br. J. Urol.*, 1993, 71, 516.

46. **Scurr, D.S. and Robertson, W.G.**, Modifiers of calcium oxalate crystallisation found in urine. III. Studies on the role of Tamm-Horsfall mucoprotein and of ionic strength, *J. Urol.*, 1986, 136, 505.

47. **Scurr, D.S. and Robertson, W.G.**, Modifiers of calcium oxalate crystallisation found in urine. II. Studies on their mode of action in an artificial urine, *J. Urol.*, 1986, 136, 128.

48. **Cao, L.C., Boevé, E.R., Schröder, F.H., Robertson, W.G., Ketelaars, G.A.M. and de Bruijn, W.C.**, The effect of two new semi-synthetic glycosaminoglycans (G871, G872) on the zeta potential of calcium oxalate crystals and on growth and agglomeration, *J. Urol.*, 1992, 147, 1643.

49. **Boevé, E.R., Cao, L.C., de Bruijn, W.C., Robertson, W.G., Romijn, J.C. and Schröder, F.H.**, Zeta potential distribution of calcium oxalate crystal and Tamm-Horsfall protein surface analysed with Doppler electrophoretic light scattering, *J. Urol.*, 1994, 152, 531.

50. **Akbarieh, M. & Tawashi, R.**, Surface phase transition of hydrated calcium oxalate crystal in the presence of normal and stone-formers' urine, *Scann. Microsc.*, 1990, 4, 387.

51. **Grases, F., Conte, A. and Gil, J.**, Simple method for the study of heterogeneous nucleation in calcium oxalate urolithiasis, *Br. J. Urol.*, 1988, 61, 468.

52. **Khan, S.R., Shevock, P.N. and Hackett, R.L.**, Membrane-associated crystallisation of calcium oxalate in vitro, *Calcif. Tissue Int.*, 1990, 46, 116.

53. **Campbell, A.A., Fryxell, G.E., Graff, G.L., Rieke, P.C. Tarasevich, B.J.**, The nucleation and growth of calcium oxalate monohydrate on self-assembled monolayers (SAMS), *Scann.*

Microsc., 1993, 7, 423.

54. **Geider, S., Dussol, B., Dupuy, P., Lilova, A.,, Berland, Y., Dagorn, J.-C. and Verdier, J.-M.,** Evidence that albumin can induce the nucleation of calcium oxalate monohydrate (COM) crystals, in *Urolithiasis: consensus and controversies,* Rao, P.N., Kavanagh, J.P. and Tiselius, H.-G., Eds., Lithotriptor Unit, South Manchester University Hospital, Manchester, U.K., 1995, 278.

55. **Söhnel, O. and Garside, J.,** *Precipitation, basic principles and industrial applications,* Butterworth-Heinemann Ltd, Oxford, 1992

56. **Mullin, J.W.,** *Crystallization,* Butterworth-Heinemann Ltd, Oxford, 1993

57. **Grover, P.K., Ryall, R.L. and Marshall, V.R.,** Effect of urate on calcium oxalate crystallization in human urine: evidence for a promotory role of hyperuricosuria in urolithiasis, *Clin. Sci.,* 1990, 79, 9.

58. **Grover, P.K., Ryall, R.L. and Marshall, V.R.,** Dissolved urate promotes calcium oxalate crystallization: epitaxy is not the cause, *Clin. Sci.,* 1993, 85, 303.

59. **Hennequin, C., Lalanne, V., Daudon, M., Lacour, B. and Drueke, T.,** A new approach to studying inhibitors of calcium oxalate crystal growth, *Urol. Res.,* 1993, 21, 101.

60. **Briellmann, T., Seiler, H., Hering, F. and Rutihauser, G.,** The oxalate-tolerance value: a whole urine method to discriminate between calcium oxalate stone formers and others, *Urol. Res.,* 1985, 13, 291.

61. **Schnierle, P., Sialm, F., Seiler, H.G., Hering, F. and Rutihauser, G.,** Investigations on macromolecular precipitation inhibitors of calcium oxalate, *Urol. Res.,* 1992, 20, 7.

62. **Ryall, R.L., Hibberd, C.M. and Marshall, V.R.,** A method for studying inhibitory activity in whole urine, *Urol. Res.,* 1985, 13, 285.

63. **Ryall, R.L., Hibberd, C.M., Mazzachi, B.C. and Marshall, V.R.,** Inhibitory activity of whole urine: a comparison of urines from stone formers and healthy subjects, *Clin. Chim. Acta,* 1986, 154, 59.

64. **Tiselius, H.-G.,** Measurement of the risk of calcium oxalate crystallization in urine, *Urol. Res.,* 1985, 13, 297.

65. **Kohri, K., Kodama, M., Ishikawa, Y., Katayama, Y., Kataoka, K., Iguchi, M., Yachiku, S. and Kurita, T.,** Simple tests to determine urinary risk factors and calcium oxalate crystallization in the outpatient clinic, *J. Urol.,* 1991, 146, 108.

66. **Rodgers, A.L., Ball, D. and Harper, W.,** Urinary macromolecules and promoters of calcium oxalate nucleation in human urine: turbidimetric studies, *Clin. Chim. Acta,* 1993, 220, 125.

67. **Rose, M.B.,** Renal stone formation. The inhibitory effect of urine on calcium oxalate precipitation, *Invest. Urol.,* 1975, 12, 428.

68. **Sarig, S., Garti, N., Azoury, R., Wax, Y. and Perlberg, S.,** A method for discrimination between calcium oxalate kidney stone formers and normals, *J. Urol.,* 1982, 128, 645.

69. **Tiselius, H.-G., Fornander, A.M. and Nilsson, M.A.,** Inhibition of calcium oxalate crystallization in urine, *Urol. Res.,* 1987, 15, 83.

70. **Bek-Jensen, H. & Tiselius, H.-G.,** Inhibition of calcium oxalate crystallization by urinary macromolecules, *Urol. Res.,* 1991, 19, 165.

71. **Atmani, F. Lacour, B., Strecker, G., Parvy, P., Drüeke, T. & Daudon, M.,** Molecular characteristics or uronic-acid-rich-protein, a strong inhibitor of calcium oxalate crystallisation in vitro, *Biochem. Biophys. Res. Commun.,* 1993, 191, 1158.

72. **Koide, T., Takemoto, M., Itatani, H., Takaha, M. and Sonoda, T.,** Urinary macromolecular substances as natural inhibitors of calcium oxalate crystal aggregation, *Invest. Urol.,* 1981, 18, 382.

73. **Edyvane, K.A., Hibberd, C.M., Harnett, R.M., Marshall, V.R. and Ryall, R.L.,** Macromolelcules inhibit calcium oxalate crystal growth and aggregation in whole human urine, *Clin. Chim. Acta,* 1987, 167, 329.

74. **Skrtic, D., Filipovic-Vincekovic, N. and Füredi-Milhofer, H.,** Crystallisation of calcium oxalate in the presence of dodecylammonium chloride, *J. Cryst. Growth,* 1991, 114, 118.

75. **Skrtic, D., Füredi-Milhofer, H. and Markovic, M.,** Precipitation of calcium oxalates from high ionic strength solutions. V. The influence of precipitation conditions and some additives on the nucleating phase, *J. Cryst. Growth,* 1987, 80, 113.

76. **Skrtic, D., Filipovic-Vincekovic, N. and Babic-Ivancic, V.,** The effect of dodecylammonium chloride on crystal growth of calcium oxalate, *J. Cryst. Growth,* 1992, 121, 197.

77. **Grover, P.K., Marshall, V.R. and Ryall, R.L.,** Tamm-Horsfall mucoprotein reduces promotion of calcium oxalate crystal aggregation induced by urate in human urine in vitro, *Clin. Sci.,* 1994, 87, 137.

78. **Grases, F., Genestar, C., March, P. and Conte, A.,** Variations in the activity of urinary inhibitors in calcium oxalate urolithiasis, *Br. J. Urol.,* 1988, 62, 515.

79. **Grases, F. and Costa-Bauza, A.,** Study of factors affecting calcium oxalate crystalline aggregation, *Br. J. Urol.,* 1990, 66, 240.

80. **Sutor, D.J., Percival, J.M. and Doonan, S.,** Urinary inhibitors of the formation of calcium oxalate, *Br. J. Urol.,* 1979, 51, 253.

81. **Ebisuno, S., Kohjimoto, Y. Yoshida, T. and Ohkawa, T.,** Effects of urinary macromolelcules on aggregation of calcium oxalate in recurrent calcium stone formers and healthy, *Urol. Res.,* 1993, 21, 265.

82. **Füredi-Milhofer, H., Kiss, K., Kahana, F. and Sarig, S.,** New method for discriminating between calcium stone formers and healthy individuals, *Br. J. Urol.,* 1993, 71, 137.

83. **Grases, F., Masárová, L., Söhnel, O. and Costa-Bauza, A.,** Agglomeration of calcium oxalate monohydrate in synthetic urine, *Br. J. Urol.,* 1992, 70, 240.

84. **Grases, F., Millan, A. and Söhnel, O.,** Role of agglomeration in calcium oxalate monohydrate urolith development, *Nephron,* 1992, 61, 145.

85. **Lachance, H. and Tawashi, R.,** The effect of controlled diffusion of ions on the formation of hydrated calcium oxalate crystals, *Scann. Microsc.,* 1987, 1, 563.

86. **Dent, C.E. & Sutor, D.J.,** Presence or absence of inhibitor of calcium-oxalate crystal growth in urine of normals and of stone formers, *Lancet,* 1971, 2, 775.

87. **Welshman, S.G. and McGeown, M.G.,** A quantitative investigation of the effects on the growth of calcium oxalate crystals on potential inhibitors, *Br. J. Urol.,* 1972, 44, 677.

88. **Sallis, J.D. and Lumley, M.F.,** On the possible role of glycosaminoglycans as natural inhibitors of calcium oxalate stones, *Invest. Urol.,* 1979, 16, 296.

89. **Akbarieh, M. & Tawashi, R.,** Calcium oxalate crystal growth in the presence of mucin, *Scann. Microsc.,* 1991, 5, 1019.

90. **Deganello, S.,** The interactions between nephrocalcin and Tamm-Horsfall proteins with calcium oxalate dihydrate, *Scann. Microsc.,* 1993, 7, 1111.

91. **Deganello, S.,** Interaction between nephrocalcin and calcium oxalate monohydrate: a structural study, *Calcif. Tissue Int.,* 1991, 48, 421.

92. **Bowyer, R.C., Brockis, J.G. and McCulloch, R.K.,** Glycosaminoglycans as inhibitors of calcium oxalate crystal growth and aggregation, *Clin. Chim. Acta,* 1979, 95, 23.

93. **Roehrborn, C.G., Schneider, H.-J. and Rugendorff, E.W.,** Determination of stone-forming risk by measuring crystal formation in whole urine with gel model, *Invest. Urol.,* 1986, 27, 531.

94. **Achilles, W,** Microphotometric quantification of crystal growth in gels for the study of calcium oxalate urolithiasis, *Scann. Microsc.,* 1991, 5, 1001.

95. **Achilles, W., Dekanic, D., Burk, M., Schalk,, Ch., Tucak, A. & Karner, I.,** Crystal growth of calcium oxalate in urine of stone formers and normal controls, *Urol. Res.,* 1991, 19, 159.

96. **Achilles, W., Lescher, C., Burk, M. and Füredi-Milhofer, H.,** Microdetermination of crystal growth rates of calcium oxalate in gel at inverse distribution of components, *in Urolithiasis 2,* Ryall, R., Bais, R., Marshall, V.R., Rofe, A.M., Smith, L.H. and Walker, V.R., Eds., Plenum Press, New York and London, 1994, 231.

97. **Achilles, W., Koethe, R., Jöckel, U., Schalk, C. and Riedmiller, H.,** A new continuous flow microsystem of crystallization in gels as a model of urinary stone formation, in *Urolithiasis 2,* Ryall, R., Bais, R., Marshall, V.R., Rofe, A.M., Smith, L.H. and Walker, V.R., Eds., Plenum Press, New York and London, 1994, 159.

98. **Achilles, W.,** Gel crystallization methods, in *Urolithiasis: consensus and controversies,* Rao, P.N., Kavanagh, J.P. and Tiselius, H.-G., Eds., Lithotriptor Unit, South Manchester University Hospital, Manchester, U.K., 1995, 75.

99. **Deganello, S. & Di Franco, L.,** Crystal growth of calcium oxalate monohydrate and calcium carbonate from dilute solutions, *Scann. Microsc.,* 1990, 4, 171.

100. **Azoury, R., Garside, J. and Robertson, W.G.,** Habit modifiers of calcium oxalate crystals precipitated in a reverse osmosis system, *J. Cryst. Growth,* 1986, 76, 259.

101. **Azoury, R., Garside, J. and Robertson, W.G.,** Calcium oxalate precipitation in a flow system: an attempt to stimulate the early stages of stone formation in renal tubules, *J. Urol.,* 1986, 136, 150.

102. **Hallson, P.C. and Rose, G.A.,** A new urinary test for stone "activity", *Br. J. Urol.,* 1978, 50, 442.

103. **Hallson, P.C. and Rose, G.A.,** Uromucoids and urinary stone formation, *Lancet,* 1979, 1, 1000.

104. **Rose, G.A. and Sulaiman, S.,** Tamm-Horsfall mucoproteins promote calcium oxalate crystal formation in urine: quantitative studies, *J. Urol.,* 1982, 127, 177.

105. **Rodgers, A.L. and Wandt, M.A.E.,** Influence of ageing, pH and various additives on crystal formation in artificial urine, *Scann. Microsc.,* 1991, 5, 697.

106. **Gohel, M.D., Shum, D.K.Y. and Li, M.K.,** The dual effect of urinary macromolecules on the crystallization of calcium oxalate endogenous in urine, *Urol. Res.,* 1992, 20, 13.

107. **Shum, D.K.Y. and Gohel, M.D.I.,** Separate effects of urinary chondroitin sulphate and heparan sulphate on the crystallization of urinary calcium oxalate: differences between stone formers and normal control subjects, *Clin. Sci.,* 1993, 85, 33.

108. **Khan, S.R. and Opalko, F.J.,** Constant composition crystallization system, in *Urolithiasis: consensus and controversies,* Rao, P.N., Kavanagh, J.P. and Tiselius, H.-G., Eds., Lithotriptor Unit, South Manchester University Hospital, Manchester, U.K., 1995, 83.

109. **Sheehan, M.E. and Nancollas, G.H.,** Calcium oxalate crystal growth. A new constant composition method for modelling urinary stone formation, *Invest. Urol.,* 1980, 17, 446.

110. **Lanzalaco, A.C., Sheehan, M.E., White, D.J. and Nancollas, G.H.,** The mineralization inhibitory potential of urines: a constant composition approach, *J. Urol.,* 1982, 128, 845.

111. **Sheehan, M.E. and Nancollas, G.H.,** The kinetics of crystallization of calcium oxalate trihydrate, *J. Urol.,* 1984, 132, 158.

112. **Lanzalaco, A.C., Singh, R.P., Smesko, S.A., Nancollas, G.H., Sufrin, G., Binette, M. and Binette, J.P.,** The influence of urinary macromolecules on calcium oxalate monohydrate crystal growth, *J. Urol.,* 1988, 139, 190.

113. **White, D.J. and Nancollas, G.H.,** Triamterene and renal stone formation: the influence of triamterene and triamterene stones on calcium oxalate crystallization, *Calcif. Tissue Int.,* 1987, 40, 79.

114. **Campbell, A.A., Ebrahimpour, A., Perez, L., Smesko, S.A. and Nancollas, G.H.,** The dual role of polyelectrolytes and proteins as mineralization promoters and inhibitors of calcium oxalate monohydrate, *Calcif. Tissue Int.,* 1989, 45, 122.

115. **Kahn, S.R., Whalen, P.O. and Glenton, P.A.,** Heterogeneous nucleation of calcium oxalate crystals in the presence of membrane vesicles, *J. Cryst. Growth,* 1993, 134, 211.

116. **Asplin, J., DeGanello, S., Nakagawa, Y.N. & Coe, F.L.,** Evidence that nephrocalcin and urine inhibit nucleation of calcium oxalate monohydrate crystals, *Am. J. Physiol.,* 1991, 261, F824.

117. **Finlayson, B.,** The concept of a continuous crystallizer. Its theory and application to in vivo and in vitro urinary tract models, *Invest. Urol.,* 1972, 9, 258.

118. **Rodgers, A.L. and Garside, J.,** The nucleation and growth kinetics of calcium oxalate in the presence of some synthetic urine constituents, *Invest. Urol.,* 1981, 18, 484.

119. **Robertson, W.G. and Scurr D.S.,** Modifiers of calcium oxalate crystallization found in urine. I. Studies with a continuous crystallizer using an artificial urine, *J. Urol.,* 1986, 135, 1322.

120. **Springmann, K.E., Drach, G.W., Gottung, B. and Randolph, A.D.,** Effects of human urine on aggregation of calcium oxalate crystals, *J. Urol.,* 1986, 135, 69.

121. **Nishio, S. and Kavanagh, J.P.,** A small-scale continuous mixed suspension mixed product removal crystallizer, *Chem. Eng. Sci.,* 1990, 46, 709.

122. **Nishio, S., Kavanagh, J.P., Faragher, E.B., Garside, J. and Blacklock, N.J.,** Calcium oxalate crystallization kinetics and the effects of calcium and gamma-carboxyglutamic acid,

Br. J. Urol., 1990, 66, 351.

123. **Kavanagh, J.P., Nishio, S., Garside, J. and Blacklock, N.J.**, Crystallization kinetics of calcium oxalate in fresh, minimally diluted urine: comparison of recurrent stone formers and healthy controls in a continuous mixed suspension mixed product removal crystallizer, *J. Urol.*, 1993, 149, 614.

124. **Bretherton, T.A. and Rodgers, A.L.**, Design, development and testing of a non-invasive continuous crystallizer system with an on-line Malvern Particle Sizer: preliminary results, in *Urolithiasis: consensus and controversies*, Rao, P.N., Kavanagh, J.P. and Tiselius, H.-G., Eds., Lithotriptor Unit, South Manchester University Hospital, Manchester, U.K., 1995, 376.

125. **Suzuki, K., Tsugawa, R. and Ryall, R.L.**, Inhibition by sodium-potassium citrate (CG-120) of calcium oxalate crystal growth on to kidney stone fragments obtained from extracorporeal shock wave lithotripsy, *Br. J. Urol.*, 1991, 68, 132.

126. **Kavanagh, J.P.**, The kidney as a mixed suspension mixed product removal system (MSMPR) crystallization chamber, in *Urolithiasis: consensus and controversies*, Rao, P.N., Kavanagh, J.P. and Tiselius, H.-G., Eds., Lithotriptor Unit, South Manchester University Hospital, Manchester, U.K., 1995, 89.

127. **Bramley, A.S., Hounslow, M.J., Ryall, R.L. and Marshall, V.R.**, Calcium oxalate crystallization in long, thin tubes, in *Urolithiasis 2*, Ryall, R., Bais, R., Marshall, V.R., Rofe, A.M., Smith, L.H. and Walker, V.R., Eds., Plenum Press, New York and London, 1994, 167.

128. **Hounslow, M.J., Bramley, A.S. and Ryall, R.L.**, Crystallization in tubes, in *Urolithiasis: consensus and controversies*, Rao, P.N., Kavanagh, J.P. and Tiselius, H.-G., Eds., Lithotriptor Unit, South Manchester University Hospital, Manchester, U.K., 1995, 29.

129. **Söhnel, O., Grases, F. and March, J.G.**, Experimental technique simulating oxalocalcic renal stone generation, *Urol. Res.*, 1993, 21, 95.

Chapter 2

INHIBITORS OF CALCIUM OXALATE CRYSTALLIZATION

Dirk J. Kok

Department of Endocrinology, Academic Hospital Leiden, The
Netherlands

I. REGULATION OF CRYSTALLIZATION

A biomineral often shows distinct features with respect to crystal-size, -habit and -phase. In plants calcium oxalate is seen as whewellite, C.O.M., and wheddellite, C.O.D., with specific morphologies like bundles of crystal called raphides and druses, agglomerates of crystals with a roughly spherical shape[1], see chapters 8 and 10. In urine mainly C.O.D. is formed, with prominent twinning. C.O.M. is seen as biconcave ovals, dumbbells and rosettes. Also, crystalluria shows a specific bimodal size-distribution[2-5]. Inside urinary stones, however, C.O.M. is more prevalent than C.O.D.[6] In pure systems, at comparable supersaturations and calcium to oxalate ratios calcium oxalate precipitates in the typical monoclinic C.O.M. form. The question arises to what extent the biological environment dictates those features.

A biological system can provide environments which (dis-) favor formation of specific crystal-phases, using specialized vesicles[7] or by regulating the concentrations of crystal components in a biological fluid. It may also provide compounds which promote or inhibit specific parts of crystallization, so-called effectors. Some may direct crystallization by providing an organic template resembling the crystal structure[8], others may specifically inhibit parts of the crystallization process. In this chapter the mechanisms by which compounds can affect calcium oxalate crystallization will be discussed, using mainly low MW compounds as examples. A review on macromolecular compounds is given in chapter two.

A. EFFECTORS OF CRYSTALLIZATION
1. Definition of an Effector

The overall crystallization process can be divided into a thermodynamic part, the supersaturation providing the driving force and a kinetic part, the rates of nucleation, crystal growth, crystal agglomeration and phase-transition. These processes occur simultaneously or overlapping in

0-8493-7673-4/95/$0.00+$.50
© 1995 by CRC Press

time but, choosing the right conditions, can be measured separately. Usually compounds effecting one or more of these processes are referred to as inhibitors. However, this is misleading, since these compounds may inhibit or promote specific parts of crystallization. We will here refer to compounds which effect (parts of) the crystallization process as effectors.

A general feature of effectors is the presence of one or more groups with an affinity for calcium or oxalate ions, either free in solution or as part of a crystal surface. Depending on the affinity constant, effectors can decrease the driving force by complexing calcium or oxalate in solution. To interfere with nucleation, crystal growth and crystal agglomeration the effector can use its affinity to bind to the nucleus or crystal surface. In this respect, additionally the three-dimensional structure of the effector is important. Rest groups may cause steric hindrance decreasing the affinity for a crystal surface. When more then one crystal-binding group is present in a macromolecular structure, the spacial arrangement of these groups and the rigidity of the backbone also plays a role[9-11]. We will discuss these features following the theoretical division between supersaturation, nucleation, crystal growth, crystal agglomeration and phase transition.

2.   Effectors discussed

Low MW compounds present in the urine, phosphate, citrate, isocitrate, magnesium, pyrophosphate. High MW compounds, pentosanpolysulfate (Sigma, St.Louis Mo. U.S.A. MW 5000 D), chondroitinsulfate (Sigma, St. Louis Mo. U.S.A. MW 50000 D), heparin (Diosynth, Oss The Netherlands MW 16000 D). The bisphosphonates EHDP (1-hydroxyethylidene-1,1-bisphosphonate), APD (3-amino-1hydroxypropilidene-1,1-bisphosphonate), DPB (3-dimethyl-amino -1hydroxypropilidene-1,1-bisphosphonate), Cl_2MDP (dichloro-methylene-bisphosphonate), DMA_4 (4-dimethylamino-1-hydroxy butylidene-1,1-bis-phosphonate) and HYPD (1-hydroxypentan-1,1-bisphosphonate). An algal polysaccharide, AP, isolated from the calcifying algae Emiliania Huxleyi[12]. This AP is involved in the formation of $CaCO_3$ structures in this algae[7,13]. The compounds were tested in concentration ranges including the minimum and maximum effects on growth and agglomeration and (for the urinary constituents) normal urine concentrations.

B.   METHODS USED
1.   Determination of the Solubility (Equilibrium) Product

Effects on the solubility were either calculated with the speciation software Equil[14] or measured directly[15-17]. Equil calculates C.O.M.-solubility for samples of known composition, using affinity constants for all complexes which may form and correcting for ionic strength effects on the activity. For compounds with unknown affinity constants, like the bisphosphonates direct measurement of the solubility is wanted. For this, the sample is added to solutions with increasing calcium oxalate concentration products and

C.O.M. crystals are added. After a fixed incubation time at 37 °C the uptake of ^{45}Ca tracer into the crystal-mass is measured. Per definition crystal growth occurs only at concentration products above the solubility or equilibrium concentration product. Thus, the solubility is measured as the lowest concentration product where uptake due to crystal growth occurs.

2.   Measurement of Effects on Nucleation

Effects on nucleation were measured as changes in the nucleation lag-time[18,20], τ, and changes in the apparent interfacial free energy[19,20], σ. Nucleation lag-times (or induction times) were measured by rapidly mixing two buffers containing calcium and oxalate and the effector of interest in a 1-cm pathlength polystyrene cuvette. The absorbance at 530 nm was measured for 10 min. Formation of solid particles will cause scattering of the light bundle which is seen as an apparent absorption. By extrapolating the straight portion of the OD530-time plot to the starting absorbance, τ is found[20]. The starting supersaturation is kept constant by correcting for effects on the solubility. Since the earliest nuclei are too small to cause appreciable scattering, the measured lag-time reflects both the time needed for nucleation to start and for particles to grow into the detection limit. This error increases when an effector also strongly inhibits crystal growth. Therefore as a second parameter the changes in σ were calculated. For this, τ is measured in a range of relative supersaturations, RS. From the slope of the plot of $\ln(1/\tau)$ against $[\ln(RS)]^{-2}$, σ can be calculated[19,20].

3.   Measurement of Crystal Growth and Crystal Agglomeration

Effects on C.O.M. crystal growth and agglomeration were measured in a seeded crystal growth system[15-17], as two separate system independent parameters, the growth constant, k_a and the agglomeration parameter [tm]. Changes in solubility, ionic strength, calcium to oxalate ratio, seed crystal concentration, seed crystal characteristics and effects of agglomeration on the measured growth rate are all accounted for or controlled. For easier interpretation, the parameters are related to control experiments containing only calcium, oxalate, buffer and saline and expressed as %G.I., percent growth inhibition and $[tm]/[tm]_c$. For the latter, a value below 1 denotes stimulation, a value above 1 inhibition of agglomeration.

4.   Adsorption Characteristics

Adsorption characteristics can be assessed from so-called Langmuir isotherms. In our experiments, based on the assumption that growth inhibition is related to the amount of surface coverage, adsorption is described by the plot of $\ln((k_{a,c}-k_{a,inh.})/k_{a,inh.})$ against $\ln(Tinh.)$[21,22]. The plot will be linear in the case of a monolayer type of adsorption. Changes in the slope indicate changes in adsorption behaviour.

Table 1
Effects on the Nucleation.

RS	Control	Citrate	DPB	EHDP	DMA4	Cl2MDP
			τ, sec			
22	231±29	350±80	241±57	230±10	156±14	180±0
24	168±42	290±47	196±39	172±34	127±36	157±37
28	120±28	170±61	140±30	128±7	103±8	102±8
33	79±18	140±33	110±18	96±6	95±5	97±1
37	54±19	90±17	92±6	84±14	88±3	92±3
σ, erg.cm^{-2}	26.7	26.8	23.0	24.0	19.6	21.4

Citrate was tested at 3.5 mM, pH 6.50, data from[20], bisphosphonates were tested at 20 μM, pH 6.00.

5. Crystal-Face and Crystal-Phase-Specificity

Crystal face- and phase-specificity of effectors was assessed by allowing the solutions in the lag-time nucleation experiments to grow for ten minutes. They were then filtered over 0.05 or 0.22 μm nucleopore filters. The filters were placed on aluminum stubs, gold coated and analyzed by scanning electron microscopy, SEM.

C. EFFECTS ON SEPARATE CRYSTALLIZATION PROCESSES
1. Effects on the Solubility.

In urine, the C.O.M. solubility is determined for approximately 14% by the thermodynamic solubility, 20% by the ionic strength and 66% by complex-formations in the solutions[22]. Ionic strength effects are well described by the Debye-Hückel equations or modifications thereof [22-27]. The binding of calcium or oxalate in solution reflects the affinity of effectors for those ions and their concentrations. In biological fluids C.O.M. solubility is therefore determined mainly by low MW compounds like citrate, magnesium and phosphate, which combine a high affinity with concentrations in the millimolar concentration range. High MW compounds may have a high affinity for calcium but are usually present in micromolar concentrations and thus do not contribute significantly to the solubility.

2. Effects on Nucleation

Table 1 shows the values for τ at varying RS in the presence of citrate[20] and several bisphosphonates. At high RS values, all these compounds delay the appearance of particles into the detection limit. However, this is probably not caused by inhibition of the nucleation process. From the calculated values for the apparent interfacial free energy, σ, it appears that citrate does not influence the stability of the nuclei. The bisphosphonates even appear to stabilize the nuclei, thus enhancing nucleation. This is seen at low RS values as decreases in τ for DMA$_4$ and CL$_2$MDP.

Table 2
Concentration Needed for 50% Crystal Growth Inhibition.

Compound	Conc., μM	Compound	Conc., μM
AP	.012	HYPD	10
Heparin	.03	APD	10
Chondroitinsulfate	1	Cl$_2$MDP	20
Pentosanpolysulfate	2	Pyrophosphate	20
DMA$_4$	5	(Iso-)citrate	400
DPB	5	Magnesium	7000
EHDP	5	Phosphate	20000

Data obtained at pH 6.00[22,27].

3.   Effects on Crystal Growth and Crystal Agglomeration

a   Crystal Growth

Table 2 summarizes the concentrations needed to obtain 50% G.I.. All of the tested substances, except uric acid, inhibited crystal growth. The high MW compounds and the bisphosphonates are effective growth inhibitors at the micromolar level. The low MW compounds (iso-)citrate, magnesium and phosphate are effective in the millimolar range. When for AP the inhibitory power was calculated per uronic acid group, responsible for growth inhibition, it was increased almost 1000-fold as compared to e.g. citrate[21,22]. This could not be ascribed to differences in affinity constants. The calcium-uronic-acid complex in the algal polysaccharide has a stability constant of $9 \cdot 10^2$ M^{-1}, the values for calcium hydrogen citrate[28], $1.23 \cdot 10^3$ M^{-1} and tri-calcium-dicitrate[29], $7.09 \cdot 10^4$ M^{-1}, are even higher. The arrangement in a macromolecular structure appears to enhance the inhibitory power. The effectiveness of the bisphosphonates seems contradictory. However, bisphosphonates tend to form very large polynuclear complexes with calcium[22,23,30] effectively they should be considered as macromolecular structures kept together by ionic interactions.

b.   Crystal Agglomeration

While effects on crystal growth are unidirectional, inhibition, effects on crystal agglomeration vary[21-23,31], Table 3. Some compounds inhibit both growth and agglomeration, (iso-)citrate, pyrophosphate, nephrocalcin and to a lesser extent magnesium and chondroitinsulfate. Some at the same time inhibit crystal growth and stimulate crystal agglomeration, AP, EHDP, DMA$_4$. Some can both inhibit and stimulate crystal agglomeration. APD, between 5-10 μM, DPB, between 15-40 μM and THP inhibit agglomeration when present as a free molecule. APD and DPB stimulate agglomeration when their concentration is increased and they form large polynuclear complexes with calcium[22,23]. THP stimulates agglomeration when it self-aggregates at low pH, high calcium concentration or high ionic strength[31]. The dissociation between the two processes growth and agglomeration shows that they probably relate to different parts of the crystal surface.

Table 3
Effects on Crystal Agglomeration

Compound	Inhibition	No Effect	Stimulation
(iso-)citrate[27]	+		
pyrophosphate[27]	+		
magnesium[27]	±		
heparin[27]		+	
pentosanpolysulfate[27]		+	
chondroitinsulfate[27]	±		
AP[21]			+
APD[22]	+		+
uric acid[27]		+	
Cl₂MDP[22]		+	
DMA₄[22]			+
DPB[22]	+		+
EHDP[22]			+
HYPD[22]		+	
nephrocalcin[31]			+
THP[31]	+		+

Data obtained at pH 6.00.

c.   *System-Dependence of Effector Action*

The action of an effector depends on its structure and on the systemic conditions. For instance, bisphosphonates work as polynuclear-complexes with calcium. The complex-formation depends on the bisphosphonate structure, the calcium-concentration and the dissociation state of the bis-phosphonate. The pH-dependence is clearly seen in Table 4. The calcium concentration dependence is seen when the calcium to oxalate ratio is varied at pH 6.00 with a constant EHDP concentration of 10 μM[23]. Growth inhibition increases from 45% at Ca/ox of 0.1 to 95% at a Ca/ox ratio of 4 or higher. The stimulation of crystal agglomeration also increases, reaching a plateau value at a Ca/ox ratio of 2. A similar pattern was found for citrate[22]. At pH 6.00 and 0.35 mM citrate the %G.I. increased from 5% to 55% and [tm]/[tm]$_c$ increased from 1.6 to 6 with increasing the Ca/ox ratio from 0.1 to 10. The effects of citrate on crystal growth[32-34] and agglomeration[32] are also enhanced when it is complexed with trace metals[33,34] or magnesium[32].

The action of an effector can also depend on the presence of other effectors competing for the same crystal surface. This is clear when a combination of different citrate concentrations and a fixed heparin concentration was tested[22,32]. At the heparin concentration used, 0.05 μM, 67 % G.I. is found. Addition of citrate in amounts up to 1.0 mM does not add to this growth inhibition, while those concentrations of citrate by themselves yield up to 60% G.I. In contrast, heparin at this concentration did not effect agglomeration. Addition of citrate, however, yielded an inhibitory curve similar to that of citrate alone. Thus heparin prevents citrate from binding to the growth sites while leaving other parts of the surface acccessible.

Table 4
pH-Dependence of Bisphosphonate Action

Compound	%G.I.		[tm]/[tm]$_c$	
	pH 5.00	pH 7.00	pH5.00	pH7.00
APD	17	94	1.11	.14
CL2MDP	4	95	.89	1.58
DMA4	30	93	1.21	.34
DPB	61	91	1.21	.38
EHDP	1	96	.60	.47

The bisphosphonates were tested at a concentration where they yield 70% G.I. at pH 6.00, 10 μM for APD, DPB and DMA$_4$, 20 μM for EHDP and 30 μM for Cl$_2$MDP.

A clear system-dependence is also shown by THP. The free THP-molecule inhibits crystal agglomeration. High ionic strength, high calcium-low citrate-concentration and low pH induce THP self-aggregation and reversal of inhibition to stimulation[31,35]. Variation in the THP-structure may also play a role, as THP from stone formers self-aggregates more readily[31].

4.   Type of Adsorption to the Crystal Surface

The Langmuir isotherm, relating $\ln((k_{a,c}-k_{a,inh.})/k_{a,inh.})$ to $\ln$(Tinh.) was linear for pentosanpolysulfate (slope 0.24), chondroitinsulfate (slope 0.6), heparin (slope 0.24) and pyrophosphate (slope 1)[22]. This indicates a monolayer type of adsorption. In view of the low effective concentrations probably involving a simple surface-kink poisoning with a preference for the growth sites. The isotherms for phosphate, citrate and magnesium were not linear, indicating a different adsorption behavior. The isotherms were linear for CL$_2$MDP (slope 1.05) and HYPD (slope 0.70) throughout the concentration range. For APD, DMA$_4$ and DPB the graphs are biphasic with slopes changing from 1 to 0.06 (APD), 1 to 0.49 (DMA$_4$) and 1.1 to 0.27 (DPB). The changes in slopes occur at fairly discrete concentrations, 40 μM for APD, 1 μM for DMA$_4$ and 10 μM for DPB. EHDP shows a linear relationship (slope 0.78) with perhaps a deviation occurring below 2 μM.

5.   Face- and Phase-Specificity

Figure 1 demonstrates how the crystal shape changes dramatically in the presence of increasing EHDP-concentrations. The average length to width ratio of the crystals decreases from 3.4 in the control situation to 1.45 at a 5 μM EHDP concentration. Comparable changes were shown to occur at prolonged incubation times of 29 hrs in _the presence of 5 μM nephrocalcin[36]. The growth perpendicular to the ($\bar{1}$01) face is inhibited, forcing the crystal to grow in that plane. When the EHDP-concentration is increased to 10 μM at pH 7.00 no crystals are found at all after ten minutes. At pH 6.00, increasing the bisphosphonate concentration results in increasing formation of calcium oxalate dihydrate crystals. Precipitation of C.O.D. was also found with nephrocalcin[36], urine and several other additives[37].

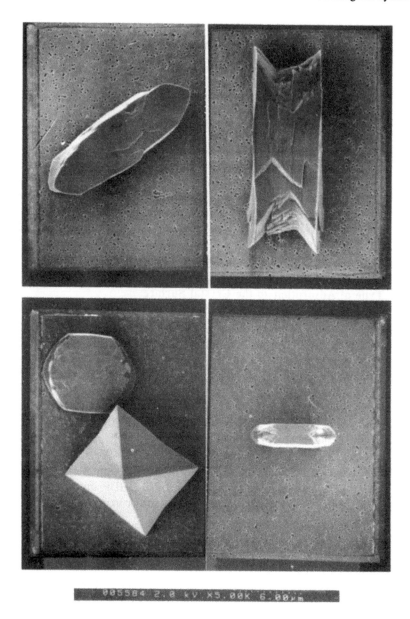

Figure 1. Crystals nucleated in equimolar calcium/oxalate solutions of RS 37, without, top and with 50 μM Cl$_2$MDP. Collected after 10 min, τ was 30 sec. for the control and 130 sec. for the EHDP experiment. Magnification used was 5000.

II. BIO-REGULATION OF CRYSTALLIZATION

Whether a crystallization process is wanted or unwanted, an organism may try to control it. The organism can direct the crystal-morphology by allowing the crystallization process to proceed inside a vesicle. In such a small closed space the kinetics of nucleation are altered[38] . A critical crystal size then exists, where crystals are in stable equilibrium with a super-saturated solution. The rate and direction of crystallization can then be controlled by adding calcium and oxalate, by restricting the space, thus also providing a physical barrier, or by providing a template. The latter is a compound which binds calcium and or oxalate ions in an arrangement resembling the crystal structure, see chapters 1 and 8. Most biominerals are formed in conditions which do not allow homogeneous nucleation. The nucleation therefore must be initiated by epitaxial growth on other mineral phases, e.g. calcium-apatite-like material in urine, or on organic structures.

If the crystallization is unwanted, the organism has several options for controlling the process. The simplest approach is to prevent supersaturation, by removal of calcium and/or oxalate or by addition of compounds which increase the solubility, like citrate and magnesium. In many instances, however, this approach will not work, because the influx of calcium and especially oxalate is just too big. A more subtle way is then possible. The organism can make use of the nucleation lag time by providing compounds which destabilize the nuclei. Manipulation of the lag-time is especially useful in situations where a time-restriction exists, like urinary stone formation. In the kidney the passage time of urine through the narrowest part of the urinary tract, the nephron, is approximately three minutes [39,40]. Increasing the lag-time may allow the crystals formed to remain small enough during that time to be excreted freely into the pelvic area, where size-restrictions are much less strict.

Strict inhibition of nucleation is rarely found. Maybe macromolecules like heparin and hyaluronic acid inhibit C.O.M.-nucleation[41]. Most compounds, however, also enhance the nucleation lag-time by inhibiting the outgrowth of the nuclei. In the example of urinary stone formation this will do also, since next to the time restriction there is a size-restriction. Particles will only remain in the urinary tract either by adherence[39] to the epithelium or by reaching a size too large to be passed freely[40]. When the particles are kept small enough, they can only be retented by adhesion. Interestingly, prolongation of the nucleation process can in this respect also be helpfull. When the nucleation process is prolonged, because crystal growth inhibition causes supersaturation to decrease slower, the total precipitable amount will be distributed over more particles which consequently on average remain smaller[20,41]. This mechanism may actually play a role in urine[41]. Mixed suspension mixed product removal experiments with urines from non-stone forming subjects showed nucleation rates were increased and average crystal sizes decreased compared to the results obtained with stone-former urines[41].

A direct way of regulating particle size is by effecting the rate of crystal growth. As was shown above, inhibition of calcium oxalate crystal growth is not a very specific process. All compounds capable of binding to the crystal surface will inhibit the crystal growth. They need for this either the ability to replace calcium, like magnesium, or the ability to bind to calcium on or in the crystal surface. A compound having neither ability, like uric acid, does not directly influence the crystal growth.

The effectiviness of growth inhibition depends partly on the strength with which the compound can bind calcium. Groups containing oxygen and nitrogen with free electron pairs available are the most effective[42,43] in this respect. Calcium prefers an eight-fold coordination, therefore compounds providing multiple free electron pairs in an arrangement which can fit a crystal-surface calcium ion will show the highest affinity. This was demonstrated using a series of bisphosphonates. The growth inhibitory power was increased 10- to 100-fold by changing the group between the two phosphonate groups from nothing, to hydrogen, to chloride to hydroxide[44].

However, the strength of adsorption is less determining for the growth inhibitory power as is the site of adsorption. The strongest effect can be sorted when an effector binds at the growth site of the crystal. In this respect arrangement of calcium binding groups in a macromolecular structures seems to add to the growth inhibitory potential. This is exemplified by the low effective concentrations of such structures, like the algal polysaccharide, heparin[45] and nephrocalcin[31]. When calculated per calcium binding site the effectiveness of the algal polysaccharide was 1000 times that of citrate[21].

An interesting feature of preferential binding to growth sites is that this may leave other parts of the surface accesible. This allows for manipulation of other surface related processes, like crystal agglomeration, with crystals which have their growth blocked. Such an effect could be seen when heparin was tested together with citrate. While the heparin obviously binds to the crystal surface, growth is inhibited strongly, it does leave room on the surface for citrate to effect crystal agglomeration. An explanation for the site preference may be that a molecule like citrate, containing one or two calcium binding groups will bind equally strong to any surface presenting one or two calcium ions to bind to. A compound like heparin, containing multiple calcium binding groups, binds stronger the more groups are used. It will bind preferentially at sites with a high density of accessible calcium-ions, the crystal growth sites.

With multiple binding sites, the fit between the binding groups in the effector and the calcium ions in the surface is also important. This latter point was clearly demonstrated for a different salt, barium sulfate[46]. When the binding groups of a bisphosphonate were placed at increasing distances along a carbon-hydrate back bone, a maximum effective separation distance of 5.7 Å was found, closely matching the distance between sulfates in the (011) barium sulfate crystal face to which they specifically adsorb, 5.6 Å.

A restraint on the adsorption specific for macromolecular structures will be caused by the rigidity of the backbone. The more rigid the backbone is, the more the polyanions will form loops or bind head-on[9-11,47]. Also a crowding-out effect may occur at increasing concentrations of the macromolecule. The algal polysaccharide was shown to cover more crystal-surface per molecule at the lowest concentrations[21]. At increasing concentrations the surface coverage per AP molecule decreases. At the same time crystal-agglomeration becomes stimulated. These data suggest that the AP stimulates agglomeration through viscous binding. It binds head-on to the crystal surface of different crystals, actively binding these crystals together. Its peculiar rod-like shape[48] may play an additional role. The binding is strong enough to overcome the dispersive force of the increased negative zetapotential upon binding of the AP. A similar effect is seen with the bisphosphonates. Binding of the bisphosphonates cause increasingly negative zetapotentials, but at the same time crystal agglomeration is stimulated. This is also reflected in the Langmuir adsorption isotherms. Changes in slopes, indicating less surface coverage per bisphosphonate molecule adsorbed, coincide with the start of agglomeration stimulation[44]. An explanation is that large polynuclear calcium bisphosphonate-complexes are being formed in solution which cause the viscous binding between crystals. When steric hindrance by large rest-groups on the bisphosphonate molecule is introduced, this abolishes the ability to stimulate crystal agglomeration[44].

It is thus clear that the crystal surface does not act as a uniform entity in the crystallization process. The processes of crystal growth and crystal agglomeration clearly can be effected seperately and even in opposing directions. Compounds which inhibit crystal growth can at the same time inhibit, have no effect or even stimulate crystal agglomeration. The combination experiments of citrate and heparin showed that effectors may bind preferentially to specific sites of the crystal surface. This site-specificity is exemplified in the experiments where calcium-oxalate is nucleated in the presence of various effectors. Citrate at 3.5 mM, nephrocalcin and several bisphosphonates at 5 μM cause dramatic changes in the crystal shape. The crystals become wider and flatter. When the concentrations are increased further, increasing amounts of C.O.D. are formed. With the bisphosphonates, at pH 6.00, eventually 100% C.O.D. is formed. _ This may be explained by preferential binding of the compounds to the   101 face, thereby preventing growth perpendicular to that plane. As their concentration increases further, they will also bind to the other faces, where calcium is less accessible to them. Eventually all faces will be blocked, calcium oxalate cannot grow in its monohydrate form anymore and C.O.D. starts to appear. Additionally the effectors may adsorb better to C.O.M. then to C.O.D., as was shown for polyphosphate[49].

III. SUMMARY

Effectors can determine the outcome of a calcium oxalate crystallization process in various ways. As basic feature they need a similarity to calcium or the presence of anionic groups. The latter need to contain oxygen or nitrogen providing free electron pairs in close proximity to eachother. For regulating crystal growth it is most effective to arrange the anionic groups in a macromolecular structure which can bind to several calcium ions in the crystal surface and thus use the increased density of available calcium ions at growth sites. The organism can also effect only specific parts of the crystallization process, thereby steering the outcome to a specific crystal-phase and morphology. Particle size can be controlled effectively by effecting the crystal agglomeration process even in directions opposite to accompanying effects on crystal growth.

Acknowledgements.

The author is recipient of a research fellowship from the Royal Dutch Academy of Sciences. Crystal growth experiments were also performed by I. Que.

REFERENCES

1   Arnott, H. J., Three systems of biomineralization in plants, *Biological Mineralization and Demineralization, Dahlem Konferenzen*, Nancollas, G.H. Ed., Springer-Verlag Berlin, 1982, chap. 10.

2   Werness, P.G., Bergert, J.H., Smith, L.H., Crystalluria, *J. Cryst. Growth*, 53, 166, 1981.

3   Robertson, W.G., Peacock, M., Nordin, B.E.C., Calcium crystalluria in recurrent renal-stone formers, *The Lancet*, ii, 21, 1969.

4   Elliot, J.S., Rabinowitz, I.N, Silvert, M., Calcium oxalate crystalluria, *J. Urol.*, 116, 773, 1976.

5   Elliot, J. S., Rabinowitz, I.N., Calcium oxalate crystalluria: crystal size in urine, *J. Urol.*, 123, 324, 1980.

6   Elliot, J.S., Structure and composition of urinary calculi, *J. Urol*, 109, 82, 1973.

7   Westbroek, P., de Jong, E.W., vd Wal, P., Borman, A.H., de Vrind, J.P.M., Kok, D.J., de Bruyn, W.C., Parker, S.B., Mechanism of calcification in the marine alga Emiliania Huxleyi, *Phil. Trans. Roy. Soc. London. B*, 304, 435, 1984.

8   Drach, G.W., Randolph, A.D., Miller, J.D., Inhibition of calcium oxalate dihydrate crystallization by chemical modifiers: 1 Pyrophosphate and methylene blue.,*J. Urol.*, 119, 99, 1978.

9   Koopal, L.K., Ralston, J., Chain length effects in the adsorption of surfactants at aqueous interfaces: comparison of existing adsorption models with a new model, *J. Coll. Int. Sci.*, 112 (2), 362 1986.

10 vd Schee, H.A., Lyklema, J., A lattice theory of polyelectrolyte adsorption, *J. Phys. Chem.*, 6661, 1984.

11 Scheutjens, J.M.H.M., Fleer, G.J., Statistical theory of the adsorption of interacting chain molecules. 2. Train, loop and tail size distribution, *J. Phys. Chem.*, 84, 178, 980.

12   de Jong, E.W., Bosch, L., Westbroek, P., Isolation and characterization of a Ca^{2+}-binding polysaccharide associated with coccoliths of Emiliania Huxleyi (Lohmann) Kamptner. *Eur. J. Biochem*, 70, 611, 1976.

13   Borman, A.H., de Jong, E.W., Huizinga, M., Kok, D.J., Westbroek, P. and Bosch, L., The role in $CaCO_3$ crystallization of an acid Ca^{2+}-binding polysaccharide associated with coccoliths of Emiliania Huxleyi, *Eur. J. Biochem.*, 129, 179, 1982.

14   Werness, P.G., Brown, C.M., Smith, L.H., Finlayson, B., Equil2: a basic computer program for the calculation of urinary saturation., *J. Urol*, 134, 1242, 1985.

15   Will, E.J., Bijvoet, O.L.M., Blomen, L.J.M.J., vd Linden, H., Growth kinetics of calcium oxalate monohydrate 1, *J. Cryst. Growth*, 64, 297, 1983.

16   Blomen, L.J.M.J., Will, E.J., Bijvoet, O.L.M. and vd Linden, H., Growth kinetics of calcium oxalate monohydrate 2, *J. Cryst. Growth*, 64, 306, 1983.

17   Bijvoet, O.L.M., Blomen, L.J.M.J., Will, E.J. and vd Linden, H., Growth kinetics of calcium oxalate monohydrate 3, *J. Cryst. Growth*, 64, 316, 1983.

18   Christiansen, J.A., Nielsen, A.E., On the kinetics of formation of sparingly soluble salts, *Acta Cim. Scand.*, 5, 673, 1951.

19   Garside, J., Nucleation, in *Biological Mineralization and Demineralization, Dahlem Konferenzen*, Nancollas, G.H. Ed., Springer-Verlag Berlin, 1982, chap. 3.

20   Antinozzi, P., Brown, C.M., Purich, D.L., Calcium oxalate monohydrate crystallization: citrate inhibition of nucleation and growth steps, *J.Cryst. Growth*, 125, 215, 1992.

21   Kok, D.J., Blomen, L.J.M.J., Westbroek, P., Bijvoet, O.L.M., Polysaccharide from coccoliths (CaCO3 biomineral), *Eur. J. Biochem.*, 158, 167, 1986.

22   Kok, D.J., The role of crystallization processes in calcium oxalate urolithiasis., *Thesis Leiden University*, Leiden 1991.

23   Blomen, L.J.M.J., Growth and agglomeration of calcium oxalate monohydrate, *Thesis Leiden University*, Leiden 1982.

24   Daniele, P.G., Sonego, S., Ronzani, M. and Marangella, M., Ionic strength dependence of formation constants part 8 Solubility of calcium oxalate monohydrate and calcium hydrogenphosphate dihydrate in aqueous solution, at 37°C and different ionic strengths, *Ann. Chim.*, 75(5-6), 245, 1983.

25   Debije, P. Hückel, E., Zur Theorie der Elektrolyte I. Gefrierpunktsniedrigung und verwandte Erscheinungen, *Physik Z*, 24, 185, 1923.

26   Debije, P. Hückel, E., Zur Theorie der Elektrolyte I. Das Grenzgesetz für die Elektrische Leitfähigkeit, *Physik Z*, 24, 305, 1923.

27   Kok, D.J., Papapoulos, S.E., Bijvoet, O.L.M., Modulation of calcium oxalate monohydrate growth kinetics in vitro, *Kidney Int.*, 34, 346, 1988.

28   Davies, C.W., Hogle, B., *J. Chem. Soc.*, 1028, 1955.

29   Walker, M., *J. Phys. Chem.*, 65: 159, 1961.

30   Wiers, B.H., Polynuclear complex formation in solutions of calcium ion and Ethane-1-hydroxy-1,1-diphosphonic acid II. Light scattering. Sedimentation, Mobility and Dialysis Measurements, *J. Phys. Chem.*, 75, 682, 1971.

31   Hess, B., The role of Tamm-Horsfall glycoprotein and nephrocalcin in calcium oxalate monohydrate crystallization processes, *Scanning Microsc.*, 5, 689, 1991.

32   Kok, D.J., Papapoulos, S.E., Bijvoet, O.L.M., The effects of low and high molecular weight substances on the citrate induced changes of growth and agglomeration of calcium oxalate, *Fortschr. Urol. u. Nephrol.*, 27, 159, 1987.

33   Meyer, J.L., Thomas, W.C., Trace metal-citric acid complexes as inhibitors of calcification and crystal growth 1, *J. Urol.*, 128, 1372, 1982.

34   Meyer, J.L., Thomas, W.C., Trace metal-citric acid complexes as inhibitors of calcification and crystal growth 2, *J. Urol.*, 128, 1376, 1982.

35   Hess, B., Zipperle L. Jaeger Ph., Citrate and calcium effects on Tamm-Horsfall glycoprotein as a modifier of calcium oxalate crystal aggregation, *Am. J. Physiol.*, 265, F784, 1993.

36   **Deganello, S.,** Interaction between Nephrocalcin and Calcium Oxalate Monohydrate: A Structural Study, *Calcif. Tiss. Int.,* 48, 421, 1991.

37   **Martin, X., Smith, L.H., Werness, P.G.,** Effect of additives and whole urine on calcium oxalate dihydrate formation, *Urol. Res.,* 12, 86, 1984.

38   **Simkiss, K.,** Mechanisms of mineralization (normal), group report in *Biological Mineralization and Demineralization, Dahlem Konferenzen,* Nancollas, G.H. Ed., Springer-Verlag Berlin, 1982, chap. 19.

39   **Finlayson, B., Reid, F.,** The expectation of free and fixed particles in urinary stone disease, *Inv. Urol.,* 15 (6), 442, 1978.

40   **Kok, D.J., Khan, S.R.,** Calcium oxalate nephrolithiasis, a free or fixed particle disease, *Kidney Int.,* 46, 847, 1994.

41   **Kavanagh, J.P.,** Methods for the study of calcium oxalate crystallisation and their application to urolithiasis research. *Scann. Microsc.,* 6(3), 685, 1992.

42   **Pearson, R.G.,** Hard and soft acids and bases, HSAB, part 1, *J. Chem. Educ.,* 45 (9), 581, 1968.

43   **Pearson, R.G.,** Hard and soft acids and bases, HSAB, part 2, *J. Chem. Educ.,* 45 (10), 643, 1968.

44   **Kok, D.J.,** Bisphosphonate action on the agglomeration and growth of crystals, in *Bisphosphonate therapy in Acute and Chronic Bone loss,* Bijvoet, O.L.M., Canfield, R.E., Fleisch, H., Russell, R.G.G., Elseviers Sciencse Publishers, Amsterdam, in press.

45   **Martin, X., Werness, W.G., Bergert, J.H., Smith, L.H.,** Pentosanpolysulfate as inhibitor of calcium oxalate crystal growth, *J. Urol.,* 132, 786, 1984.

46   **Davey, R.J., Black, S.N., Bromley, L.A., Cottier, D., Dobbs, B., Rout, J.E.,** Molecular design based on recognition at inorganic surfaces, *Nature,* 353, 549, 1991.

47   **Papenhuijzen, J., vd Schee, H.A., Fleer, G.J.,** Polyelectrolyte adsorption, *J. Coll. Interf. Sci.,* 104 (2), 540, 1985.

48   **Borman, A.H., Kok, D.J., de Jong, E.W., Westbroek, P., Varkevisser, F.A., Bloys van Treslong, C.J. and Bosch, L.,** Molar mass determination of the polysaccharide associated with coccoliths of Emiliania Huxleyi, *Eur. Polymer. J.,* 22, 521, 1986.

49   **Nancollas, G.H..,** Phase transformation during precipitation of calcium salts, in *Biological Mineralization and Demineralization, Dahlem Konferenzen,* Nancollas, G.H. Ed., Springer-Verlag Berlin, 1982, chap. 5.

Chapter 3

MECHANISMS OF CALCIUM OXALATE AGGREGATION IN THE BIOPHYSICAL ENVIRONMENT

James H. Adair, Richard V. Linhart, Craig F. Habeger, and Saeed R. Khan

I. INTRODUCTION

Kidney stone disease has been called an opportunistic disease because it is thought that it takes an unfortunate, simultaneous convergence of factors that may be achieved by several pathways for a stone to form.[1-4] Thus, injury, supersaturation (particularly hyperoxaluria), nucleation and growth mechanisms all play a role in contributing to stone formation. Even so, urolithiasis is idiopathic with no one risk factor providing an obvious cause of stones in chronic stone formers. However, both the clinical *in-vivo* and *in-vitro* evidence indicates that the presence of particles is a necessary but insufficient condition for stone formation. Most normal individuals, at certain times, have crystalluria yet do not have a stone incident. For a stone to be created from freely flowing particles, either a particle must grow to a large enough size to occlude the tubule or aggregation of multiple particles must occur. Furthermore, if aggregation is responsible for stone formation, the attractive energy or energies holding the primary particles together in an aggregate must be great enough to withstand the hydrodynamic shear forces due to the flow of fluid through the nephron.

The possibility that the growth of a large crystal leads to a stone has been shown to be unlikely by several teams of investigators.[5,6] Finlayson and Reid[5] predicted that there was not enough time for a freely flowing particle to grow to a size sufficient to occlude the tubule based upon the hydrodynamics in the nephron and measured rates of crystal growth for calcium oxalate monohydrate. It was also suggested by Finlayson and Reid that aggregation of freely flowing particles was not likely due to the relatively low concentration of particles within the nephron at any given point in time. However, a recent analysis of the hydrodynamic aspects of transit time through the nephron by Kok and Khan,[6] based on better estimates of the diameter of different regions in the nephron, indicate that aggregation of freely flowing particles is possible. Regardless, attachment by calcium oxalate particles to the epithelial nephron wall has been demonstrated in a rat animal model,[3,5] as shown in Figure 1. Thus, the concept of fixed particles at the epithelial wall has become a fundamental principle in the formation of stones. In more recent work,[7-10] it has been demonstrated that particle attachment can be induced by epithelial wall damage in which intercellular species, such as phospholipids, may play a role.

The objective of this chapter is to examine the dynamics of aggregation for particles and interactions of the particles and/or aggregates with biological surfaces such as the interluminal wall of the nephron. The nature

0-8493-7673-4/95/$0.00+$.50
© 1995 by CRC Press

of the forces among particles and surfaces will be discussed with respect to the hydrodynamic flow scheme found in the nephron.

Figure 1. COM particle aggregate attached to the wall of the proximal tubule in the nephron of the rat with one end of the constituent crystals joined together (near the arrow) and the other end free (bar = 5 μm). Note also the organic layer present on the surface of the particles. Photomicrograph by S.R. Khan.[11]

II.   AGGREGATION MECHANISMS AMONG PARTICLES AND PARTICLES AT SURFACES

Numerous studies[2,3,5,10,12-29] have shown that the aggregation and/or adhesion of COM particles potentially can lead to the ultrastructures determined for kidney stones. However, the actual mechanism(s) of aggregation within the biophysical environment of the human kidney have not been examined in detail. There are at least six aggregation mechanisms, as shown in Figure 2, acting alone or in concert that may contribute to stone formation.

To establish what dispersion techniques may work in stone prevention, first it is necessary to determine which of the possible aggregation mechanisms may be most important with respect to interparticle strength. There are a number of possible mechanisms for aggregation of COM within the human kidney. In addition to minimum double layer interactions, other interactions include secondary minimum coagulation, heterocoagulation, polymer bridging flocculation, aggregation by secondary nucleation and

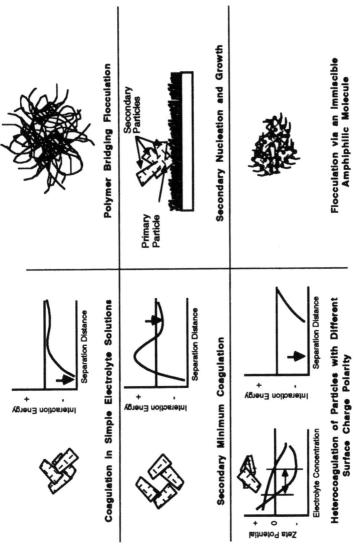

Figure 2. Possible aggregation mechanisms for particles in the nephron.

growth, and immiscible liquid flocculation. The physical nature of these aggregation mechanisms is shown in Figure 2. There are also hydrodynamic factors that need to be considered including the size of the tubule and the shear rate associated with fluid flow within the nephron, as discussed by Finlayson and Reid,[5] and Kok and Khan.[6] The concentration of the COM particles also has a role whether one is considering fixed stone disease or aggregation of the particles in a freely flowing state. Finally, the flow patterns associated with peristaltic vis-a-vis continuous flow within the nephron have not been addressed in past studies but should be to realistically assess aggregate formation in the kidney. Each of the mechanisms will be discussed with respect to their likelihood in the human nephron.

A. PARTICLE AGGREGATION IN SIMPLE ELECTROLYTES

The aggregation of COM particles has traditionally been attributed to the absence of charge on COM particle surfaces and/or the high ionic strength of urine. However, neither of these aggregation mechanisms accommodates the observation that COM particles observed *in-vivo* invariably have organic matter on their surfaces, as shown in Figure 1. Based on the electrostatic model for aggregation, in the simplest case, the repulsive contributions between particles in solution arise from the interaction of ionic clouds surrounding the particles. Generally, particles in aqueous solution have a surface charge created by a combination of several different charging mechanisms. In the case of COM, Curreri et al.[30-32] showed that surface charge is due to incongruent dissolution of the constituent Ca^{2+} and $C_2O_4^{2-}$. These and other species from solution then can adsorb (in their hydrated form) into an adsorbed ion layer known as the Stern layer. The electroneutrality of the system, composed of the surface and Stern charges and surrounding solution, is achieved by the charge in the diffuse cloud of ions that only are attracted electrostatically toward the surface.

Adair and Linhart[33] have developed a computer program, STABIL©, that calculates the repulsive contributions for solutions of arbitrary complexity such that the distance of separation, D, for interacting particles is given in its most general form by:

$$D = \int_{\psi_{D/2}}^{\psi_\delta} \frac{2d\psi}{\left[\frac{8\pi kT}{\varepsilon_3}\right]^{\frac{1}{2}} \left[\sum_i^\infty n_i \left(e^{-\frac{z_i e\psi(x)}{kT}} - e^{-\frac{z_i e\psi_{D/2}}{kT}}\right)\right]^{\frac{1}{2}}} \tag{1}$$

where ψ_δ and $\psi_{D/2}$ are the Stern potential and potential at one-half of the separation distance between two identical interacting particles and n_i is the ionic concentration of the ith species. Equation (1) is solved numerically to give $\psi_{D/2}$ as a function of D. These values are then put into the relevant generalized relationship to provide a calculation of the interaction energy for particles in solutions of arbitrary complexity to predict whether aggregation is likely in complex, multispecies solutions such as urine.

B.   SECONDARY MINIMUM COAGULATION

As shown in Figure 2, there is a minimum at relatively large distances of separation in the interaction energy for COM particles. As the ionic strength increases, magnitude of this secondary minimum increases. The secondary minimum in the interaction energy curve is a consequence of the longer range of the van der Waals attractive forces than the electrostatic repulsive interactions.[34,35] Thus, even when the charge at the surfaces of the interacting particles is great enough to produce an energy barrier at intermediate separation distances, secondary minimum coagulation may take place because there is no energy barrier for this mechanism of aggregation. The strength of the interparticle bonds for secondary minimum interactions has been evaluated theoretically and experimentally by Chan and Halle.[36] It was demonstrated that the mean lifetime for secondary minimum aggregates increased with increasing ionic strength of the suspension containing the model polystyrene spherical particles.

In the only known study on secondary minimum aggregation for COM, Adair[12] showed that COM suspensions composed of relatively coarse primary particles (~5 μm equivalent spherical diameter) aggregate over a wide range of solution and surface charge conditions. Secondary minimum aggregation was implicated in conditions where primary minimum aggregation was minimized because of low ionic strength and relatively high zeta potential. However, the results were ambiguous because only the thermodynamic aspects of the coagulation process were addressed in Adair's study. The strength of the proposed secondary minimum interaction was not determined by analyzing the hydrodynamic shear forces required to promote breakup of the aggregates. Aggregate bond strength measurements as a function of particle size were suggested since the magnitude of the secondary minimum increases as a function of the radii of the interacting particles. The increased likelihood of secondary minimum coagulation with increasing particle size may have important implications to stone disease since it has been reported by Robertson[37] that stone forming individuals have particles significantly larger in size than non-stone formers.

C.   HETEROCOAGULATION

Heterocoagulation is the aggregation among particles of different materials. We are not aware of any investigators that have addressed this potential mechanism for stone formation. However, there have been a number of investigators[38-41] within the colloid chemistry community that have developed the theoretical and practical framework for a study of heterocoagulation. The basis for heterocoagulation is that particles of different materials may have different polarity for their surface charge. Thus, a positively charged COM particle may be electrostatically (as well as through the van der Waals interactions) attracted to a negatively charged hydroxyapatite (HAP) particle. However, even if the surface charge on particles of dissimilar materials are the same, the van der Waals attractive forces may be strong enough to promote heterocoagulation (as well as homocoagulation).

Within the human nephron there are a variety of potential combinations that can lead to heterocoagulation. These include: COM-HAP, COM-HU (uric acid), COM-NH4U (ammonium urate), COM-COD (calcium oxalate dihydrate), COM-E.coli and other bacteria, and COM with various macromolecules. Heterocoagulation will be reversible only when the sign of the surface charge is made the same for all particles. Thus, the addition of strongly charged species such as citrate should clinically minimize stone disease in patients whose stones are formed via heterocoagulation. The sign of the surface charge and corresponding solution conditions that will promote or inhibit heterocoagulation need to be evaluated for each system. It has been predicted that epitaxy of COM and HAP is unlikely because of incoherent crystal structures.[42] However, heterocoagulation can explain the presence of HAP with COM in a stone.

Heretofore, heterocoagulation has not been addressed for material systems relevant to urolithiasis. Preliminary experiments have been conducted in our laboratory to determine the conditions in which particles composed of various materials will be likely to form heterocoagules. An obvious starting point in this initial evaluation is to examine the effect of the polarity of the zeta potential for the various materials as a function of relevant solution conditions. The zeta potential data for COM and HU are summarized in Figure 3A as a function of Ca^{2+} and $C_2O_4^{2-}$ concentration. The COM data are from Curreri et al.[31] and incorporate the solubility product for COM in the Ca^{2+} concentrations. The uric acid data from Adair et al.[43] are given as a function of Ca^{2+} or $C_2O_4^{2-}$. COM has a point of zero charge (pzc) at pCa = 5.2 (Ca^{2+} = 6.3×10^{-6} M) with COM having negatively charged surfaces above this pCa and positively charged surfaces at higher Ca^{2+} concentrations. Adair et al. showed that uric acid is negatively charged over a wide range of pH and Ca^{2+} and $C_2O_4^{2-}$ concentrations.

Thus, mixing COM and HU particles where pCa is less than 5.2 should promote heterocoagulation between the negatively charged HU and the positively charged COM. As shown in Figure 3B and 3C, fine COM particles produced by the dimethyl oxalate decomposition adhere to the larger, prismatic HU particles when pCa = 4. When COM and HU particles in saturated COM solutions are mixed with pCa = 5, the fine COM particles have a greater affinity toward one another than the HU particles and heterocoagulation does not occur, as shown in Figure 3C.

D.   FLOCCULATION

This mechanism takes place when there is insufficient polymer (or macromolecule if the polymer is self-organizing) for full surface coverage on particles. This is the only aggregation mechanism that explains some ultrastructural observations of Boyce, Khan et al., and others[44-50] on the role of matrix macromolecules in the microstructure and ultrastructure of the mature stone. Maximum flocculation takes place when one-half of the surface of a particle is covered by polymer.[35,51] However, flocculation takes place anywhere between about one-tenth surface coverage to greater

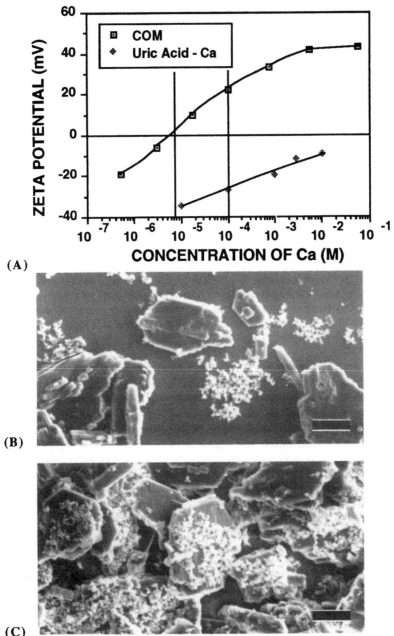

(A)

(B)

(C)

Figure 3. Heterocoagulation of COM with HU is predicted based on: (a) zeta potential determinations, (b) mixing of HU and COM at pCa ≤ 5 where COM and HU are both negatively charged, and (c) mixing at pCa = 4 where COM and HU are both oppositely charged (bar = 20 μm).

than 75 percent coverage. We are not aware of any studies relevant to urolithiasis that have addressed the role of flocculation in detail. Kok et al., Finlayson, and Robertson and Peacock[2,20,21] have discussed this mechanism with other possible mechanisms of aggregation but the research emphasis has been on the prevention of aggregates by employing large concentrations of macromolecules or polymers.[13,14,22,25,27,28,52,53] Regardless, this mechanism can also explain the conflicting reports of inhibition versus promotion.[15,54,55] Thus, macromolecular species such as uropontin, Tamm-Horsfall mucoprotein, and nephrocalcin may play a dual role: at sufficiently low concentration, aggregation is promoted through flocculation while at higher surface coverages, dispersion of particles is achieved through the protective colloidal effect of the macromolecular coating.[34,35]

Although there have been limited studies on flocculation with respect to urolithiasis, there have been a number of investigations on floc formation and hydrodynamic breakup because of its importance in wastewater treatment and other technologies.[56-60] These studies provide a basis for evaluating the role of flocculation as an aggregation mechanism for COM and other relevant particles in urine. For example, an excellent starting point is to determine the macromolecular or polymer dosage for relevant urinary species required to achieve the maximum flocculation. The critical flocculation concentrations (CFC) for a particular macromolecule will indicate whether this mechanism is likely by comparison with its concentration range in urine.

In preliminary studies, we have examined a flocculant commonly used in mineral recovery. Polyethyleneimine (PEI) is a positively charged highly branched polymer molecule used by Pelton and Allen[61] to produce positive charge on glass surfaces in their particle adhesion studies. It has been shown in preliminary studies, using the apparatus developed by Eisenlauer and Horn,[57] to evaluate the aggregation of freely flowing particles in suspension. This device has the advantage that flow rate and mode of flow (i.e., continuous versus peristaltic) can be varied. The commercial analogue to Eisenlauer and Horn's device, known as a photometric dispersion analyzer (Rank Brothers, Cambridge, UK), has been used extensively to monitor the flocculation of model systems. Initial experiments have demonstrated that fine COM particles flocculate at intermediate dosages of PEI. The relative degree of flocculation varies as a function of charge (as dictated by the concentrations of Ca^{2+} and $C_2O_4^{2-}$). Urinary macromolecules are currently being evaluated for their ability to flocculate COM particles in freely flowing systems as a function of shear rate.

E. ADHESION OF COM PARTICLES AT SURFACES

Determination of the adhesion strength for particles adhering at surfaces is of fundamental importance in deducing whether the fixed particle mechanism for stone formation proposed by Finlayson and Reid[5] is reasonable within the hydrodynamic system of the nephron.

It has been demonstrated by numerous investigators that counting the particles adhering to a surface as a function of time at a particular shear rate can be used to estimate the interparticle bond strength. Mandel and coworkers[8-10,29] have shown that COM particles adhere specifically to certain surface irregularities on rat papillary collecting tubule (RPCT) cell cultures. The irregularities consist of "clumped" cells thought to be either rearranged cell surface molecules or basement membrane remnants bound and exposed to the cell surface. While providing exciting insight into the nature of COM binding on epithelial surfaces of the nephron, an obvious extension to the current level of understanding is to evaluate the kinetic aspects and the strength with which COM particles adhere to the specific binding sites on the RPCT cell culture.

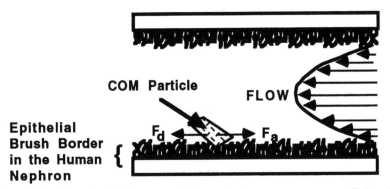

Figure 4. Scenario depicting the balance of hydrodynamic forces (F_d) and adhesive forces (F_a) acting on a COM particle bound to the brush border in the human nephron.

Another approach in evaluating the bond strength of particles adhering to a surface is to determine the hydrodynamic shear force required to remove particles. Pelton, Busscher, Matijevic, Owens, and others[40,41,61-64] have used this approach to evaluate thermodynamic models for the attachment of particles to surfaces. The balance of forces proposed for a COM particle attached to the epithelium is shown schematically in Figure 4. The shear stress at the wall (τ_w) is given by:[64,65]

$$\tau_w = \frac{dP}{dl} b \qquad (2)$$

where τ_w is the shear stress at the wall, P is the hydrostatic pressure drop across the conduit of length, l, and wall separation, b.

Using the relationship in Equation (2) and a specially adapted adhesion cell, Habeger et al.[66] have recently demonstrated that COM particles adhere to a hard surface (fused silica) at shear rates well above those associated with the human nephron. Specificity of the adhesion strength of the COM particles on the silica surface with respect to crystallographic habit of the COM was also observed, as shown in Figure 5. This finding was consistent with earlier observations by Mandel and co-workers[67] that there was a specificity depending on whether the COM particles are oriented with respect to either the (010) or the ($10\bar{1}$) face on the COM particles.

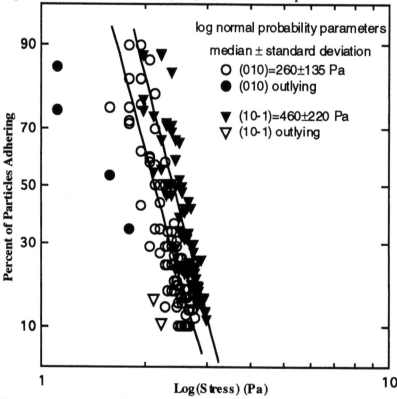

Figure 5. Percent of COM particles adhering to a model inorganic, negatively charged, surface (silica) as a function of hydrodynamically induced shear stress. Note difference in adhesive strength for particles oriented with the oxalate-rich (010) COM face vis-a-vis the calcium-rich ($10\bar{1}$) face.[66]

F. FLOCCULATION OF COM PARTICLES BY PHOSPHOLIPIDS AND OTHER INTERCELLULAR SUBSTANCES

Work by Khan, and Mandel and co-workers[67,68] indicates that interaction of COM particles with phospholipids at either the epithelial surface of cells or with phospholipids liberated into the bulk solution by cell injury are important features in the formation of aggregates in the interluminal channel of the nephron. Flocculation by sparingly soluble, amphiphilic molecules such as the phospholipids forming a major component of the epithelial cell membrane have not been studied with respect to aggregation, but Mandel et al. clearly show that adhesion of COM particles is important. The ability of phospholipids to adsorb to the surfaces of sparingly soluble inorganic particles and surfaces is well established as is the forces arising from the interaction of phospholipid monolayers on mica and similar surfaces based on force balance work by Israelachivili and others.[69] Thus, one would expect phospholipids to have an effect similar to macromolecules or polymers capable of promoting flocculation. However, the interparticle strength of particles flocculated in either freely flowing suspensions or at surfaces containing phospholipids (i.e., cell membranes) would be expected to depend on the concentration of phospholipids and the efficiency of adsorption of such species to COM surfaces.

III. CONCLUSIONS

The variety of possible aggregation mechanisms for COM in the biophysical environment of the human nephron reflect the opportunistic nature of urinary stone disease. Whether a particular aggregation mechanism results in persistent aggregates depends on the interparticle adhesive forces among the primary particles making up the aggregates balanced against dynamic forces in the nephron generated by the hydrodynamic shear of urinary flow. It has also been experimentally confirmed that the strength of adhesion for COM particles to a model inorganic surface is dependent upon the crystallographic orientation of the adherent COM surface. Several heretofore unrecognized aggregation mechanisms have been proposed whose validity may be easily established by carefully designed experiments.

ACKNOWLEDGMENTS

The work reported herein is sponsored by the National Institutes of Health grant number POG 5P01 DK20586-17. The authors also wish to thank personnel in the Major Analytical Instrumentation Center at the University of Florida for help with the electron microscopy and Mrs. Pam Howell for help with manuscript preparation.

REFERENCES

1.   **Conyers, R.A.J., Bals, R., and Rofe, A.M.**, The relation of clinical catastrophes, endogenous oxalate production, and urolithiasis, *Clin. Chem.*, **36**[10], 1717, 1990.
2.   **Finlayson, B.**, Physicochemical aspects of urolithiasis, *Kidney Int.*, **13**, 344, 1978.
3.   **Finlayson, B., Khan, S.R., and Hackett, R.L.**, Mechanisms of stone formation - An overview, *Scan. Elec. Microsc.*, III, 1419, 1984.
4.   **Robertson, W.G.**, Pathophysiology of stone formation, *Urol. Int.*, **41**, 329, 1986.
5.   **Finlayson, B. and Reid, F.**, The expectation of free and fixed particles in urinary stone disease, *Invest. Urol.*, **15**[6], 442, 1978.
6.   **Kok, D.J. and Khan, S.R.**, Calcium oxalate nephrolithiasis, a free or fixed particle disease, *Kidney Int.*, **46**, 847, 1994.

7.   **Khan, S.R., Cockrell, C.A., Finlayson, B., and Hackett, R.L.**, Crystal retention by injured urothelium of the rat urinary bladder, *J. Urol.*, **132**, 153, 1984.

8.   **Mandel, N.S., Mandel G.S., and Hasegawa, A.T.**, The effect of some urinary stone inhibitors on membrane interaction potentials of stone crystals, *J. Urol.*, **138**, 557, 1987.

9.   **Mandel, N. and Riese, R.**, Crystal-cell interactions: Crystal binding to rat renal papillary tip collecting duct cells in culture, *Am. J. Kidney Diseases*, **17**[4], 402, 1991.

10. **Riese, R.J., Riese, R.W., Kleinman, J.G., Wiessner, J.H., Mandel, G.S., and Mandel, N.S.**, Specificity in calcium oxalate adherence to papillary epithelial cells in culture, *Am J. Physiol.*, **225**, F1025, 1988.

11. **Khan, S.R. and Hackett, R.L.**, Retention of calcium oxalate in renal tubules, *Scann. Micros.*, **5**[3], 707, 1991.

12. **Adair, J.H.**, Coagulation of Calcium Oxalate Monohydrate, Ph.D. dissertation, University of Florida, Gainesville, FL, 1981.

13. **Coe, F.L., Nakagawa, Y., and Parks, J.H.**, Inhibitors within the nephron, *Am. J. Kidney Diseases*, **17**[4], 407, 1991.

14. **Deganello, S.**, Interaction between nephrocalcin and calcium oxalate monohydrate: A structural study, *Calcif. Tissue Int.*, **48**, 421, 1991.

15. **Edyvane, K.A., Ryall, R.L., and Marshall, V.R.**, Macromolecules inhibit calcium oxalate crystal growth and aggregation in whole human urine, *Clin. Chim. Acta.*, **156**, 81, 1986.

16. **Edyvane, K.A., Ryall, R.L., Mazzaachi, R.D., and Marshall, V.R.**, The effect of serum on the crystallization of calcium oxalate in whole human urine: Inhibition disguised as apparent promotion, *Urol. Res.*, **15**, 87, 1987.

17. **Edyvane, K.A., Hibberd, C.M., Harnett, R.M., Marshall, V.R,. and Ryall, R.L.**, The Influence of serum and serum proteins on calcium oxalate crystal growth and aggregation, *Clin. Chim. Acta.*, **167**, 329, 1987.

18. **Hartel, R.W., Gottung, B.E., Randolph, A.D., and Drach, G.W.**, Mechanisms and kinetic modeling of calcium oxalate crystal aggregation in a urinelike liquor: Part I, Mechanisms, *AICHE Journal* , **32**[7], 1176, 1986.

19. **Hartel, R.W. and Randolph, A.D.** Mechanisms and kinetic modeling of calcium oxalate crystal aggregation in a urinelike liquor: Part II, Kinetic modeling, *AICHE Journal*, 1186, 1986.

20. **Kok, D.J., Papapoulos, S.E., and Bijvoet, O.L.M.**, Crystal agglomeration is a major element in calcium oxalate urinary stone formation, *Kidney Int.*, **37**, 51, 1990.

21. **Robertson, W.G. and Peacock, M.**, Pathogenesis of urolithiasis, in *Pathogenesis of Urolithiasis in Urolithiasis Etiology Diagnosis*, Springer-Verlag, New York, 185, 1985.

22. **Ryall, R., Harnett, R.M., and Marshall, V.R.**, The effect of urine, pyrophosphate, citrate, magnesium and glycosaminoglycans on the growth and aggregation of calcium oxalate crystals, *In Vitro*, *Clin. Chim. Acta.*, **112**, 349, 1981.

23. **Ryall, R., Ryall, R.G., and Marshall, V.R.**, Interpretation of particle growth and aggregation patterns obtained from the Coulter counter: A simple theoretical model, *Invest. Urol.*, **18**[5], 396, 1981.

24. **Ryall, R.L., Bagley, C.J., and Marshall, V.R.**, Independent assessment of the growth and aggregation of calcium oxalate crystals using the Coulter counter, *Invest. Urol.*, **18**[5], 401, 1981.

25. **Ryall, R.L. and Marshall, V.R.**, The relationship between urinary inhibitory activity and endogenous concentrations of glycosaminoglycans and uric acid: Comparison of urines from stone-formers and normal subjects, *Clin. Chim. Acta*, **141**, 197, 1984.

26. **Ryall, R., Harnett, R.M., and Marshall, V.R.**, The effect of monosodium urate on the capacity of urine, chondroitin sulfate and heparin to inhibit calcium oxalate growth and aggregation, *J. Urol.*, **135**, 174, 1986.

27. **Scurr, D.S. and Robertson, W.G.**, Modifiers of calcium oxalate crystallization found in urine: II. Studies on their mode of action in an artificial urine, *J. Urol.*, **136**, 128, 1986.

28. **Scurr, D.S. and Robertson, W.G.**, Modifiers of calcium oxalate crystallization found in urine: III. Studies on the role of Tamm-Horsfall mucoprotein and of ionic strength, *J. Urol.*, **136**, 505, 1986.

29. **Wiessner, J.H., Kleinman, J.G., Blumenthal, S.S., Garancis, J.C., Mandel, G.S., and Mandel, N.S.**, Calcium oxalate interaction with rat renal inner papillary collecting tubule cells, *J. Urol.*, **138**, 640, 1987.

30. **Curreri, P.A.**, The Electrokinetic Properties of Calcium Oxalate Monohydrate, Ph.D. dissertation, University of Florida, Gainesville, FL, 1979.

31. **Curreri, P., Onoda, Y., Jr., and Finlayson, B.**, An electrophoretic study of calcium oxalate monohydrate, *J. Colloid Interface Sci.*, **69**[1], 170, 1979.

32. **Curreri, P.A., Onoda, G.Y., Jr., and Finlayson, B.**, Microelectrophoretic study of calcium oxalate monohydrate in macromolecular solutions, in *Proteins at Interfaces, Physicochemical and Biochemical Studies*, J.L. Brash and T.A. Horbett, Eds., American Chemical Society, Washington, D.C., 278, 1987.

33. **Adair, J.H. and Linhart, R.V.**, A generalized program to calculate interparticle interactions in a variety of suspension conditions, in *Handbook on Characterization Techniques for the Solid-Solution Interface*, Adair, J.A., Casey, J.A., and Venigalla, S., Eds., American Ceramic Society, Westerville, OH, 1993, 69.

34. **Hough, D.B. and White, L.R.**, The calculation of Hamaker constants from Lifshitz theory with applications to wetting phenomena, *Adv. Colloid Interface Sci.*, **14**, 3, 1980.

35. **Sonntag, H. and Strenge, K.**, *Coagulation and Stability of Disperse Systems* (Engl. trans.), Halsted Press, New York, 1972.

36. **Chan, D.Y.C. and Halle, B.**, Dissociation kinetics of secondary-minimum flocculated colloidal particles, *J. Colloid Interface Sci.*, **102**[2], 400, 1984.

37. **Robertson, W.G.**, A method for measuring calcium crystalluria, *Clin. Chim. Acta.*, **26**, 105, 1969.

38. **Barouch, E., Matijevic, E., Ring, T.A., and Finlan, J.M.**, Heterocoagulation: II. Interaction energy of two unequal spheres, *J. Colloid Interface Sci.*, **67**[1], 1, 1978.

39. **Derjaguin, B.V.**, A theory of the heterocoagulation, interaction and adhesion of dissimilar particles in solutions of electrolytes, *Disc. Faraday Soc.*, **18**, 85, 1954.

40. **Kuo, R.J. and Matijevic, E.**, Particle adhesion and removal in model systems, *J. Colloid Interface Sci.*, **78**[2], 407, 1980.

41. **Matijevic, E.**, Interactions in mixed colloidal systems (heterocoagulation, adhesion, microflotation), *Pure & Appl. Chem.*, **53**, 2167, 1981.

42. **Mandel, N.S. and Mandel, G.S.**, Epitaxis between stone-forming crystals at the atomic level, in *Urolithiasis: Clinical and Basic Research*, Plenum Press, New York, 1981, 469.

43. **Adair, J.H., Aylmore, L.A.G., Brockis, J.G., and Bowyer, R.C.**, An Electrophoretic mobility study of uric acid with special reference to kidney stone formation, *J. Colloid Interface Sci.*, **124**, 1, 1988.

44. **Boyce, W.H.**, Organic matrix of human urinary concretions, *J. Med.*, **45**, 673, 1968.

45. **Khan, S.R., Finlayson, B., and Hackett, R.L.**, Stone matrix as proteins adsorbed on crystal surfaces: A microscopic study, *Scan. Elec. Microsc.*, pt. 1, 379, 1983.

46. **Khan, S.R., Finlayson, B., and Hackett, R.L.**, Agar-embedded urinary stones: A technique useful for studying microscopic architecture, *J. Urol.*, **130**, 992, 1983.

47. **Khan, S.R. and Hackett, R.L.**, Role of the scanning electron microscopy and X-ray microanalysis in the identification of urinary crystals, *Scan. Microsc.*, 1[3], 1405, 1987.

48. **Meyer, J.L. and Thomas, W.C., Jr.**, Trace metal-citric acid complexes as inhibitors of calcification and crystal growth: I. Effects of Fe(III), Cr(III), and Al(III) complexes on calcium phosphate crystal growth, *J. Urol.*, **128**, 1372, 1982.

49. **Meyer, A.S., Finlayson, B., and DuBois L.**, Direct observation of urinary stone ultrastructure, *Br. J. Urol.* **43**, 154, 1971.

50. **Prien, E.L. and Prien, E.L.**, Composition and structure of urinary stone, *Am. J. Med.*, **45**, 654, 1986.

51.   **Hunter, R.J.**, *Foundations of Colloid Science*, Vols. I and II, Clarendon Press, Oxford, 1986.

52.   **Lanzalaco, A.C., Singh, R.P., Smesko, S.A., Nancollas, G.H., Sufrin, G., Binette, M., and Binette, J.P.**, The influence of urinary macromolecules on calcium oxalate monohydrate crystal growth, *J. Urol.*, **139**, 190, 1988.

53.   **Leal, J. and Finlayson, B.**, Adsorption of naturally occuring polymers onto calcium oxalate crystal surfaces, *Invest. Urol.*, **14**[4], 278, 1977.

54.   **Campbell, A.A., Ebrahimpour, A. Perez, L., Smesko, S.A., and Nancollas, G.H.**, The dual role of polyelectrolytes and proteins as mineralization promoters and inhibitors of calcium oxalate monohydrate, *Calcif.Tissue Int,* **45**, 122, 1989.

55.   **Grover, P.K., Ryall, R.L., and Marshall, V.R.**, Does Tamm-Horsfall mucoprotein inhibit or promote calcium oxalate crystallization in human urine, *Clin. Chim. Acta.*, **190**, 223, 1990.

56.   **Ditter, W., Eisenlauer, J., and Horn, D.**, Laser optical method for dynamic flocculation testing in flowing dispersions, in *The Effect of Polymers on Dispersion Properties*, Academic Press, New York, 1982, 323.

57.   **Eisenlauer, J. and Horn, D.**, Fibre-optic sensor technique for flocculant dose control in flowing suspensions, *Colloids & Surfaces,* **14**, 121, 1985.

58.   **Gregory, J.**, Flocculation in laminar tube flow, *Chem. Engr. Sci.*, **36**[11], 1789, 1981.

59.   **Gregory, J.**, Turbidity fluctuations in flowing suspensions, *J. Colloid Interface Sci.*, **105**[2], 357, 1985.

60.   **Ray, D.T. and Hogg, R.**, Agglomerate breakage in polymer-flocculated suspensions, *J. Colloid Interface Sci.*, **116**[1], 3256, 1987.

61.   **Pelton, R.H. and Allen, L.H.**, Factors influencing the adhesion of polystyrene spheres attached to Pyrex by polyethyleneimine in aqueous solution, *J. Colloid Interface Sci.*, **99**[2], 387, 1984.

62.   **Busscher, H.J., Weerkamp, A.H., van der Mei, H.C., van Pelt, A.W.J., de Jong, H.P., and Arends, J.**, Measurement of the surface free energy of bacterial cell surfaces and its relevance for adhesion, *J. Appl. Enviro. Microbio.*, **48**[5], 980, 1984.

63.   **Olsson, J.**, Studies on Initial Streptococcal Adherence, Ph.D. dissertation, University of Goteborg, Sweden, 1978.

64.   **Owens, N.F., Gingell, D., and Rutter, P.R.**, Inhibition of cell adhesion by a synthetic polymer adsorbed to glass shown under defined hydrodynamic stress, *J. Cell Sci.*, **87**, 667, 1987.

65.   **Eskinazi, S.**, *Principles of Fluid Mechanics* (2nd ed.), Allyn and Bacon, Inc., Boston, 1968.

66.   **Habeger, C.F., Linhart, R.V., and Adair, J.H.**, Adhesion to Model Surfaces in a Flow Through System, presented at Science,

Technology, and Applications of Colloidal Suspensions Symp., 96th Annual Meeting of the Am. Ceram. Soc., Indianapolis, IN, 1994.
67.   Mandel, N., Crystal-membrane interaction in kidney stone disease, *J. Am. Soc. Nephrol.* **5**, 537, 1994.
68.   Khan, S.R., Shevock, P.N., and Hackett, R.L., *In vitro* precipitation of calcium oxalate in the presence of whole matrix or lipid components of the urinary stones, *J. Urol.*, **139**(2), 418, 1988.
69.   Israelachvili, J.N., Intermolecular and Surface Forces, 2nd ed., Academic Press, New York, 1992.

Chapter 4

CALCIUM OXALATE FORMATION IN HIGHER PLANTS

Harry T. Horner and Bruce L. Wagner
Department of Botany, Iowa State University, Ames

I. INTRODUCTION

Calcium oxalate is of widespread occurrence in the animal and plant kingdoms and among a wide variety of microorganisms.[1,2] Its formation and presence constitute a portion of a larger field of study called biomineralization, which encompasses a diverse array of processes by which organisms are capable of converting ions in solution into solid minerals. These processes are controlled by cellular activities that cause physical and chemical changes for mineral formation and growth.

The occurrence of calcium oxalate is of differing significance, depending upon the specific organisms it occurs with and the environments in which the organisms live. Calcium oxalate is a product of environmentally derived calcium and the organism's ability to metabolically form oxalate. Calcium oxalate is the crystalline form, whereas the anionic oxalate complexed with hydrogen forms the free oxalic acid; or with sodium and potassium produces water-soluble substances; or with other cations, such as aluminum, copper, magnesium, manganese, and strontium creates other solid and crystalline substances. These various compounds of oxalate are potentially of both biological and environmental importance, particularly with respect to many plants.

Certain algae, lichens, fungi,[3] and many higher plants[4,5] are capable of producing significant quantities of the different forms of oxalate, particularly calcium oxalate, and to a much lesser degree, the calcium salts of carbonate and phosphate, and silica.[6] Unlike the cases of calcium oxalate occurring in the urinary tracts of animals and humans, calcium oxalate and the other forms of oxalate are not pathological in plants, in that their occurrence and locations within their cells or on their surfaces do not disrupt normal development and function. In fact, their ability to produce these various forms of oxalate may be helpful to them in certain phase(s) of their life cycles.

This chapter will focus on calcium oxalate specifically in the two higher plant groups, the gymnosperms (cone bearing plants) and the angiosperms (flower bearing plants) where sometimes significant amounts of calcium oxalate, and the other forms of oxalate, are produced. These two major groups of plants are important for many environmental and economic reasons since they cover large portions of the earth's land masses and provide an array of consumable products such as oxygen, wood and food.

0-8493-7673-4/95/$0.00+$.50
© 1995 by CRC Press

A. HISTORICAL PERSPECTIVES

Calcium oxalate has existed within higher plants (and the algae, lichens and fungi) longer than the existence of mankind. The first recorded instances of the presence of oxalate were made about the time of the Greeks when Pliny and Dioscorides mentioned plants (now known as *Rumex* and *Oxalis*) that had bitter or sour tastes.[1] Anton von Leewenhoek can be credited with probably first observing needle-like crystals of calcium oxalate in the plant *Arum* in the late 17th Century with his rudimentary microscope. Since his initial discovery, there have been many observations and physical and chemical identifications of calcium oxalate, and its related forms, occurring in higher plants.[4-8]

B. TAXONOMIC VALUE

Calcium oxalate has been reported in more than 215 families of angiosperms[9] and gymnosperms[10] and occurs in the wood of more than 1000 genera of 180 families.[11,12] Calcium oxalate crystals have been shown to be of limited taxonomic value in certain plant species because of their identifiable shape(s) and their specific location(s) within the plant body such as in the root, stem, leaf, flower, and seed.[7,8,13,14] The shape and location of the calcium oxalate crystals is species and tissue specific. A variety of anatomical studies, along with tissue culture, strongly suggest that calcium oxalate crystal formation and shape are under genetic control;[15] can be influenced by the environment; and, can be affected by the presence of other molecules and cations that are associated with crystal formation.[16,17]

C. ENVIRONMENTAL AND CELLULAR CALCIUM

Depending on the environment, calcium in the water and in the soil surrounding aquatic and land plants may be in much higher concentrations than the plant cells can withstand. This environmental calcium is readily taken up and is transported through the plant body in the vascular system, commonly via the transpiration stream.[4,18] This means that calcium is available to all parts of the plant and can move through the cell walls surrounding the cells. However, the level of cellular calcium cannot exceed 10^{-6} to 10^{-8} M before it becomes toxic to the cells. As a result, plant cells attempt to maintain a low level of calcium by actively pumping it from the cell cytoplasm to the wall, or storing it in their intracellular compartments, such as in mitochondria, the endoplasmic reticulum, and vacuoles.

D. CELLULAR OXALATE

The pathways of oxalate formation have been reviewed elsewhere.[5,19-21] Briefly, oxalate can be formed by enzymatic conversions of isocitrate and glycolate to glyoxylate, the latter is then enzymatically converted to oxalate. Glycolate is considered one of the major precursors of oxalate, the former being derived from photosynthesis and photorespiration. Oxaloacetate and formate also can be converted to oxalate in some systems. Ascorbic acid (vitamin C), formed in plastids, is considered to be another major precursor of oxalate. A recent study using axenic *Lemna* plants,[20] incorporated [^{14}C]-glycolic acid, -glyoxylic acid, -oxalic acid and -ascorbic acid to determine which precursors were involved in crystal formation. The experiments carried out in both light and dark, and using inhibitors of glycolate oxidase, indicated glycolic acid and glyoxylic acid were the potential substrates for oxalic acid formation, and ascorbic acid catabolism may give rise to glycolic acid and glyoxylic acid in the dark.

Other studies suggest that, whatever the precursor of oxalate, it may be induced to form because the high levels of calcium in the transpiration stream lead to oversaturation of both the available cell wall sights (apoplast) and cytoplasmic compartments (symplast; such as the mitochondria and endoplasmic reticulum). This oversaturation may induce the cell to differentiate high-capacity calcium sinks (and ultimately crystals) to form either in the cell vacuoles, or outside of the cytoplasm in the cell wall (see examples of crystal development later).

E. FUNCTIONAL CONSIDERATIONS

There are only a few higher plant families that show little or no oxalate accumulation.[9] In contrast, there are a variety of species in five angiosperm families where soluble oxalate accumulation may cause unpalatable bitter or sour tastes due to eating them (i.e., *Oxalis*, spinach, rhubarb, *Halogeton*), or the high levels of oxalate may cause various toxic reactions leading to hyperoxaluria,[1] crystal formation, and renal edema in animals and humans. In the plant system, these high levels of soluble oxalate may be involved in maintaining an ion or pH balance within the cells.[5,20]

Calcium oxalate formation, which may be formidable in terms of total dry weight of a whole plant, or one of its organs, may serve several different functions depending upon the plant. Possible roles of calcium oxalate can be summarized in the categories of detoxification,[5,22,23] protection,[13,24-26] structural strength, calcium storage,[5,27,28] and light gathering and reflection. It has been reported that some cacti contain up to 80 percent dry weight calcium oxalate. Horner (pers. commun.) has estimated that a large deciduous tree, in one growing, season may produce up to several hundred pounds of calcium oxalate in its leaves, petioles, branches and trunk.

II. HIGHER PLANT CALCIUM OXALATE

A. LOCATION

The literature shows that, in general, calcium oxalate crystals can be found associated with almost every type of higher plant cell throughout the plant body, even though in a given species their location, shape and hydration form are specific. Crystals have been described as occurring in four general locations with respect to the plant cell cytoplasm and its surrounding wall. Figure 1. depicts these four locations as being in the cell vacuole (most common location), in seed protein storage bodies,[27] somewhere in the plant cell wall, or on the surface of the cell wall. Some crystals form within the cell vacuole and later become encased in a cell wall that is produced internal to the normal cell wall. They are called Rosanoffian crystals and are commonly found in legumes and in some other families.

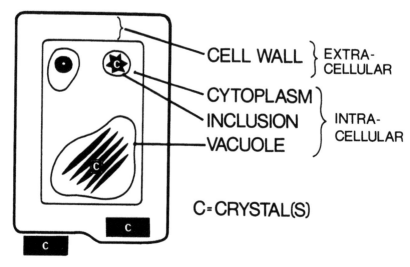

Figure 1. Diagram showing four locations of crystals associated with higher plant cells; on cell wall, in cell wall, in storage protein body, and in vacuole.

B. SHAPE AND FORM

The shape of crystals varies considerably even though five general shapes are typically described.[5,6] They are raphides (needles), druses (spherical aggregates), crystal sand, styloids, and prisms (Figure 2A-E). Raphides, for instance, vary in their cross-sectional shape depending upon the species in which they occur (Figure 3). Crystals occur as either the monohydrate (whewellite; monoclinic crystal system) or dihydrate (weddellite; tetragonal crystal system). These two hydration forms have been reviewed in higher plants.[29] For instance, the two hydration forms, as druses, were found in different taxonomic groups of cacti.[30] X-ray diffraction, infrared spectroscopy, differential interference contrast microscopy and polarizing microscopy aid in identifying the hydration forms and, along with other physical and chemical tests, the crystals can be distinguished from other kinds of inorganic substances also present in some plants.

1. Gymnosperms

The coniferous (evergreen) gymnosperms commonly display calcium oxalate crystals in or on the surfaces of cell walls (Figure 4A, B) exposed to the intercellular spaces in their needle-like leaves.[10,31] Crystals can also occur in the walls of epidermal cells, and intra- and extracellularly in the vascular

Figure 2. Five shapes of crystals commonly found in higher plants. (A) Needle-like raphide, *Ornithogalum* leaf. (B) Styloid, *Peperomia* leaf. (C) Crystal sand, *Nicotiana* leaf. (D) Druse, *Peperomia* leaf. (E) Prism, *Begonia* leaf.

bundle xylem and phloem parenchyma. Under abnormal conditions, such as acid rain and high ozone levels, the crystals in spruce needles appear different than under normal conditions.[32] Under acid conditions, the crystals are absent in the outer walls of the epidermal cells, whereas ozone exposure shows complexes of crystals surrounded by cellulose and callose extending from the

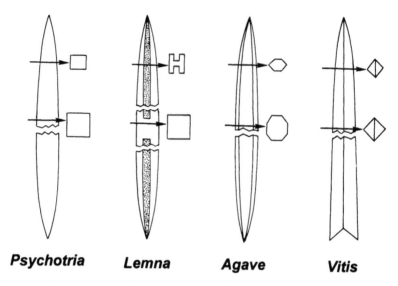

Psychotria   Lemna   Agave   Vitis

Figure 3. Diagram of four different types of needle-like raphide crystals found in higher plant cell vacuoles. Sectional views depict shapes as being four-sided (*Psychotria*), H-shaped (*Lemna*), six- and eight-sided (*Agave*), and divided (*Vitis*).

inside of the walls into the cells, thus, engorging the lumens of the epidermal and hypodermal cells. In the mesophyll cells, the crystal complexes may project from the walls into the vacuoles, or occur within the vacuoles. Dead cells display calcium oxalate, or phosphate, the latter probably due to destruction of the cytoplasm and excess calcium. It is thought that the more common apoplastic (wall) formation of the crystals is due to the ability of the conifer cells to actively pump oxalic acid from the cytoplasm into the walls where it combines with the calcium to form the crystals.[10] There is no indication as to how certain cells selectively carry out this latter process, and other cells form the crystals within the vacuoles and lumens of the cells. Certainly the former process does not require cells to develop specialized cytoplasmic compartments and vacuole machinery to deal with the intracellular transport and storage of the calcium. However, there are no developmental studies which clearly show what occurs cytoplasmically during early extra- and intracellular crystal formation in the conifers.

   Ginkgo biloba, a deciduous gymnosperm with a typically bilobed, flat leaf lamina, displays druses within the parenchyma cells surrounding the dichotomously branched veins, and between the veins in the leaf mesophyll.[4] Crystal formation seems to be intracellular, within the vacuoles, and may be similar to that observed in the angiosperms (see next section).

2. Angiosperms

   About 74 percent of the angiosperm families display calcium oxalate crystals.[9] There does not seem to be a correlation between whether the plants are C_3 or C_4, with respect to their photosynthetic system. Species containing

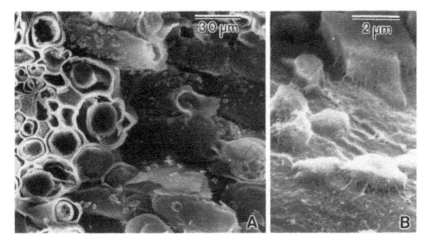

Figure 4. Crystals in outer portion of cell walls exposed to air spaces in *Tsuga* needle-like leaves. (A) Exposed walls of cells coated with crystals. (B) Close-up of crystals partially covered with wall material.

crystals show the widest variety of crystal shapes, forms, number, and locations. A particular species of plant may contain only one shape of crystal in a specific cell type or tissue within an organ, whereas another plant species may display different shapes and/or forms of crystals in the same or different tissues and organs. The important fact is that, whatever the situation, the location(s) is/are species specific. Examples of the former are prismatic crystals in special paired epidermal cells on *Canavalia* leaves[33] and crystal sand in leaves of *Beta*;[34] an example of the latter are the four kinds of crystals (druses, prisms, spherulites, and crystal sand) in the anther connective tissue of sweet pepper[35] and the prisms and druses in leaves of *Begonia*.[36]

The majority of studies dealing with higher plants crystals depict their mature stage, either within plant cells (typically in the vacuoles) or, in a few instances, within their surrounding walls. There are a relatively few studies that deal with the development of angiosperm crystals and the specialized cells in which they form.

III. CELLS FORMING CRYSTALS

A. GENERAL CONSIDERATIONS

In instances where crystals are associated with the walls of cells exposed to air spaces or to the environment, no special name is given to the cells other than that they are usually parenchymatous. There are other instances where crystals form in cells that, in other plants, appear similar but without the crystals. These cells do not have special names either, other than those already defined by their anatomy and location (i.e., photosynthetic palisade parenchyma of *Peperomia* species;[5] wood ray parenchyma of many woody species;[11,12] parenchymatous tissues in many aroids[25]). However, some researchers refer to all cells containing crystals as crystal idioblasts (see next section).

B. SPECIALIZED CELLS

Their are many cells which produce crystals where, prior to and during early crystal development, the cells may enlarge and take on shapes and sizes that are quite different from the non-crystal-forming cells of the same tissue.[5] These former cells are called crystal idioblasts.[6,37] Most crystal idioblasts develop from young cells and as these cells enlarge, changes occur within the cytoplasm and vacuole which lead to crystal formation. Mature cells may also produce crystals under special conditons.[38,39] In many idioblasts, the nucleus enlarges and the DNA increases, probably due to DNA amplification.[40] At the same time, unique structures appear in the vacuole and changes occur in the cytoplasm such as the development of specialized plastids (crystal plastids), cytoplasmic channels contiguous with the vacuole, increased endoplasmic reticulum, and elevated levels of ribosomal RNA. It is assumed these latter changes are involved with the structures that appear in the vacuoles, as well as locations for either the synthesis or transport of substances necessary for crystal formation.

C. CRYSTAL FORMATION IN VACUOLES

Formation of crystals in vacuoles is most common in the angiosperms. Vacuoles are fluid-filled cytoplasmic organelles surrounded by a single membrane called a tonoplast. The high turgor pressure of the vacuole fluid provides cells with their rigidity and many dissolved substances and ions are present.[41] Crystal idioblasts typically form from young meristematic cells which have a few, very small vacuoles. As these cells enlarge and differentiate, the small vacuoles coalesce, forming a large central vacuole that takes up most of the volume of the cell. In the few developmental studies that have been conducted on crystal idioblasts, the vacuoles develop in a similar manner as just described. In addition, vacuole structures associated with crystal formation appear as the vacuoles coalesce and enlarge. These vacuole structures vary, depending on the species. At the present time, two general systems can be described, based on the presence or absence of membranes associated with crystal development.

1. System I.

This system is represented by the druses in *Capsicum* (sweet pepper, Solanaceae)[42] and in *Vitis* (grape, Vitaceae),[43,44] the raphides in *Psychotria* leaves (leaf-nodulated genus, Rubiaceae),[45] and crystal sand in *Beta* leaves (sugar beet, Chenopodiaceae),[34] all dicotyledonous plants. This system is characterized by having such structures as cytoplasmic spherosomes, vacuole organic paracrystalline bodies, membrane complexes, plasmalemmasomes, and crystal chambers whose walls are membrane-like (Figure 5). Individual

Capsicum ,Psychotria CRYSTAL IDIOBLASTS

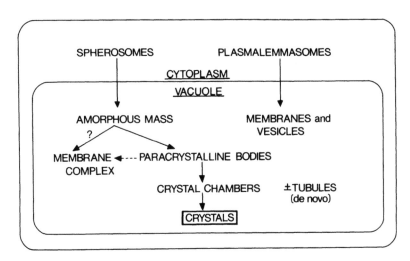

Figure 5. General diagram of a **System I.** crystal idioblast, as represented by *Capsicum* and *Psychotria*, showing special cytoplasmic organelles and vacuole structures.

tubular elements and vesicles also may be present in the vacuole. The paracrystalline bodies of varying shapes are unique structures which are organic and may display subunits with large periodicity. These bodies are distinct in sweet pepper and *Psychotria*, and more diffuse in sugar beet. The bodies are clearly associated with a network of membranes that are linked to, or may give rise to, the crystal chambers. Preliminary indications are that these bodies may consist of calcium-rich glycoproteins and may serve as the nucleation or initiation sights for crystal and/or crystal chamber formation. In *Psychotria* (Figure 6A-D), there are numerous bodies, with each giving rise to many chambers [2140 crystals (chambers) were counted in one crystal idioblast in *Eichhornia*[4]]. In sweet pepper stomium cells (Figure 7A-D), there are also numerous bodies, each of which serves as the origin and core for many crystal chambers resulting in a druse.[42,46] In sugar beet, the bodies are more diffuse, and the resulting small, individual sand crystals and their chambers are loosely bound to each other.[34] In *Vitis*, the organic matrix associated with druses has been analyzed.[43,44]

Tubules 10-13 nm are found around the developing crystal chambers in *Psychotria*, but not in sweet pepper, sugar beet or *Vitis*. It is presumed these tubules may help to orient and align the elongating crystal chambers that contain the needle-like raphides. These tubules would not be necessary for the formation and orientation of the druses in sweet pepper or the crystal sand in sugar beet.

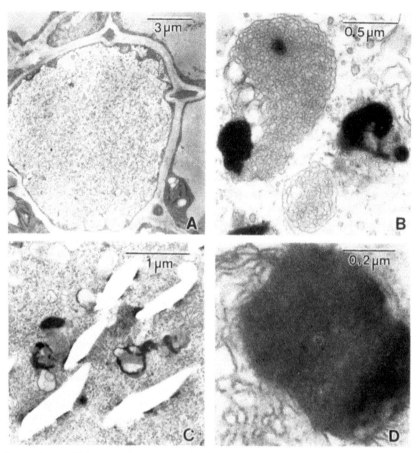

Figure 6. Special features of a *Psychotria* leaf **System I.** crystal idioblast. (A) Young idioblast without crystals in vacuole. (B) Portion of cytoplasm and vacuole containing paracrystalline bodies, membrane complexes, vesicles, and tubules. (C) Center of vacuole showing same structures as in (B), as well as crystal chambers. (D) Paracrystalline body and associated membrane complex.

2. System II.

This system is represented by the raphide crystal idioblasts in *Typha* (cattail, Typhaceae),[47] *Vanilla* (Orchidaceae),[48] and *Yucca* (Agavaceae),[49] all monocotyledonous plants. This system differs from **System I.** in certain significant ways (Figure 8). There are no membrane complexes found in the vacuoles (Figure 9A-D). Paracrystalline bodies are present and display closely spaced subunits. However, flocculate material, possibly emanating from the bodies, is prevalent in the vacuole. Crystal chamber walls seem to form as interface aggregations of the flocculate material. In addition, filaments (not tubules) may form, in some species, and aggregate to produce cables of filaments that anchor together the ends of the raphide crystals.[47,49]

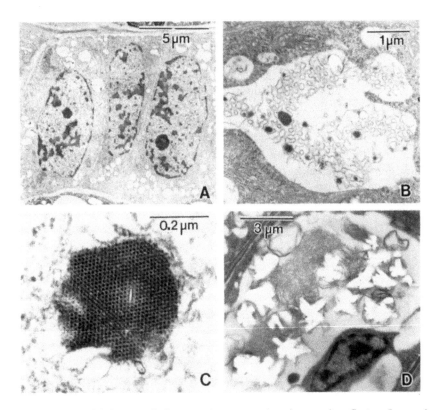

Figure 7. Special features of *Capsicum* (sweet pepper) anther stomium **System I.** crystal idioblasts. (A) Young idoblasts with small vacuoles. (B) Small vacuole with membrane complex and many, dense paracrystalline bodies. (C) Paracrystalline body showing several crystal chambers and associated membrane complex. (D) Vacuole containing druses associated with membrane complex. Paracrystalline bodies are surrounded by crystals and are not visible.

In *Yucca*, as well as in some other species containing **System II.**, a mucilage-like material appears around the developing crystal chambers. Lamellae arise in the mucilage around the crystal chamber walls (Fig. 10A, B). As these lamellae form, the raphide crystals change shape from being four-sided in cross-section to being six- and eight-sided.[28,49] Specialized plastids, called crystal plastids, may be present in the cytoplasm. The significance of these events and structures, if any, is not known.

3. Other Systems

Sweet pepper anthers display druses, prismatic crystals, spherulites and crystal sand in its connective tissue and stomium.[35] Druses[42,46] and crystal sand[34] have already been described as to their development. The prismatic crystals and spherulites may form separately in different connective cells or they may form together in the same cell and, physically in association with each other. Images of these developing cells shed little information on the development of the prismatics other than they seem to develop from accumulated masses of vacuolar calcium-rich glycoprotein material (Figure 11A, B). In contrast, starburst structures, consisting of central non-crystalline cores and radiating tubules, also appear in the vacuoles (Figure 11C). The tubules seem to attract the calcium-rich material similar to that associated with the prismatic crystals. This material surrounds the core and builds up along the tubules producing a spherical mass of dense material and crystals, collectively called a spherulite (Figure 11D). No membranes or other strucutures mentioned in **Systems I.** and **II.** are visible. These two examples of crystal formation are different from those described for the two systems, and suggest that, as more crystal-forming systems are studied in other plants, additional systems will be identified.

Typha, Vanilla, Yucca CRYSTAL IDIOBLASTS

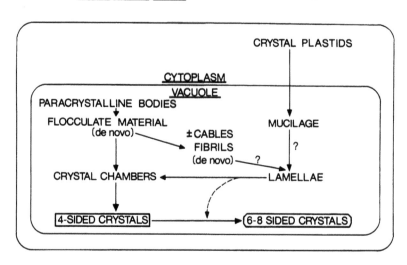

Figure 8. General diagram of a **System II.** crystal idioblast, as represented by *Typha, Yucca,* and *Vanilla,* showing special cytoplasmic organelles and vacuole structures. Vacuole flocculate material, paracrystalline bodies, and non-membrane-bound crystal chambers are major features.

4. Other Special Features

Previous sections have identified organelles (specialized plastids, endoplasmic reticulum) in the cytoplasm and structures (vacuole channels, vesicles, paracrystalline bodies, tubules, fibrils, membrane complexes, flocculate material, crystal chambers) within the vacuole that seem to be related to calcium oxalate crystal formation. All of this data is circumstantial evidence, based on ultrastucture information.

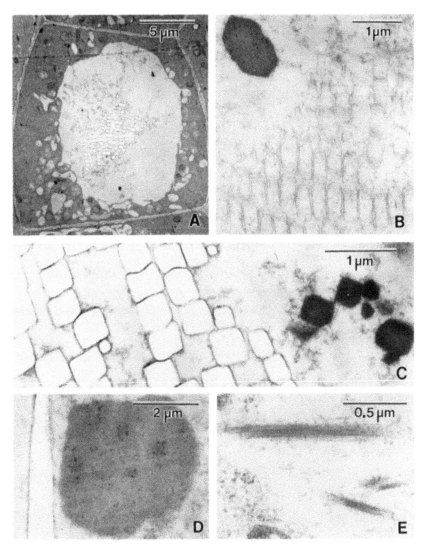

Figure 9. Special features of **System II.** raphide crystal idioblast from *Yucca.* (A) Young idioblast with flocculate material, paracrystalline bodies, and chambers in vacuole. (B) Paracyrstalline body, flocculate material, and young crystal chambers. (C) Several paracrystalline bodies, flocculate material, and older crystal chambers. (D) Paracrystalline body near crystal chamber. (E) Portions of cables consisting of vacuole filaments.

In water lettuce crystal idioblasts,[50] a calsequestrinlike protein has been immunologically localized in the calcium-enriched endoplasmic reticulum. It was concluded that these crystal idioblasts have a buffering system using this protein to bulk control calcium in plant cells. In addition, subcellular regions

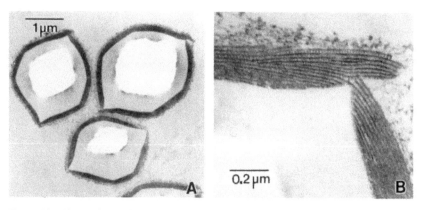

Figure 10. Six-sided crystal chambers in **System II.** *Yucca* raphide crystal idioblast vacuole showing sheath of lamellae. (A) Crystal chambers surrounded by lamellae. Holes represent crystal ghosts. (B) Portion of multi-layered lamellar sheath showing spacing elements between lamellae. Hole is crystal ghost.

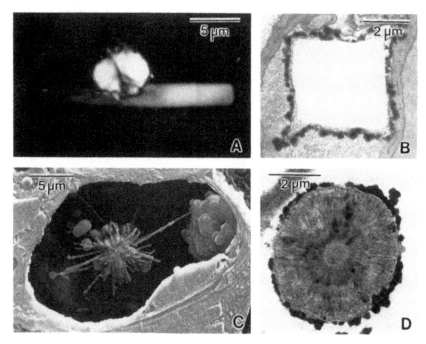

Figure 11. Crystals in connective tissue of sweet pepper anthers. (A) Polarizing microscopy showing a prism and spherulite bound together in one cell. (B) Calcium-rich glycoprotein (black) in vacuole surrounding crystal. (C) Starburst body in vacuole consists of central core, radiating tubules, and black material. (D) Section through a spherulite showing organic core, radiating tubules, and interspersed black material.

around crystals in the stomium and connective tissues of anthers in tobacco have been immunologically localized.[51] These regions have been identified histochemically as a calcium-rich glycoprotein in sweet pepper,[35] and possibly may be associated with the paracrystalline bodies, membranes, and dense material previously described. Various laboratories are trying to localize similar substrates in other plant idioblasts that will aid in better understanding the nucleation process and eventually the mechanism(s) controlling crystal formation. Figure 12. summarizes the cytoplasmic and vacuole structures found in crystal idioblasts, in general.

D. CRYSTAL FORMATION IN WALLS

Unlike the coniferous gymnosperms, only a few angiosperm species have been identified where prismatic crystals are either embedded in the cell walls or on the surface of exposed cell walls. Examples are the special astrosclereids and associated exposed parenchyma in *Nymphaea* leaves (water lilly, Nymphacaeae),[4,5,52] and the epidermal cell walls of branchlets and scale leaves in species of three genera of casuarina trees (Casuarinaceae). [53] In *Nymphaea* the crystals are sandwiched between the primary and secondary walls of the astrosclereid cell arms, and on the wall surfaces of nearby parenchyma (Figure 13A, B). There is no indication as to how these crystals form in this species. However, the crystals appear during both the early development of the astrosclereids and the formation of the air chambers. In Casuarinaceae species, epidermal cells forming the crystals show a distinct increase in the amount of cytoplasmic endoplasmic reticulum. Crystals first appear between the cell membrane and the primary wall. These crystals continue to enlarge as they become encased in wall material during further wall deposition. It was further suggested that the endoplasmic reticulum contains large quantities of calcium which, for unknown reasons, is transported to the outside of the cell, instead of into the vacuole.[53] The presence of the many crystals in these walls may serve as a defense mechanism against predators, particularly insects.

IV. EVOLUTIONARY CONSIDERATIONS

Calcium oxalate crystals are of widespread occurrence among extant animals and microorganisms, and in the algae, lichens, fungi, and higher plants. Calcium oxalate formation is part of the overall process of biomineralization, the latter being summarized as having its ancient origins explained by several hypotheses.[2] Whatever the origin of biomineralization, calcium oxalate formation is a highly specialized process involving living cells. In higher plants this process is not detrimental. It seems its main purpose is to regulate the presence of high concentrations of environmental and apoplastic calcium, by either sequestering it in intracellular compartments, or by secreting it back to the apoplast, in a form that is non-toxic, stable, and non-diffusible. Even though there is little proof regarding other functions the crystals and crystal idioblasts supposedly serve in higher plants, it is difficult to not envision that some of these plants have further evolved their ability to use these specialized cells and the calcium oxalate crystals for other specific purposes. Further research should help to clarify this suggestion.

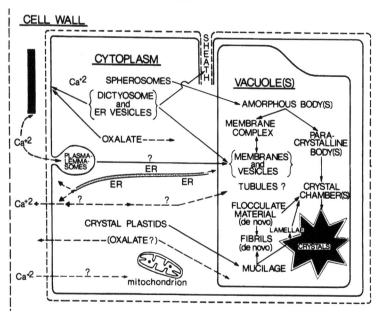

Figure 12. Summary diagram depicts cellular features of crystal idioblasts in higher plant cells, including cell wall.

V. FUTURE RESEARCH DIRECTIONS

The ability of higher plants to produce calcium oxalate crystals in association with cells provides unique opportunities for further study related to a number of different areas. 1. Crystal idioblasts may provide a better understanding as to how cells, in general, are able to concentrate, transport and sequester large amounts of calcium; 2. Understanding the breakdown of the crystals of lichens, fungi, and higher plants in the environment may provide insight about calcium and oxalate cycling and their roles in soil genesis and soil fertility; 3. Changes in number, location, shape, and/or chemical structure of crystals in higher plants may serve as physiological indicators, as well as pollution indicators of detoxification of certain environmental pollutants; 4. Understanding the metabolism of oxalate, and its association with calcium, may provide a means for either reducing levels of oxalate in certain plants to

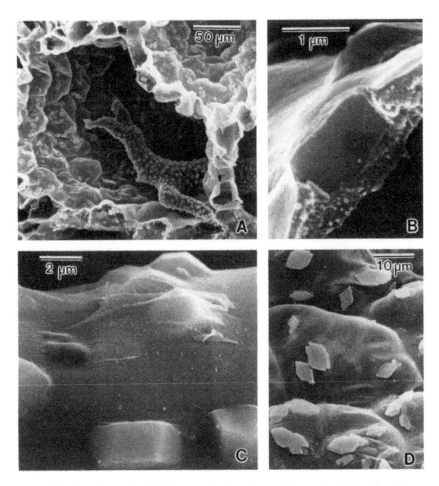

Figure 13. Crystals associated with astrosclereids and parenchyma cells of *Nymphaea* leaves. (A) Portions of astrosclereid arms studded with crystals; airspace parenchyma cell surfaces display crystals also. (B) Section through arm showing cell lumen, thick inner secondary wall and thin outer primary wall with a crystal sanwiched between them. (C) Surface view of astrosclereid showing crystals covered by primary wall. (D) Crystals on outside wall surfaces of airspace cells.

make them more edible, or to increase the levels of oxalate in some agricultural plants, to allow them to grow on tropical soils high in cations toxic to plants; 5. Understanding the genetics and molecular biology of higher plant crystal nucleation sights (organic bodies) as a way of controlling crystal induction and growth; 6. Exploiting information in 5. to determine whether a similar system is operational in animals and humans to reduce or eliminate kidney and bladder stone formation.

Acknowledgments: The authors thank the Department of Botany and the Bessey Microscopy Facility, a Life Sciences and Biotechnology Center, for partial support and use of their facilities for preparing this chapter. We also thank Joanne Nystrom for formatting and preparing the manuscript for publication and Dr. Elisabeth Zindler-Frank for her suggestions.

REFERENCES

1. Hodgkinson, A., *Oxalic Acid in Biology and Medicine*, Academic Press, New York, 1977.
2. Simkiss, K. and Wilbur K. M., *Biomineralization: Cell Biology and Mineral Deposition*, Academic Press, New York, 1989.
3. Horner, H. T., Tiffany, L. H. and Knaphus, G., Oak-leaf-litter rhizomorphs from Iowa and Texas: calcium oxalate producers, *Mycologia* 87(1), 34, 1995.
4. Arnott, H. J. and Pautard, F. G. E., Calcification in plants, in *Biological Calcification*, Schraer, H., Ed., Appelton-Century-Crofts, New York, 1970, 375.
5. Franceschi, V. R. and Horner, H. T., Jr., Calcium oxalate crystals in plants, *Bot. Rev.*, 46(4), 361, 1980.
6. Arnott, H. J., Three systems of biomineralization in plants with comments on the associated organic matrix, in *Biological Mineralization and Demineralization*, Dahlem Konferenzen, Nancollas, G. H., Ed., Springer-Verlag, New York, 1982, 199.
7. Metcalfe, C. R. and Chalk, L., *Anatomy of the Dicotyledons, I., Systematic anatomy of leaf and stem, with a brief history of the subject*, 2nd ed., Clarendon Prress, Oxford, 1979.
8. Metcalfe, C. R. and Chalk, L., *Anatomy of the Dicotyledons, II., Wood structure and conclusion of the general introduction*, 2nd ed., Clarendon Press, Oxford, 1983.
9. Zindler-Frank, E., Oxalate biosynthesis in relation to photosynthetic pathway and plant productivity -- a survey, *Z. Pflanzenphysiol.*, 80(S), 1, 1976.
10. Fink, S., Comparative microscopical studies on the patterns of distribution in the needles of various conifer species, *Bot. Acta*, 104(4), 306, 1991.
11. Chattaway, M. M., Crystals in woody tissue, Part I., *Tropical Woods*, 102, 55, 1955.
12. Chattaway, M. M., Crystals in woody tissue, Part II., *Tropical Woods*, 104, 100, 1956.
13. Sunell, L. A. and Healey, P. L., Distribution of calcium oxalate crystal idioblasts in leaves of taro (*Colocasia esculenta*), *Am. J. Bot.*, 72(1), 1854, 1985.
14. Zindler-Frank, E., Calcium oxalate crystals in legumes, in *Advances in Legume Systematics*, Part 3, Stirton, E., Ed., Royal Botanic Gardens, Kew, 1987, 279.
15. Kausch, A. P. and Horner, H. T., A comparison of calcium oxalate crystals isolated from callus cultures and their explant sources, *Scan. Electron Microsc.*, II, 211, 1982.
16. Cody, A. and Horner, H. T., Crystallographic analysis of crystal images in scanning electron micrographs and their application to phytocrystalline studies, *Scan. Electron Microsc.*, III, 1451, 1984.
17. Zindler-Frank, E., Calcium oxalate crystal formation and growth in two legume species as altered by strontium, *Bot. Acta*, 104(3), 229, 1991.
18. Kinzel, H., Calcium in the vacuoles and cell walls of plant tissues, *Flora*, 182(1), 99, 1989.
19. Libert, B. and Franceschi, V. R., Oxalate in crop plants, *J. Agric. Food Chem.*, 35(6), 926, 1987.
20. Franceschi, V. R., Oxalic metabolism and calcium oxalate formation in *Lemna minor* L., *Plant, Cell Environ.*, 10(5), 397, 1987.
21. Kausch, A. P. and Horner, H. T., Absence of $CeCl_3$-detectable peroxisomal glycolate-oxidase activity in developing raphide crystal idioblasts in leaves of *Psychotria punctata* Vatke and roots of *Yucca torreyi* L., *Planta*, 164(1), 35, 1985.
22. Borchert, R., Functional anatomy of the calcium-excreting system of *Gleditsia triacanthos* L., *Bot. Gaz.*, 145(4), 474, 1984.

23. **Borchert, R.**, Calcium-induced patterns of calcium-oxalate crystals in isolated leaflets of *Gleditsia triacanthos* L. and *Albizia julibrissin* Durazz, *Planta*, 165(3), 301, 1985.

24. **Thurston, E. L.**, Morphlogy, fine structure, and ontogeny of the stinging emergence of *Tragia ramosa* and *T. saxicola* (Euphorbiaceae). *Am. J. Bot.*, 63(6), 710, 1976.

25. **Genua, J. M. and Hillson, C. J.**, The occurrence, type and location of calcium oxalate crystals in the leaves of fourteen species of Araceae, *Ann. Bot.*, 56(3), 351, 1985.

26. **Gardner, D.G.** , Injury to the oral mucous membranes caused by the common houseplant, dieffenbachia. *Oral Surg. Oral Med. Oral Pathol.*, 78(5), 631, 1994.

27. **Webb, M. A. and Arnott, H. J.**, A survey of calcium oxalate crystals and other mineral inclusions in seeds, *Scan. Electron Microsc.*, III, 1109, 1982.

28. **Tilton, V. R. and Horner, H. T., Jr.**, Calcium oxalate raphide crystals and crystalliferous idioblasts in the carpels of *Ornithogalum caudatum*, *Ann. Bot.*, 46(5), 533, 1980.

29. **Frey-Wyssling, A.**, Crystallography of the two hydrates of crystalline calcium oxalate in plants, *Am. J. Bot.*, 68(1), 130, 1981.

30. **Rivera, E. R. and Smith, B. N.**, Crystal morphology and ^{13}carbon/^{12}carbon composition of solid oxalate in cacti, *Plant Physiol.*, 64(6) 966, 1979.

31. **Fink, S.**, The micromorphological distribution of bound calcium in needles of Norway spruce [*Picea abies* (L.) Karst.], *New Phytol.*, 119(1), 33, 1991.

32. **Fink, S.**, Unusual patterns in the distribution of calcium oxalate in spruce needles and their possible relationships to the impact of pollutants, *New Phytol.*, 119(1), 41, 1991.

33. **Zindler-Frank, E.**, Zu Entstehung und Lokalisation von Kristallidioblasten in der Höheren Pflanze, *Ber. Deutsch. Bot. Ges.*, 89(S), 269, 1976.

34. **Franceschi, V. R.**, Developmental features of calcium oxalate crystal sand deposition in *Beta vulgaris* L. leaves, *Protoplasma*, 120(3), 216, 1984.

35. **Horner, H. T. and Wagner, B. L.**, Association of four different calcium crystals in the anther connective tissue and hypodermal stomium of *Capsicum annuum* (Solanaceae) during microsporogenesis, *Am. J. Bot.*, 79(5), 531, 1992.

36. **Horner, H. T. and Zindler-Frank, E.**, Histochemical, spectroscopic, and X-ray diffraction identifications of the two hydration forms of calcium oxalate crystals in three legumes and *Begonia*, *Can.. J. Bot.*, 60(6), 1021, 1982.

37. **Foster, A. S.**, Plant idioblasts: remarkable examples of cell specialization, *Protoplasma*, 46(2), 184, 1956.

38. **Borchert, R.**, Calcium acetate induces calcium uptake and formation of calcium-oxalate crystals in isolated leaflets of *Gleditsia triacanthos* L., *Planta*, 168(4), 571, 1986.

39. **Borchert, R.**, Ca^{2+} as developmental signal in the formation of Ca-oxalate crystal spacing patterns during leaf development in *Carya ovata*, *Planta*, 182(3), 339, 1990.

40. **Kausch, A. P. and Horner, H. T.**, Increased nuclear DNA content in raphide crystal idioblasts during development in *Vanilla planifolia* L. (Orchidaceae), *Europ. J. Cell Biol.*, 33(1), 7, 1984.

41. **Wagner, G. J.**, Compartmentation in plant cells: the role of the vacuole, in *Cellular and Subcellular Localization in Plant Metabolism*, Vol. 16, Creasy, L. L. and Hrazdina, G., Eds., Plenum Press, New York, 1982, 1.

42. **Horner, H. T. and Wagner, B. L.**, The association of druse crystals with the developing stomium of *Capsicum annuum* (Solanaceae) anthers, *Am. J. Bot.*, 67(9), 1347, 1980.

43. **Webb, M. A. and Arnott, H. J.**, Inside plant crystals: a study of the noncrystalline core in druses of *Vitis vinifera* endosperm, *Scan. Electron Microsc.*, IV, 1759, 1983.

44. **Webb, M A.**, Analysis of the organic matrix associated with calcium oxalate crystals in *Vitis mustangensis*, in *Mechanisms and Phylogeny of Mineralization in Biological Systems*, Suga, S. and Nakahara, H., Eds., Springer-Verlag, Tokyo, 1991, 117.

45. **Horner, H. T. and Whitmoyer, R. E.**, Raphide crystal cell development in leaves of *Psychotria punctata* (Rubiaceae), *J. Cell Sci.*, 11(2), 339, 1972.

46. **Wagner, B. L.**, Genesis of the vacuolar apparatus responsible for druse formation in *Capsicum annuum* L. (Solanaceae) anthers, *Scan. Electron Microsc.*, II, 905, 1983.

47. **Kausch, A. P. and Horner, H. T.**, The development of mucilaginous raphide crystal idioblasts in young leaves of *Typha angustifolia* L. (Typhaceae), *Am. J. Bot.*, 70(5), 691, 1983.

48. **Kausch, A. P. and Horner, H. T.**, Development of synctial raphide crystal idioblsasts in the cortex of adventitious roots of *Vanilla planifolia* L. (Orchidaceae), *Scan. Electron Microsc.*, II, 893, 1982.

49. **Kausch, A. P. and Horner, H. T.**, Differentiation of raphide crystal idioblasts in isolated root cultures of *Yucca torreyi* (Agavaceae), *Can. J. Bot.*, 62(7), 1474, 1984.

50. **Franceschi, V. R., Li, X, Zhang, D. and Okita, T. W.**, Calsequestrinlike calcium-binding protein is exprressed in calcium accumulating cells of *Pistia stratiotes*, *Proc. Natl. Acad. Sci.*, 90(15), 6986, 1993.

51. **Trull, M. C., Holaway, B. L., Friedman, W. E. and Malmberg, R. L.**, Developmentally regulated antigen associated with calcium crystals in tobacco anthers, *Planta*, 186(1), 13, 1991.

52. **Kuo-Huang, L.-L.**, Calcium oxalate crystals in the leaves of *Nelumbo nucifera* and *Nymphaea tetragona*, *Taiwania*, 35(3), 178, 1990.

53. **Berg, R. H.**, A calcium oxalate-secreting tissue in branchlets of the Casuarinaceae, *Protoplasma*, 183(1), 29, 1994.

Chapter 5

CALCIUM OXALATE IN FUNGI

Howard J. Arnott

I. INTRODUCTION

A. General.

Calcium oxalate crystals are commonly found associated with fungi; in fact, over a hundred years ago Anton de Bary (1887) stated that "calcium oxalate is a substance so generally found in the fungi that it is quite unnecessary to enumerate instances of its occurrence." In his monograph, De Bary (1887) also pointed out that the walls of *Psalliota campestris* and other species were so encrusted that the mycelia appeared chalky. He illustrated the occurrence of calcium oxalate as irregular crystal aggregates on the outer surface of *Phallus caninus* and needle-like crystals on the hyphae of *P. campestris*. Hein (1930) also illustrates the needle-like crystals of *Agaricus campestris* (*P. campestris*) grown on compost with 30-45% water content. In describing his culture of this commercial mushroom, he said, "Mycelia present a diffused appearance because of the more or less complete covering of needle-like and other crystals." He reported the crystals were from 1 to 20 μm in length and 1-3 μm in thickness, and that "all [crystals] react to the usual tests for calcium oxalate." Heckman et al. (1989) called hyphae with needle-like crystals of calcium oxalate "ornamented hyphae." They found that ornamented hyphae appear after only a few days of growth.

Similar needle-like crystals have been observed by scanning electron microscopy (SEM) (Arnott and Fryar, 1984) on the hyphae of fungi associated with compost litter and were studied on the hyphae of *Agaricus bisporus* (Arnott and Whitney, 1987), on a variety of gymnosperm litter samples from five southwestern states (Blackmon, 1992) and on oak leaf litter from Texas and Iowa (Horner et al., 1995).

This paper is fashioned to provide an introduction to the literature of calcium oxalate in fungi, to consider some of the research literature and to examine some previously unpublished electron microscopic studies. It is not intended to be an inclusive discussion of calcium oxalate in the fungi and other microorganisms.

Foster (1949) says "calcium oxalate is deposited as granules or minute crystals in the membranes of many common fleshy fungi: *Agaricus, Lactarius, Russula, Cantharellus, Boletus, Polyporus, Fistularia, Lycoperdon, Lectia and Peziza.*" Many types of calcium oxalate crystals, on or near the surface of the hyphae, have been reported to be present in association with different groups of fungi, slime molds, lichens and other microorganisms (Arnott, 1982a, 1984, 1987; Arnott and Webb, 1983; Arnott and Whitney, 1986a,b, 1987, 1988; Benny and Khan, 1988; Cairney, 1990; Cairney and Clipson, 1991; Farley et al., 1985; Foster, 1949; Hamlet and Plowright, 1877; Heckman et al., 1989; Hintikka et al., 1979; Horner et al., 1985 a, b, 1995; Jackson, 1981; Keller, 1985; Krisai and

0-8493-7673-4/95/$0.00+$.50
© 1995 by CRC Press

Mrazek, 1986; Lapeyrie et al., 1990; Powell and Arnott, 1985; Schoknecht and Keller, 1977; Sunhede, 1990; Timdal, 1984; Waldsten and Moberg, 1985).

B. Calcium oxalate in systematics.

Systematists report (or sometimes only illustrate) calcium oxalate in their work, however, calcium oxalate crystal structure does not usually seem to be of systematic importance. Notable examples of such work are Krisai and Mrazek (1986), Sunhede (1990) on *Geastrum* and Keller (1985) on the Aphyllophorales. Sunhede's magnificent monograph on Northern European species of *Geastrum* illustrates the exact location and character of calcium oxalate crystals in 14 of the 26 species he studied (Table 1). In a careful SEM study, Krisai and Mrazek

Table 1. Characteristics of peridial calcium oxalate crystals in *Geastrum*.

Species name	Hydration and Types	Size	Author(s)
G. minus	COD, small and large bipyramids; COM, rosettes	0.5 to 95 µm	Horner, et al. (1985)
G. saccatum	COD, bipyramidal druses, large flat eroded	5 to 20 µm	Whitney & Arnott (1986)
G. minimum	COD, many flat tetragonal bipyramids, short prisms topped by pyramids, sometimes druse like; endoperidium and exoperidium	40 to 120µm 5 to >100µm	Krisai & Mrazek[1] (1986) Sunhede (1990) present author
G. quadrifidum	COD, pyramids, single or in loose aggregates	11 to 30 µm	Krisai & Mrazek (1986) Sunhede (1990)
G. elegans (*G. badium*)	COD, prisms, bipyramids, compact druses	1.3 to 4 µm	same authors
G. campestre (*G. pedicellatum*)	COD, single bipyramids	30 to 70 µm	same authors
G. schmidelii (*G. nanum*)	COD, single or crowded, flat bipyramids	6 µm	same authors
G. pectinatum	COD, few prismatic-bipyramidal	6 µm	same authors
G. berkeleyi	COD, perfect bipyramids	4.5 to 10 µm	same authors
G. striatum	COD, druses composed of bipyramids	0.2 to 2 µm	same authors
G. coronatum	CO, crystalline matter	tiny	Sunhede (1990)
G. pouzarii	CO, crystalline matter	2 to 5 µm	same author
G. leptospermum	CO, crystals small	2 to 12 µm	same author
G. triplex	CO, fine crystalline pruina	2 to 3 µm	same author
G. fornicatum	CO, many patches of crystalline matter, over 200 µm in breadth	2 to 8 µm	same author
G. hungaricum	CO, crystalline grains	5 to 40 µm	same author

[1]Characteristics are integrated when reported by more than one author (except for *G. minimum*). The three taxa in parentheses are synonyms of Sunhede's (1990) taxa as used by Krisai & Mrazek (1986). They are employed here without taxonomic prejudice.

(1986) showed that 8 out of 14 *Geastrum* species produced calcium oxalate crystals (Table 1). From a taxonomic viewpoint, while they found crystal shape constant within species, "distinct systematic significance" is doubtful; however, the total absence of calcium oxalate in certain species "seems to be more significant." Keller (1985) studied the cystidia of 60 species of Aphyllophorales using SEM and x-ray diffraction. His paper is replete with excellent SEM micrographs of cystidia. Using them he was able to identify ten different types of crystals which he classified into four major groups: tetragonal bipyramids, prisms, tablets and needles (Fig. 1). However, he found that the crystal morphology was not of great systematic use even though the crystal structure was usually consistent

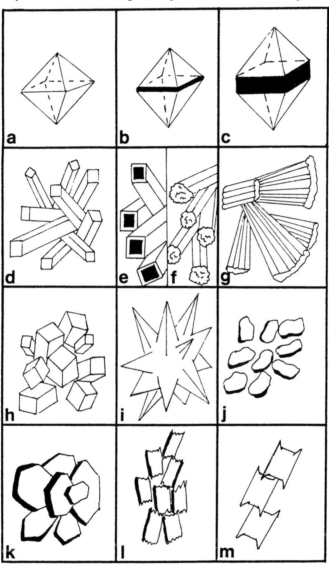

Figure 1. Types of calcium oxalate crystals found on the cystidia of the Aphyllophorales. Redrawn from Keller, J., Les Cystides Cristalliferes des Aphyllophorales, Mycologia Helvetica, 5, 1985, 277.

within collections of the same species. All the species of Aphyllophorales that he studied had crystals of calcium oxalate as determined by x-ray analysis. Crystals found on two species Keller studied are worth noting; in *Bulbillomyces farinosus* (collected in Sweden) he found rod-shaped square crystals with square holes and in *Tubulicium clematidis* (from France) he found what I would call, "a terminal nest of rod-shaped crystals resting on a bed of organic material (Fig. 1 e, f). Both represent unique structures. Timdal (1984) used the presence of calcium oxalate in the hypothecium as a delimiting character for the lichen genus *Psora*. However, when considering the relationship between *Psora, Eremastrella* and *Lecidea pulcherrima* he says, "it is more probable that this character [the presence of calcium oxalate] has developed independently in the three taxa."

In an ambitious SEM survey, Blackmon (1992) studied litter samples from 28 sites in Colorado, New Mexico, Oklahoma, Texas and Utah. From a series of several hundred micrographs, she chose 80 to document the nature of the crystal complements of litter samples found in a variety of different sites. Most of the litter was derived from gymnosperms (Blackmon, 1992). Fourteen (50%) of the sites had only raphide-like crystals, ten (36 %) of the sites had only druse-like crystals while four (14%) had bumps and blisters associated with the litter hyphae. None of the 28 sites had mixtures of druse-like and raphide-like crystals, however, several sites showed positive evidence of recrystallization. Raphide-like crystals were the prevailing type associated with litter from Junipers (*Juniperus sps.*) and Firs (*Abies sps.*) while druse-like crystals prevailed in Pine (*Pinus sps.*) litter. Blackmon (1992) attempted to correlate basic soil types found in the collection sites with the type of crystals produced on the litter but she was unable to discover any significant trends.

C. Mycorrhizal involvement with calcium oxalate.

Hysterangium crassa is believed to be a mycorrhizal fungus often found associated in deep mats of forest litter (Cromack et al., 1979). *H. crassa* forms calcium oxalate very strongly and the crystals have been implicated with the retention of calcium and the freedom of certain elements in the soil (Graustein et al., 1977, Cromack et al., 1979). The formation of weddellite by the mycorrhizal fungus, *Paxillus involutus*, has been reported by Lapeyrie et al. (1984) and in *Monotropa uniflora* by Snetselaar and Whitney (1990). In the former case, using SEM, massive crystals of COD were found on the surface of hyphae grown *in vitro*, however, more recently Lapeyrie et al. (1990) studied the same system using TEM and believe they have found examples of amorphus calcium oxalate associated with hyphae grown in pure culture. The evidence for this was stated as follows: "At high magnification, they [the "crystals"] appear heterogeneous, made of spherical vesicles...These vesicles varied in size and density. Some crystals have geometrical shapes." Later they say, "These inclusions are protoplasmic...or vacuolar." In general, their pictures appear to be similar to those of Arnott and Webb (1983) or Figs. 4, 15, and 35 in the present report. An alternate explanation for their findings in the non-contrasted sections might be that the crystalline calcium oxalate has undergone beam damage similar to that shown by Arnott and Pautard (1970) in Fig. 36. It is clear that calcium oxalate is easily damaged by an

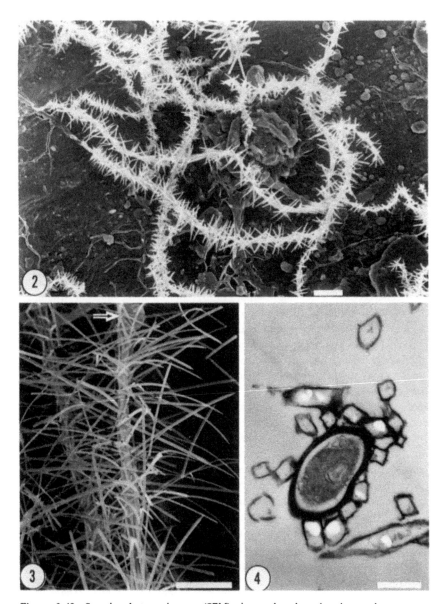

Figures 2-49. Scanning electron microcope (SEM) micrographs unless otherwise stated.

Figure 2. Crystalliferous hyphae bearing sharp-pointed crystals of calcium oxalate on the surface of an aspen leaf. Hyphae without crystals are also seen. Note that crystals are absent on the leaf surface. Bar = 10 μm. (From Arnott, H. J., Calcium oxalate (weddellite) crystals in forest litter, Scanning Electron Microscopy 1982, 1141. With permission.)

Figure 3. Basidiomycete hyphae bearing acicular crystals of calcium oxalate from Arlington, Texas compost. Note sharp ends of crystals extending from within the hyphae; arrow points to clamp connection. Bar = 5.0 μm. (From Arnott, H. J., and Fryar, A., Raphide-like fungal crystals from Arlington, Texas compost, Scanning Electron Microscopy, 1984, 1745. With permission.)

Figure 4. Cross section of crystalliferous hypha bearing raphide-like crystals as seen with TEM. Specimen from Arlington, Texas compost. Bar = 0.5 μm.

electron beam and this damage could produce the "protoplasmic" or "vacuolar" type structures reported in their paper. Recently Arnott et al. (1992) have shown the rapid time course in electron beam damage of calcium oxalate crystals. In their first paper, Lapeyrie et al. (1984) showed clearly by x-ray diffraction and SEM that crystalline deposits occured. Referring to their earlier paper, Lapeyrie et al. (1990) say "the external depositions have been firstly considered as mainly calcium oxalate crystals, however, while some depositions gave an electron microdiffraction spectrum characteristic of crystalline compounds, most of them are amorphous." They continue, "This is more in agreement with the 'vesicular structure' of these external depositions observed in TEM." In saying this, they ignore their previous x-ray diffraction study and rely on electron diffraction. As has already been mentioned, calcium oxalate is electron beam sensitive and this is especially so with the concentrated beam used for electron diffraction where rapid breakdown of the crystalline calcium oxalate can occur. The results reported by Lapeyrie et al. (1990) may also be explained as beam artifacts.

D. Rhizomorphs.

Currently research workers are looking at the internal structure of rhizomorphs or mycelial cords. For example, Cairney (1990) reported on the internal structure of mycelial cords in *Agaricus carminescens* where he found cords with a substantial degree of internal differentiation. Loosely interwoven hyphae formed the outer layer of the cortex and were thin-walled and crystal encrusted (crystalline material not identified). The medulla, which constituted approximately 70% of the cord's cross-sectional area contained "large diameter 'vessel' hyphae" that sometimes had crystalline deposits. In another study, Cairney and Clipson (1991) state, "The internal structure of mature rhizomorphs appears to be intraspecifically consistent...in *Agaricus* species at least, there appear to be strong structural similarities at the generic level" (Cairney, 1990). Cairney et al. (1988) reported on the internal structure of "linear mycelial organs" (rhizomorphs) of four basidiomycete species. *Mutinus caninus*, *Tricholomopsis platyphylla* and *Steccherinum fimbriatum* exhibited crystal-encrusted hyphae in their outer layers and crystals in some internal cells while *Lycoperdon pyriforme* did not produce crystals. The crystals were not identified in this study, however, *Mutinus caninus* has been reported to form calcium oxalate crystals (see above). While it is not in the general scope of this paper to discuss rhizomorph structure, it is clear that some taxonomic insight to the fungi that populate litter may be gleaned by understanding the rhizomorph structure. Papers by Cairney and his coworkers provide an entry into this literature which begins as far back as Hein (1930).

Figures 5-9. Stages in the development of raphide-like crystals on the hyphae of compost litter from Arlington, Texas. (From Arnott, H. J., and Fryar, A., Raphide-like fungal crystals from Arlington, Texas compost, Scanning Electron Microscopy, 1984, 1745. With permission.)

Figure 5. Very early stage in crystallogenesis, note how the crystals extend outward from inside of the hypha. Bar = 2.0 μm.

Figure 6. Early stage in the production of raphide crystal by a hypha. Bar = 1.0 μm.

Figure 7. Elongating crystals which still show evidence of wall stretching. Bar = 1.0 μm.

Figure 8. Hypha bearing numerous short (low length/width ratio) twin crystals. Bar = 1.0 μm.

Figure 9. Hypha with low length/width ratio twin crystals. Reentrant angles can be seen on many of the crystals. Bar = 0.5 μm.

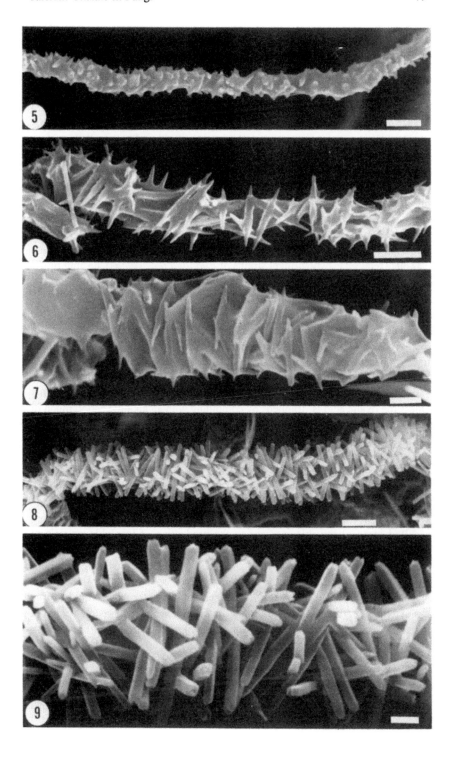

E. Plant diseases and calcium oxalate.

Calcium oxalate formation in association with fungi that produce plant diseases has been the subject of several reports (McCarroll and Thor, 1978; Noyes and Hancock, 1981; Punja and Jenkins, 1984; Havir and Anagnostakis, 1985; Punja, 1987; Bennett, 1989; Bennett and Hindal, 1990; Punja et al., 1995). *Rhizoctonia carotae* and *Sclerotium rolfsii* are disease organisms which produce calcium oxalate in or around infection sites (See Punja, 1987, Fig. 5d). It is believed that during infection these fungi secrete both oxalic acid and cell wall degrading enzymes (especially polygalacturonase). As the middle lamella is broken down, calcium (normally utilized to bond pectins together) is released and is then precipitated with oxalic acid to form crystals of calcium oxalate in association with the fungal hyphae. A similar precipitation of calcium oxalate crystals has been noted in cultures of *S. rolfsii* where Punja and Jenkins (1984) found that both changes in medium composition and pH affected the amount of oxalic acid produced. Havir and Anagnostakis (1985)found that the ability to produce oxalate was quite reduced in hypovirulent strains of *Endothia parasitica* (Chestnut blight organism) as compared to virulent strains.

F. Oxalic acid secretion and crystal formation.

According to Foster (1949) "oxalic acid occupies a unique situation in the history of mold metabolism since it was the first oxidation product resulting from the aerobic breakdown of carbohydrate found to accumulate in fungus cultures. It was the forerunner of a great variety of oxidation products since discovered..." It is interesting to point out that very high yields of oxalate may be formed from various substrates (acetate, tartrate, malate, citrate, glycerol, peptone, etc., and that unless the oxalic acid is "trapped" by combining with an alkaline cation much of it will be decomposed by the fungus (Foster, 1949)). The presence of cations in the medium allows the determination of just how much oxalic acid is formed by various fungi. Foster (1949) reports that a 10 day old culture of *Aspergillus niger* grown on peptone substrate produced 0.5 g of oxalic acid while the dry weight of the fungus itself amounted to only 0.23 g. He also reports on other fungi which secrete oxalic acid into the medium. In cultures of *Sclerotium rolfsii*, Punja and Jenkins (1984) reported that the addition of calcium to the growth medium resulted in an observable increase in the production of insoluble crystals of calcium oxalate.

Given the proposition that oxalic acid is secreted by fungi growing in culture and is then trapped in the medium surrounding it by combining with calcium ions, it is no surprise that Graustein et al. (1977) assumed that the same would be true in soil fungi like *Hysterangium crassum*. In a stunning 1977 Christmas cover of Science, Graustein et al. showed an amazing SEM view of this soil fungus with

Figures 10-12. *Armillariella tabescens* in a laboratory culture.
Figure 10. Early development of crystals at the hyphal tip. Note that the tip is free of crystals but that very close to the actual tip many small crystals are already in stages of development. Bar = 0.5 μm.
Figure 11. Hypha a short distance from the tip showing the development of the crystals from within (see arrow) and near the surface. Bar = 0.5 μm.
Figure 12. Hyphal tip with more rapid development of crystals just below the tip. Individual crystals appear to be pressed outward and into the surface wall material. Bar = 1.0 μm.

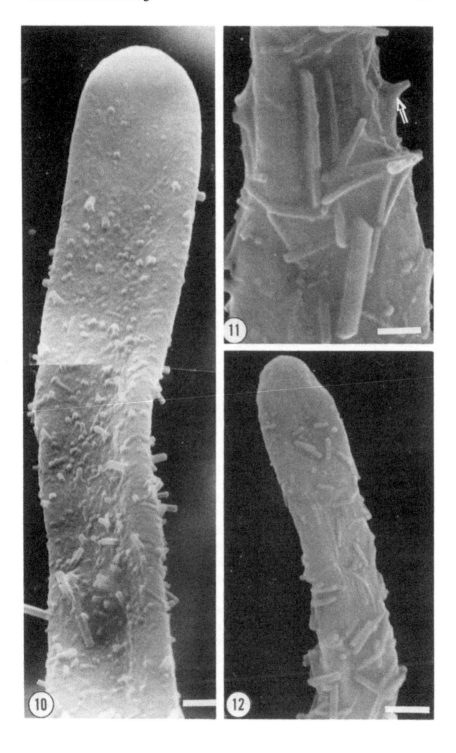

abundant star-like calcium oxalate crystals attached to its hyphae at relatively regular intervals. They believed that the fungus secreted oxalic acid and that crystals of calcium oxalate precipitated on the surface of the fungus when the oxalic acid interacted with calcium ions in the soil. However, the regularity of the deposits could have been a hint that it was not a simple precipitation. The investigations of Arnott (1982), Arnott and Webb (1983) and Arnott and Fryar (1984) showed that it was clear that the development of calcium oxalate crystals comes from within the hyphae, and that, in general, a surface precipitation is not the condition one finds in the case of calcium oxalate crystals associated with fungal hyphae.

G. Crystallinity and hydration state of calcium oxalate.

Most calcium oxalate deposits in the fungi are crystalline, and often the crystals are large enough to visualize by light or polarization microscopy. The morphology of the crystals forming the deposits is extremely variable: the individual crystals may be separate and completely independent of any other crystal, or they may be arranged in a variety of different arrays (druse-like, fan-like, etc.). Many of the arrays are obviously the result of one or more twinning operations. In shape, the crystals may be elongate, needle-like, plate-like, spine-like, large or small rhombohedrals, or pyramidal (Figs. 1, 38). In some cases the crystal shape appears to be the result of a biofabrication, as a result there are no obvious crystal facets. Whatever their shape, most fungal crystals are found directly on the hyphal surface, however, fungi growing on agar cultures can produce calcium oxalate crystals in the medium away from the hyphae (Foster, 1949; Punja, 1987). When fungal calcium oxalate deposits are studied by x-ray diffraction they are often identified as calcium oxalate dihydrate (COD) weddellite (Arnott, 1982; Arnott and Fryar, 1984; Cromack et al., 1979; Graustein et al., 1977; Horner et al., 1985b; Lapeyrie et al., 1984; Punja and Jenkins, 1984; Whitney and Arnott, 1987). Calcium oxalate monohydrate (COM) whewellite crystals have also been reported to occur (Graustein et al., 1977; Verrecchia et al., 1993; Horner et al., 1995). The two hydration states can sometimes be determined on the basis of birefringence; COM crystals are strongly birefringent while COD crystals have only a "pale" level of birefringence. They sometimes can be distinguished because of their crystal morphology as COM and COD crystals are in different crystal systems; COM crystals belong to the monoclinic system while COD crystals belong to the hexagonal system.

Some authors believe that when COM crystals are found associated with fungal deposits they are the result of an alteration to COD crystals previously produced by the fungus. Through some kind of diagenetic process the monohydrate crystals result from a recrystallization process (Verrecchia , 1993). It is well established that the monohydate is the more stable form of the two common hydration states of calcium oxalate and that diagensis would normally lead from COD to COM which is the lower free energy state. However, Horner et al. (1995) believe that the opposite is true in the oak rhizomorphs which they studied. Using x-ray diffraction and polarization microscopy they found that the youngest portions of the oak rhizomorphs, collected from Iowa and Texas, have COM crystals while

Figures 13-14. Cross sectional fractures through the hyphae of *Armillariella tabscens* in a laboratory culture fixed with glutaraldehyde, frozen in liquid nitrogen, fractured and critial point dried.
Figure 13. Fractures showing hyphae with cytoplasm and bearing elongate crystals or bubble-like structures (right). Bar = 1.0 μm.
Figure 14. Fracture showing thick-walled, apparently nonliving, hyphae bearing numerous crystals. Note that the crystals extend perpendicularly from the hyphal wall and that some "crystals" have bulbous or expanded areas (arrows). Bar = 1.0 μm.

the older parts of the rhizomorphs have COD crystals; likewise, a change in the shape of the crystals from young to older parts of the hyphae was depicted in their micrographs. They interpret this data to indicate that the first crystals formed on the hyphae are COM and that these are later "converted" into COD crystals and that the transformation in hydration form is accompanied by a concomitant alteration in the shape of the crystals in young and older regions of the rhizomorphs. Presumably the energy required for these changes could be derived from the metabolic activities of the fungus or from some other as yet unidentified source.

Currently, when calcium oxalate is found associated with fungi it is necessary to consider that the deposits may be either the monohydrate or dihydrate or that both may be present. The present state of knowledge also indicates that the hydration state may be the result of diagenesis and/ or biological processes which lead to changes in crystal morphology and/or hydration state; furthermore, we must consider the possibility that changes may occur either from COD to COM or from COM to COD.

II. LITTER CRYSTALS

A. General.

My first contact with calcium oxalate in litter came from a micrograph that Susan Pratt showed me after I had just given a paper on calcium oxalate crystals in higher plants. What was clear was that the structures in her micrographs of forest litter from Utah had a great deal in common with calcium oxalate raphide crystals in higher plants. Using some samples she sent me I began to investigate the structure of fungi associated with litter. Natural litter is composed of all types of organic material (leaves, stems, bark, wood, cones, flower parts, animal fragments, insects, dung, etc.) found under and in association with plants. Litter may be especially thick, 50 cm, and occupy many square meters in some forest situations (for example, the dense forests of Oregon or Washington) or it may be extremely thin, perhaps only 1 or 2 mm in thickness and less than a meter in diameter when seen in association with pinion pines and junipers in arid areas of the Great Basin. Still, both sites will almost certainly have fungal hyphae associated with their litter fragments. SEM examination of white or gray litter fragments often show many hyphae and/or rhizomorphs growing across their surface (Figs. 2, 24). Commonly the hyphae are decorated with crystals of calcium oxalate (Figs. 2, 3, 5-9, 24-26). Often it is possible to separate the fungus, with its crystals of calcium oxalate, from the substrate and these can be examined by x-ray diffraction. However, when doing x-ray diffraction it is important to remove any higher plant material from the sample as calcium oxalate may be present in almost any plant parts commonly found in litter.

Figure 15. Raphide-like and tabular or plate-like crystals on a hypha of *Agaricus bisporus* in laboratory culture. Bar = 1.0 μm. (From Whitney and Arnott, Calcium oxalate crystal morphology and development in *Agaricus bisporus*, Mycologia, 1987, 180. With permission.)
Figure 16. TEM view of plate-like crystals associated with a hypha of *Agaricus bisporus* in laboratory culture. Bar = 1.0 μm.

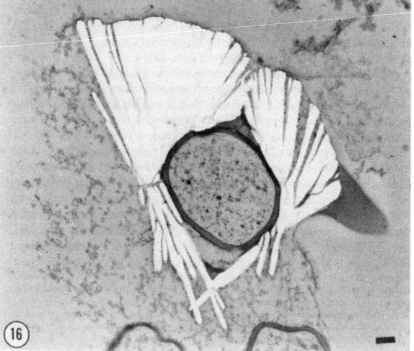

Recently Horner, et al. (1995) studied unidentified fungi producing calcium oxalate crystals on oak-leaf-litter. They found dagger-like COM cyrstals, and thin, plate-like crystals as well as COD crystals with pyramidal ends in Texas litter. The "oak-leaf-litter" from Iowa had a mixture of COD and COM, being predominantly short or elongated styloid-like crystals growing on leaves.

There is a discussion in the literature concerning the usefulness of studying unidentified litter organisms. In 1983 H. T. Horner and A. M. Cody expressed concern over this subject in the published review of Arnott and Webb (1983), saying, "This is the third paper on a fungus producing Ca oxalate where the fungus has not been taxonomically characterized...future workers will not be able to discern which organism was studied..." In answering this criticism Arnott and Webb (1983) said, "it is important to examine organisms in nature, to see what happens in natural systems and to use electron microscopy as a tool in these investigations... Faced with a decision as to whether to study organisms as they are in nature, providing as much information about their taxonomy as one can and a careful discussion of the location and site of collection or to spend our time on culturing organisms, we chose the former." The concept of studying calcium oxalate-producing organisms in nature still seems both "ethically and scientifically correct" as there are many other investigators who specialize in the study of fungi in aseptic laboratory cultures.

B. Needle-like crystals.

Investigation of litter from Utah showed many hyphae growing on the surface of litter fragments all bearing numerous elongate needle-like structures extending from the hyphae like the bristles of a bottle brush (Figs. 2, 3, 13, 14). Examination of these structures does not immediately call to mind a series of crystals. The structures are long, with a length/width ratio approaching 20/1. Many are flexible in the electron beam and when isolated or attached the structures are often strongly curved (Fig. 3). However, examination of these materials with x-ray diffraction clarified the situation and indicated that calcium oxalate crystals were present. Confirmation of their character was also obtained by energy dispersive x-ray analysis which showed that the long narrow structures contained only calcium, carbon and oxygen (Arnott and Webb, 1983). Careful enumeration of the number of crystals per hyphal length and their distance apart in the Arlington, Texas compost litter provided data that show that as many as 3×10^9 crystals can be found per cm^2 and that the crystals included in a cm^3 sample would have a surface area of $450 \ cm^2$ (Arnott and Fryar, 1984). In her study of litter from 28 sites in the Southwestern U.S., Blackmon (1992) found that 14 contained needle-like crystals associated with the fungal hyphae. TEM views of sectioned hyphae bearing raphide-like crystals show that each crystal is covered by a membrane-like

Figures 17-19. Litter samples collected in the mountains near Provo, Utah (Arnott # 270), fixed and freeze etched at Brigham Young University courtsey of Dr. W. Hess. The surfaces of hyphae and the raphide-like crystals are covered with rodlets.

Figure 17. Raphide-like crystal attached to the surface of a hypha. Note the difference between the rodlets on the surface of the hypha which form a herringbone pattern and those on the surface of the crystal which are orientated along its long axis. Bar = 100 nm.

Figure 18. Surface view of rodlets on a raphide-like fungal crystal. Note that the majority of the rodlets are oriented parallel to the long axis of the raphide crystal. Bar = 100 nm.

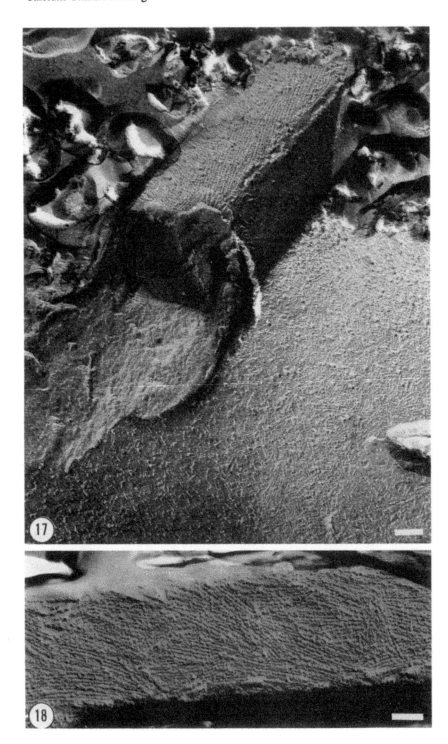

sheath which is attached to the hyphal wall at some point. Little distinction between the hyphal wall and the crystal "membrane" can be made (Fig 4.), however, it is clear that this "membrane" has the same sharp angles that would be consistent with the facets of a crystal.

Examination of these and other samples showed that considerable variation in the length of the crystals was possible. For example, some were relatively short and thick (Figs. 2, 8, 9). In some samples it was almost impossible to still see the central hypha around which the crystals cluster; however, in broken hyphae it is possible to show the relationship between the hyphal wall and the crystals. The central lumen is maintained in size and position and the crystals appear to be attached to the periphery of the hyphal cell wall. The short crystals which decorate these hyphae often appear to have reentrant angles characteristic of twins (Fig. 9). It is now well established that the raphide crystals of higher plants, *Vitis* for example, are twins (Fig. 9). It appears that these raphide-like crystals in the litter fungi are also sometimes twins. Such is not the case for the longer crystals mentioned above in which both ends are pointed and no reentrant angle can be seen (Fig. 3).

The development of raphide-like crystals in litter samples was studied using the SEM by Arnott (1982) and again by Arnott and Fryar (1984). Raphide-like crystals arise within the hyphae and grow outward (Figs. 5-7). As the crystals grow and extend, a separate hyphal wall often can be seen about the individual crystals (Figs. 6, 7). This developmental style appears to be true for both elongate and short needle-like crystal systems (Figs. 3, 8, 9, 22e, 22g). Careful observation of hyphae in which crystals are developing shows sharp pointed crystals that develop in 2 directions, their development stretching the hyphal cell wall as they grow from the hyphal axis (Figs. 6, 7).

In laboratory cultures of *Armillariella tabescens* views of developing hyphae show that hyphal tips are free of crystals, however, only a few μm from the tips crystals are already developing on the surface of the hyphal wall (Figs. 10,12). As these crystals develop they begin to orient in a radial direction (Figs. 10-12) so that when the hyphae mature elongate crystals extend radially in all directions from the hyphal surface, sometimes from extremely thick cell walls (Figs. 13,14). In views where the inside of the hyphae can be seen no examples of crystals can be seen internally. They seem to be completely "external," that is, they develop and are a "part" of the hyphal wall. In this case the crystals do not develop in two directions similar to that shown in the litter examples above, but rather grow in one radially oriented direction, a substantially different pattern of development. In a few cases it was possible to see developing crystals which clearly have an inflated (wall) material around them (Fig 13). It also appears that additional crystals can develop from "blisters" such as those seen on the hyphal surface (Fig. 13).

Some litter fungi have hyphae with "crystals" distributed in a more or less uniform manner over their surface (Figs. 20-23). These "crystals" may be angular and flat (Fig. 20), blister-like (Fig. 21), sharp and triangular in profile (Fig. 22), or "post-like" (Fig. 23); all give x-ray spectra with a strong calcium peak.

Figure 19. Fractured raphide-like crystal showing rodlets covering the surface. End views of the rodlets can be seen at the fracture plane. Each rodlet appears to be formed from a "twisted double strand." Bar = 50 nm.

C. Druse-like crystals.

Druse-like crystals of calcium oxalate were found in a regular distribution on the mycelium of *Hysterangium crassum* by Graustein et al. (1977). In a Douglas fir stand Cromack et al. (1979) found that 9.6% of the upper 10 cm of soil was occupied by fungal mats. In addition to the druse-like crystals, rhombohedral and irregular COM crystals were also seen. The authors believed that the crystals found on the surface of the fungi were the result of oxalate secretion and precipitation on the hyphal surface. Graustein et al. (1977) also called attention to the potential ecological effects of oxalate crystals in calcium retention and the importance of oxalate in the solubility of various elements necessary for plant growth. Druse-like crystals were also found associated with the mycelium attached to the lower surface of the basidiocarps of *Geastrum minus* (Horner et al., 1985). Crystals of this morphology are widely distributed in litter. Blackmon (1992) found that 36% of the litter samples from the Southwestern United States that she examined had druse-like crystals attached to their hyphae. Arnott and Webb (1983) found numerous examples of druse-like calcium oxalate crystals associated with a pine wood rot fungus. These usually showed regular distribution and size along a particular hypha (Figs. 24, 25, 31). The individual deposit appeared to be a multiinterpenetrant twin in which the individual crystals had the shape of a steep four-sided pyramid, a typical tetragonal pattern exhibited by COD crystals (Fig. 26). In some cases the individual crystals are almost plate-like in that certain parameters of their growth are exaggerated (Figs. 26, 31).

In the pine associated litter Arnott and Webb (1983) were able to find developmental stages which clearly showed that the druse-like crystals arise internally. A series of stages was found extending from very slight bumps on the surface of the hypha (Figs. 27, 28) through intermediate stages (Figs. 29, 30) to mature stages (Figs. 26, 31). Two things were noted in studying developmental stages. First, throughout development, whether young or old, the crystals or their precursors were spaced more or less evenly along the hypha. Secondly, all the structures, crystals or their precursors, appear to be at the same stage of maturation; if the crystals were large all were large, if the crystals were just beginning to develop then all were at that stage (Figs. 27-31). In a few cases, however, the arrangement and the shape of the crystals was not consistent (Fig. 25). Here, we found that not only was the distribution uneven but also that a variety of different crystal development stages were present on the same hypha.

D.   Terminal crystal arrays. Terminal crystal arrays were found in the bark beetle excavations of *Pinus ponderosa* bark. These arrays were similar to the druse-like crystals (see above) except they were on side branches and each side branch was capped with a terminal druse-like crystal array (Figs. 32-34, 38h). Note that the stalks shown in Figs. 32 and 33 are entirely free of crystals. The terminal crystal arrays are also like some cystidia reported by Keller (1985). It appears that in this system certain parts of the hypha are designated to produce the crystal   deposits while other areas are designed to be free of the crystals. The crystals making up these arrays were similar to the druse-like arrays mentioned above and appear to represent some sort of interpenetrant twinning process. In a few cases circular druse-like deposits were found. These consisted of calcium oxalate crystals arranged in a ring around a limited area of a hypha (Fig. 31, 38h). TEM observations of thin-sectioned druse-like crystals seem to show that each crystal is contained within an individual membrane-bound compartment. These compartments are irregular near the center of the druse-like array but become pointed and have sharp angles at the periphery of the arrays (Fig. 35). When the membrane bound compartments are seen associated with a hypha, little distinction can be made between the "membrane" and the cell wall.

Figures 20-23. Hyphae found growing on mixed broad leaf and gymnosperm litter collected at Beaver's Bend, Broken Bow, Oklahoma, collected by W. McDonald.

Figure 20. Numerous small angular crystals often arranged in groups. Bar = 1.0 μm.

Figure 21. Round to angular crystals closely packed on the surface. Energy dispersive x-ray examination showed that the "crystals" contained only calcium . Bar = 1.0 μm.

Figure 22. Irregular, randomly oriented, kinked or triangle-shaped crystals extending in random directions from the hyphal surface. Bar = 1.0 μm.

Figure 23. Post-like crystal appendages with dimensions of about 1 x 0.5 μm extending from the surface of the hypha in radial directions and distributed in a regular pattern. Bar = 1.0 μm.

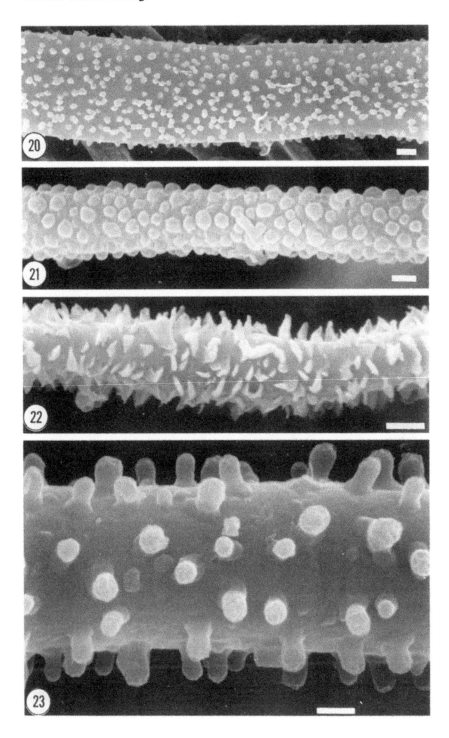

III. RODLET STRUCTURE AND DISTRIBUTION
A. Rodlets on spores.

The surface of certain fungal spores and hyphae is covered with rodlets. These fungi include: *Penicillium* (Hess et al., 1968), *Aspergillus* (Hess and Stocks, 1969; Cole et al., 1982; Claverie-Martin et al., 1986), *Neurospora* (Beaver and Dempsey, 1979) and *Tilletia indica* (Gardner et al., 1983). Similar rodlets have been found on the surface of bacterial spores (Holt and Leadbetter, 1969). Rodlets are reported to be short or long rod-shaped structures about 7-100 nm in diameter (Holt and Leadbetter, 1969; Dempsey and Beever, 1979; Gardner et al., 1983; Arnott, 1984; Claverie-Martin et al., 1986), sometimes hollow (Dempsey and Beever, 1979; Claverie-Martin et al., 1986), usually found in fascicles and, in most fungi, are found only on the external cell surface. Rodlet alignment changes during ontogeny so that different parts of a conidiospore may have rodlet fascicles with patterns which are representative of the stresses occurring during development (Cole, 1973). The rodlet surface of conidia may break up into patches during germination (Hobot and Gull, 1981).

The rodlets of *Neurospora crassa* conidia have been chemically characterized and consist of 91% (or more) protein, 2 % carbohydrate, and 0.9% phospholipid (Beever et al., 1979); those of *Trichophyton mentagrophytes* microconidial walls were 80 to 85 % protein and 7 to 10% glucomannans (Hashimoto et al., 1976); and, the conidia of *Aspergillus nidulans* 41.7% protein, 40.2% melanin and 17.5% carbohydrate in the wild type, and 46% protein, 3.8% melanin and 14.1% carbohydrate in a white mutant (Claverie-Martin et al., 1986) . A melanin-like pigment was associated with *Trichophyton* while a β-carotene was associated with the *Neurospora* rodlets (Beever et al., 1979). The rodlet layer of *Agaricus bisporus* (Rast and Hollenstein, 1977) and *Schizophyllum commune* (Wessels et al., 1972)is composed of β-$(1\rightarrow3)$-glucan. Several authors have suggested that the rodlet layer is hydrophobic and is probably responsible for the difficulty of wetting conidia with water and is hence an aid to dispersion. The relative amounts of amino acids present in rodlet proteins are shown in Table 2. In *Neurospora crassa* over 57% of the constituent amino acids present can be characterized as non-polar hydrophobic amino acids; in *Trichophyton mentagrophytes* and in *Aspergillus nidulans* over 40% of the constituents are non-polar hydrophobic amino acids. Certainly the characteristics of these amino acids must play an important role in the hydrophobicity demonstrated in conidia.

B. Rodlets on calcium oxalate crystals.

Rodlets are also associated with the surface of both druse- and raphide-like calcium oxalate crystals as well as with the vegetative hyphae that bear the crystals (Arnott, 1984). Freeze-etch preparations of gymnosperm litter viewed with TEM show small rod-shaped structures (rodlets) on both the surface of hyphae and crystals (Figs. 17-19, 36, 37). The surfaces of hyphae form an intricate herringbone pattern, often associated with a rough or irregular surface contour (Fig. 17). On the cell wall the rodlets are present in groups of 4-7 in parallel fascicles which collectively form the herringbone pattern, a pattern similar to that

Table 2. Amino acid composition of rodlet layers in fungi.

Amino Acid	Mole %		% Dry weight
	*N. crassa**	*T. mentagrophytes*	A. nidulans
Lysine	4.98	7.86	1.07
Histidine	0.39	1.26	4.48
Arginine	0.57	1.11	0.41
Aspartic acid	13.96	15.67	3.66
Threonine	4.76	5.10	1.35
Serine	9.19	3.99	1.74
Glutamic acid	4.22	8.26	3.03
Proline	5.81	3.38	1.53
Glycine	10.68	9.02	2.33
Alanine	8.85	6.97	2.11
Cysteine	8.09	1.90	
Valine	8.03	7.47	1.53
Methionine	1.64	0.14	
Isoleucine	6.39	3.23	1.05
Leucine	7.38	7.40	1.62
Phenylalanine	1.84	2.69	1.06
Tryptophan	0.21		
Tyrosine	3.01	1.29	1.63
Cysteic acid		0.07	

*Rodlet layers of *Neurospora crassa* conidia (Beever et al., 1979), *Trichophyton mentagrophytes* microconidial wall (Hashimoto et al., 1976) and *Aspergillus nidulans* conidia (Claverie-Martin et al., 1986).

seen on the surface of the conidia of several species. In the initial report of these crystal rodlets Arnott (1984) listed their diameter as ca. 100 nm, however, current measurements indicate their diameter is 5 nm with a 7 nm center-to-center distribution (Fig. 19).

There is a distinct difference in the orientation of the rodlets, or rodlet fascicles,

Figures 24-31. Wood rotting fungus from the wood of *Pinus ponderosa* collected in Pagosa Springs, Colorado. (From Arnott, H. J., and Webb, M.A., The structure and formation of calcium oxalate crystal deposits on the hyphae of a wood rot fungus, Scanning Electron Microscopy, 1983, 1747. With permission.)

Figure 24. Mycelium with hyphae producing druse-like crystals growing over the surface of decaying wood. Many druse-like deposits of various sizes can be seen. Note the multilayered nature of the mycelium. Bar = 50 μm.

Figure 25. Area of aberrant hyphae in which druses-like deposits and other crystal types are developing asynchronously. Bar = 10 μm.

Figure 26. Isolated strand of hyphae with druses-like deposits on a glass background. Note the common occurrence of twinning in the formation of these bodies. Bar = 5.0 μm.

Figures 27-31. Series of hyphae showing developmental stages in the formation of druses.

Figure 27. Almost smooth hypha is just beginning to form small protrusions which will develop into druse-like deposits. Bar = 1.0 μm.

Figure 28. Hypha with "young" crystal deposits just beginning to develop. Note crystal development at arrow and angular shape of deposit on the lower right. Bar = 1.0 μm.

Figure 29. Hypha with druse deposits developing in an asynchronous manner. Bar = 1.0 μm.

Figure 30. Intermediate stage in druse development, hypha growing close to the surface of wood elements (seen in background). Crystalline deposits arise on every side of the strand and are developing in a uniform and synchronous manner. Bar = 1.0 μm.

Figure 31. Older intermediate stage in druse development. Note obvious twinning and partially collapsed hypha on the right end. Bar = 1.0 μm.

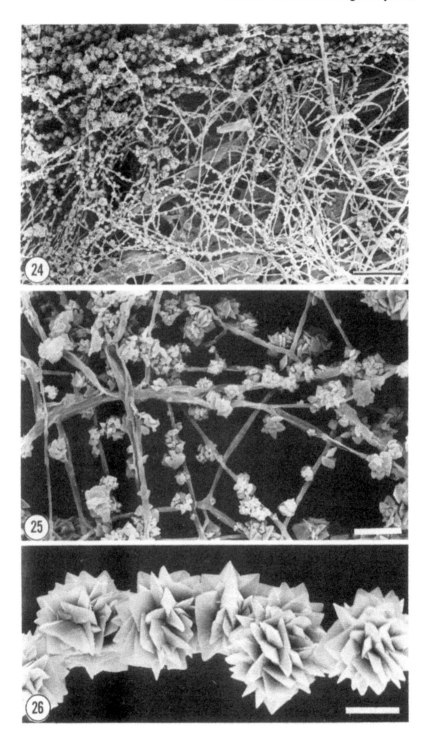

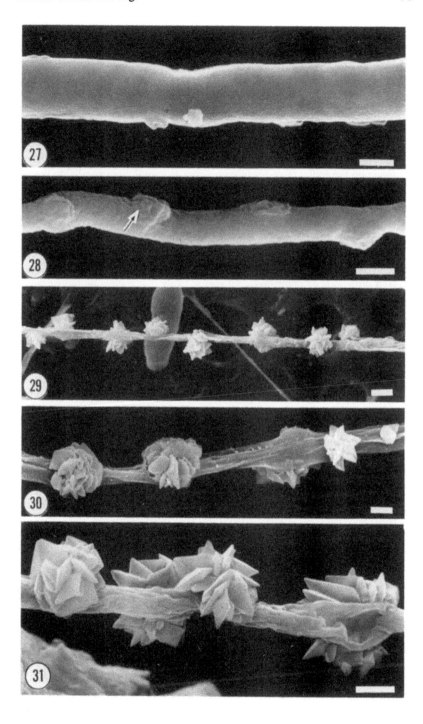

on the raphide-like crystal surfaces as compared to those of druse-like crystals (Figs 17-19). The rodlets are primarily oriented parallel to the long axis of the raphide-like crystals and the herringbone pattern characteristic of the vegetative hyphal surface is much less obvious. On the other hand the rodlet pattern seen on druse-like crystals does not show this type of fascicle orientation and is patterned much more like that of the vegetative hyphae (compare Figs. 17-19 with 36 and 37). Morphogenesis of the raphide-like crystals may account for this difference as it does in conidial development. The primary axis of growth in the raphide-like crystals is the long axis; growth in this direction would have the effect of placing a torsion on the rodlet fascicles in such a way as to pull them parallel to the axis of growth. Examination of the rodlet patterns on the surface of druse-like crystals shows little "growth" orientation (Fig. 36). In some cases, however, where sharp angles exist the rodlets are obviously "bent" around the edge. The chemical structure of the crystal rodlets is not known, however, it is clear from my own experience that the dry hyphae and/or rhizomorphs of litter samples which have calcium oxalate crystals are extremely difficulty to wet with water, but are easily wetted with oils or other lipids. It seems likely that rodlets on the surface of hyphae and crystals provide a hydrophobic surface like that of conidia. The rodlet layer impedes both the breakdown of calcium oxalate and of the litter fungi thus providing a relatively long term stabilizing mechanism within the litter. Because the rodlets are a part of the cell wall and thus produced by the fungus, this is conclusive evidence that the crystals were in fact produced inside the cell wall of the hypha bearing them (Arnott, 1984).

There are numerous types of calcium oxalate crystal deposits associated with litter fungi. The major types are shown in Figure 38, however, other types of crystals may occur. The more important consideration, however, is the wide occurrence of calcium oxalate crystals on the surface of litter fungi hyphae.

IV. CALCIUM OXALATE ON NON-LITTER ORGANISMS
A. Calcium oxalate crystals in *Geastrum*.

The genus *Geastrum* has about 50 species and is world wide in distribution (Krisai and Mrazek, 1986). Crystals associated with the surface of the endoperidium were first reported by Hollós in 1904 and have been reported by many authors since. The crystals are described as white, powdery, glistening "crystallin" or stipitate and in certain species can easily be seen by eye or with a hand lens. In *G. minimum* the crystals may have a size of more than 120 μm (Table 1). The crystals were first identified as calcium oxalate by Lohwag (1934, 1941). Investigators agree that most crystals are weddellite, COD (Horner et al., 1985; Krisai and Mrazek, 1986; and Whitney and Arnott, 1986), however, COM crystals were found in weathered specimens of *G. minus* by Horner et al. In the latter case the authors were unsure as to their origin and suggest they might be the result of recrystallization. The phenomenon of recrystallization is always a big bugaboo in the comprehension of the hydration state of calcium oxalate in soil organisms. Fifteen species of *Geastrum* are reported to have crystals on the endo- or exoperidium; for the most part these are believed to be calcium oxalate (Table

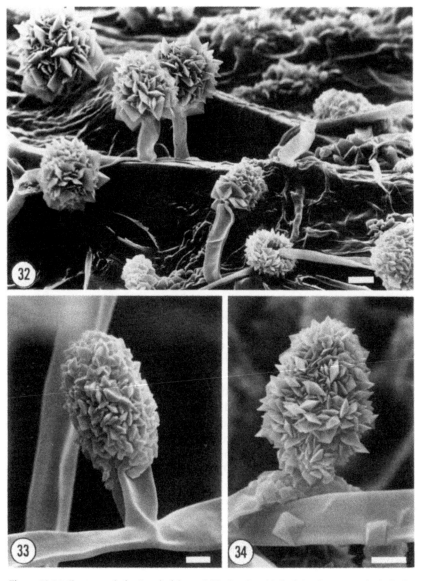

Figures 32-34. Fungus producing "terminal druses." Hyphae found in bark beetle passages in the bark of Pinus ponderosa collected in Pagosa Springs, Colorado.

Figure 32. This micrograph shows several terminal druses which are vertically oriented and extending upward as branches from hyphae following the surface. Note that the crystals appear only on the ends of the upright hyphal strands. Bar = 2.0 μm.

Figure 33. Hyacinth shaped terminal druse, note the complete absence of crystals on the lower stalk and the supporting horizontal hypha. Bar = 1.0 μm.

Figure 34. Terminal druse in which the stalk and the hypha from which the druse arises are all covered with crystals. Hypha in lower foreground has two developed and one developing pyramidal crystals. Bar = 2.0 μm.

1). The crystals range in diameter from 0.5 to 120 μm and are represented by a variety of shapes with both COD and COM having been reported to be the hydration states found in nature (Table 1).

Horner et al. (1985) found five different types of crystals on the surface of *G. minus*, which they characterized as follows: form One, small bipyramid crystals; form Two, crystals with split tips; form Three, crystals with oriented overgrowths; form Four, large bipyramid crystals; and form Five, rosettes of plate-like crystals. All except form Five were COD. These were made up of COM crystals. They also characterized the crystals present as to the stage of development. In unopened basidiocarps where exo- and endoperidia had not separated there were no crystals seen. In unopened basidiocarps in which the exo- and endoperidia had separated and in just fully opened basidiocarps they found forms One and Two. In exposed (weathered) peridia of fully open basidiocarps they found forms Three, Four and Five. As weathering progressed only forms Four and Five were found. At the soil level pale white mycelia surround the basidiocarps of *G. minus*. These had individual hyphae which displayed small druse-like clusters of crystals.

I have collected many specimens of *Geastrum* in the field. They often have crystals associated with either or both the endo- and the exoperidium. In Texas, among other species, *G. minimum, G. coronatum* and *G. saccatum* are found. All have crystals associated with the peridium. In sections of *G. coronatum* the adhering hyphae of the subtending mycelium have raphide-like crystals attached. Sunhede (1990) showed that it is possible to "cultivate" basidiocarps in the laboratory by bringing in suitable substrate from nature and maintaining it in a moist condition. The sporocarps shown in Figures 39-46 were grown from pine litter collected near Taos, New Mexico and cultivated in my laboratory. The individual sporocarps were collected as soon as they formed and thus were not subjected to any weathering. They developed from litter without any "eggs" (sporocarp primordia). In these specimens both the endo- and exoperidium are covered with relatively large crystals, most of which would be classified as "forms Three and Four" by the Horner et al. (1985) system. They have low birefringence and are tetragonal indicating that they are COD. Many of these crystals reach a diameter of 100+ μm, but most are of a somewhat smaller size. In this species the crystals jacket the entire endoperidium, including the peristome, but are not found on the stalk. The peridial surface has the appearance of sandpaper. Face view SEM examination of the peridial surfaces using the secondary electron detector is less than satisfactory for demonstrating the crystal density or distribution (Fig. 41). Using the backscattered electron detector, much better results were obtained and it is possible to clearly distinguish the large number and the area which is covered by the COD crystals (Figs 42, 43). Areas in the endoperidium without crystals appear to be matched by areas of the exoperidium which have crystals. This

Figures 36-37. Litter sample collected in Gilpin County, Colorado (Arnott # 264).

Figure 36. Freeze-etch preparation of rodlets on the surface of a druse. Irregular shadowing results from the high profiles shown by druses. Litter sample collected in Gilpin County, Colorado (Arnott # 264). Bar = 50 nm. See Figure legend 17-19.

Figure 37. Small portion of a hyphal surface exhibiting the normal herringbone pattern. Compare with the irregular pattern shown on the druse surface (Fig. 36). Bar = 50 nm.

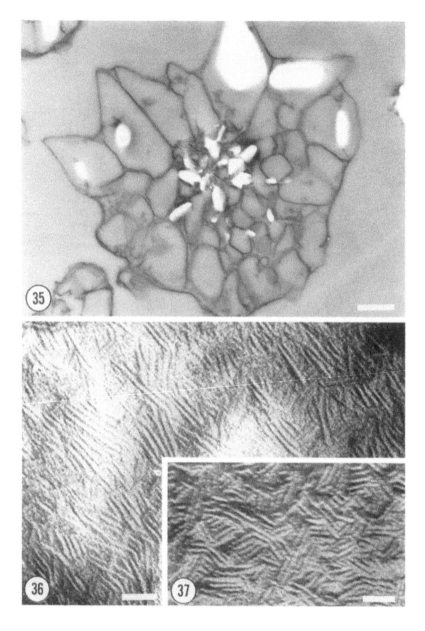

Figure 35. TEM section from an old rotting pine stem from the Kaibab National Forest, Arizona (Arnott 249) in which the plane of section does not run through the hypha. Over 50 crystals can be seen in this section. Most of the crystals were extracted by bulk uranyl acetate staining. Bar = 0.5 μm. (From Arnott, H. J., and Webb, M.A., The structure and formation of calcium oxalate crystal deposits on the hyphae of a wood rot fungus, Scanning Electron Microscopy, 1983, 1747. With permission.)

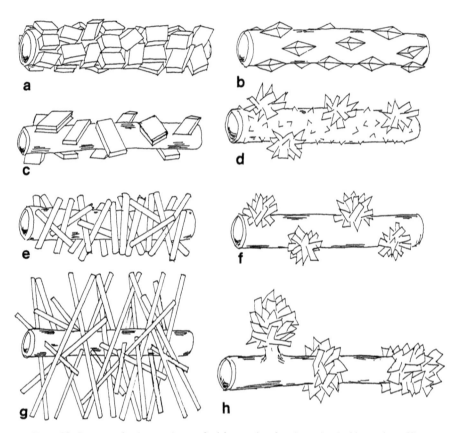

Figure 38. Summary of major crystal types of calcium oxalate found associated with a variety of litter fungi.

distribution of crystals shows that almost the entire surface between the endo- and exoperidium formed a solid layer of crystals just before splitting, and that some crystals end up on the exoperidium and the rest on the endoperidium.

Perhaps the best demonstration of the crystal distribution and morphology can be seen in tilted views of the endoperidium (Fig. 44). Here we can see another feature of the crystals found on the surface of the peridium, namely, that they often have hyphae associated with them in such a manner as to keep them attached to the surface (Figs 45, 46).

When the surface of the peridium, well endowed with crystals, is viewed under the dissecting microscope while a gentle stream of air is blown on it, many crystals will rise and move slightly in the air stream, however, they do not blow away. This

Figure 39-46. *Geastrum minimum* sporocarp produced in the laboratory from litter collected at the Fort Bergwin Field Station of Southern Methodist University, near Taos, New Mexico (Arnott # 467).
Figure 39. Endoperidium covered with granular calcium oxalate crystals. Although not clearly shown the exoperidium also has granular crystals on its surface. Bar = 2.0 mm.
Figure 40. Top view of endoperidium covered with granules, some of which are found on the peristome. Bar = 2.0 mm.

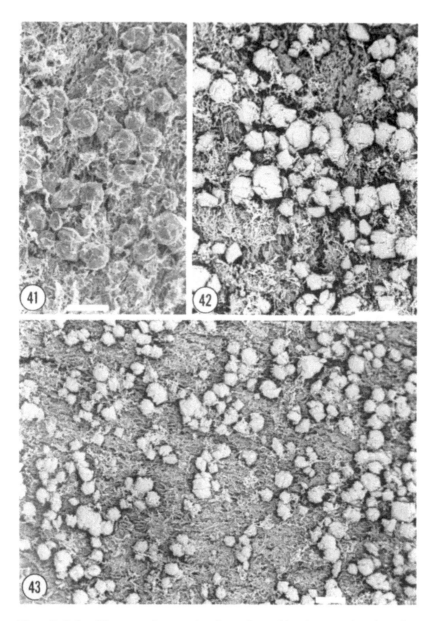

Figure 41. Endoperidium as seen in a secondary electron image. Many large crystals on the surface are somewhat difficult to interpret. Bar = 100 μm.

Figure 42. Endoperidium as imaged using the backscattered electron detector. A comparison of Figs. 41 and 42 clearly shows the advantage of the latter system in terms of viewing the calcium oxalate crystals. Bar = 100 μm.

Figure 43. Exoperidial surface as imaged with backscatter. Note the large number of crystals, many of which are over 50 and some near 100 μm in diameter. Bar = 100 μm.

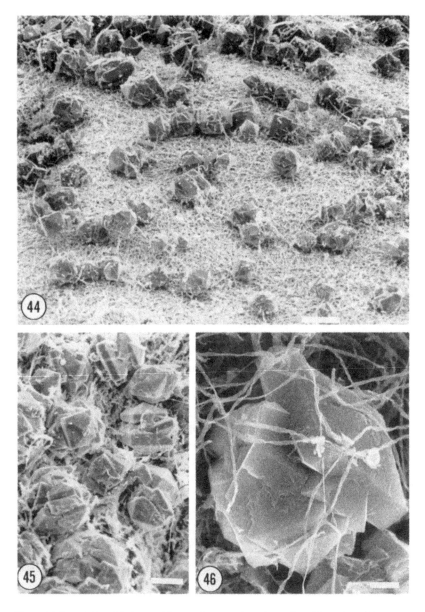

Figure 44. Lateral view of the large calcium oxalate crystals on the surface of the endoperidium. Bar = 100 μm.

Figure 45. Large multi-interpenetrant druse-like crystals on the surface of the endoperidium. Note the pyramidal or prismatic characteristics of many of the weddellite (COD) crystals seen on the surface. Bar = 30 μm.

Figure 46. Large multi-interpenetrant weddellite crystal on the surface of the endoperidium. Note the many hyphae which "tie" the crystal to the surface of the endoperidium. Bar = 10 μm.

kind of observation shows clearly that most of the crystals are actually attached to the surface of the peridium. Figure 46 is an especially graphic display of the many hyphae which "attach" the crystals to the surface by forming a network over their surface. Hyphae can also be observed that actually enter into the crystals. Sometimes several enter a single crystal at various points and occasionally they can be observed coming out on a different side; these hyphae are functional equivalent tie-down straps. It is likely that this situation comes about through the crystals forming about hyphae during crystallization. Whatever the method, it seems clear that attachment of the hyphae to the surface and to the crystals thus insures that the crystals will be maintained for some time on the surface of the sporocarp. The selective advantage of such a crystal tie-down apparatus is not clear.

B. Calcium oxalate crystals in the Mucorales.

Calcium oxalate crystals are found as the sporangial spines of *Mucor plumbeus* and *Cunninghamella echinulata* (Jones et al., 1976), the zygospores and sporangia of *M. mucedo* (Urbanus et al., 1978), the sporangia of *M. hiemalis* and *Rhizopus oryzae* (Powell and Arnott, 1985) and on the sporangia of *Gilbertella persicaria* (Whitney and Arnott, 1986). They were able to identify the first 2 as COD crystals. In *G. persicaria* Whitney and Arnott (1985) found four types of crystals. The first type was elongate plates. These were flattened crystals found along most of the sporangiophore. They are sometimes further subdivided into simple (no upright components) and complex (with upright appendages) plates. The second type is simple spines which consist of a flattened, polygonal base plate and a single, short, upright column. The third type was called complex spines which are comprised of three parts; a polygonal base plate, an upright, angular column and a flattened, hexagonal cap. The final type was patches of elongate bipyramidal crystals occurring along the length of the sporangiophore and called crystal plaques. The first three types lie within the walls of the sporangia and sporangiophores, whereas the crystal plaques appear to be produced outside the cell wall after the sporangia have developed and may represent recrystallization.

Figure 47. COD crystals on unfixed, fresh specimen of *Pyxine caesiopruinosa* showing matrix coating on crystals (arrows). Bar =1.0 µm. (from Jackson, D. W.., An SEM study of lichen pruina crystal morphology, Scanning Electron Microscopy III, 279, 1981. With permission.)

Figures 48-50. *Umbilicaria papulosa* (Ach.) Nyl. collected on large granite boulders in the JEOL Woods, Essex County, Mass. (Arnott # 323).

Figure 48. Three large prismatic weddellite crystals on the surface of a raised pustule on the thallus. In the foreground a large interpenetrant druse can be seen, all the crystals are partly embedded in the thallus surface. Bar = 5.0 µm.

Figure 49. Calcium dot map for same area as seen in Fig.50. Crystals give positive spectrum for calcium and because of their shape and birefringince they are presumed to be calcium oxalate dihydrate, weddellite. Bar =10 µm.

Figure 50. Area on upper part of the thallus containing many large prisms with pyramidal ends; same area as seen in the dot map of Fig. 49. Bar = 10 µm.

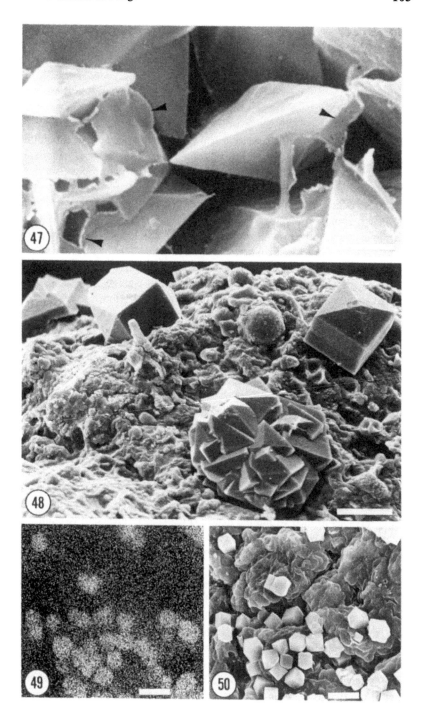

In a second study, Whitney and Arnott (1988) showed that *Gilbertella persicaria* when grown on a liquid synthetic medium without calcium soon lost the ability to form crystals of calcium oxalate. The addition of calcium to the medium restores normal form and crystal development. When no calcium is added to the medium, significantly more fungal dry weight is produced and the aerial portions of the colony increase in size dramatically. They concluded that the environmental calcium levels can modify the growth and development of *G. persicaria*. They state, "Further, the ability of *G. persicaria* to accumulate, transport and immobilize calcium in aerial portions of the mycelium suggests that calcium oxalate biomineralization in this fungus is potentially a detoxifying process, serving to reduce calcium ion concentration in the environment."

Crystals of calcium oxalate are formed by *Rhizopus oryzae* and *Mucor hiemalis* on their sporangia (Powell and Arnott, 1985). They vary in morphology from simple crystals, consisting of single spines in *M. hiemalis*, to complex crystals with twin spines. Each spine is sometimes three-parted. The spines are attached to a common base in *R. oryzae*. Spines in both species are at first covered by wall material, and while the spines of *M. hiemalis* rapidly emerge, those of *R. oryzae* appear to remain covered with a layer of wall material. Both species also have spines of a similar nature on their sporangiophores, however, the relative number of crystals (spines) is much less in *M. hiemalis* than in *R. oryzae* where a "felty" mass of spines is seen on the wall of the sporangiophore.

C. Calcium oxalate crystals in lichens.

Calcium oxalate is commonly associated with lichens (Jackson, 1981; Timdal, 1984; Wadsten and Moberg, 1985). Jackson (1981) studied the pruina, a surface layer of calcium oxalate crystals, in the lichens, *Pyxine caesiopruinosa, Physica aipolia, Heterodermia albicans* and *Physica americana*. He found a pruina made up of bipyramidal octahedra or dodechahedra crystals on the first three species and using x-ray diffraction identified the crystals as COD (Fig. 47). In *Physica americana* he found bipyramidal octahedrons which were identified as COD but also found flat plate-like crystals which he thought were COM. Jackson (1981) presented evidence that a matrix was secreted by the hyphae, and that most of the initial crystals formed within the matrix (Fig. 47). Evidence of twinning, and multiple twinning was very evident. Twins arise from preexisting crystals which nucleate further crystal growth in the form of twins. Both twins and the "original" crystals were contained in a matrix.

Arnott (present communication) has investigated the structure of *Umbilicaria papulosa* and *U. pennsylvanica* and found large rhombohedral crystals of calcium oxalate commonly associated with the thallus of these lichens. The most common type of crystals is large bipyramidial octahedra which approach 10 μm in length, however, a developed pruina does not form. Multiinterpenetrant twins still larger in size, are less common, but are often seen. In *Umbilicaria* the crystals are on both surfaces of the thallus (Figs. 48, 50). They also can be found in the central part of the thallus between loose strands of hyphae. In a few cases the crystals appear inside the hyphae. Examination of the surface crystals by SEM and energy dispersive x-ray analysis shows that the large rhombohedral crystals have a high calcium content (compare Figs. 49, 50).

V. CONCLUSIONS

The above discussion makes it clear that calcium oxalate is broadly distributed in the fungi. Certainly, all fungi do not produce calcium oxalate crystals or even oxalic acid, but in groups like the basidiomycetes and Mucorales calcium oxalate crystal formation is to be expected. De Bary's (1887) statement about the frequent occurrence of calcium oxalate is certainly authenticated by these observations. However, it is essential to recognize that **it is this very frequency** of calcium oxalate's association with fungi, that makes calcium oxalate significant. When considering the number of organisms that produce it and the rate at which it is produced, it seems likely that the fabrication of this compound provides a selective advantage to the organisms that produce it. Each instance of calcium oxalate occurrence is important in determining the absolute importance of its production. Furthermore, each instance is important in understanding the relationship between the fungi and their environment..

It also seems clear that fungi can produce both calcium oxalate crystals and secrete oxalic acid. When secreted, the latter may precipitate as calcium oxalate in the growth medium, with the number of crystals being dependent on the amount of calcium in the medium. Calcium oxalate crystals in nature may be produced through the secretion of oxalic acid, as in plant diseases, but more generally the crystals are formed in wall "chambers." which have specific sites of origin and are controlled by the organism bearing them.

Both hydration states, COD and COM, are found associated with fungi in nature. The production of COD crystals is more common than the formation of COM crystals. Clearly some very elaborate crystal types are formed which are not always representative of the hydration state present. Long needle-shaped crystals are not the normal morphology for COD crystals, however, they occur commonly in litter fungi.

Rodlets seem to be an important feature of some fungal crystals. The rodlets provide a "protective" cover over the crystals and alter their interaction with the environment around them. Such a protective mechanism may have considerable ecological consequence and certainly represents an interesting area for future research. The chemical nature of rodlets is of considerable interest in that it may give some indications of systematic relationships. The substances that rodlets are formed from, protein, carbohydrate, or other, and the pigments that they are combined with have potential systematic (taxonomic) importance.

It is of interest to speculate as to the importance of the calcium oxalate crystals. In some of the Mucorales, and perhaps in other groups, when large masses of calcium oxalate are produced on the hyphal surface these crystals may provide support for the cells they surround. In a sense, the rodlets are a kind of exoskeleton. Such a coating around calcium oxalate crystals may even provide protection from grazing. However, the physiological role played by calcium oxalate is more important from an evolutionary viewpoint. Oxalate can be used to control the level of Ca^{2+} to which cells are exposed. Once crystallized, the calcium in calcium oxalate is essentially removed from the physiological system and represents a calcium sink. Similar suggestions have been made for calcium oxalate in higher plants. Because Ca^{2+} is abundant in the earth's crust all organisms must deal with it. In this regard future research should examine the consequence fungal mineralization has, not only on the individual fungus or hypha, but also the effect on surrounding organisms. In mycorrhiza or other fungi, it is important to know what the elimination of calcium from the environment by crystal formation will do. It is also of interest to know what happens when calcium and oxalate ions are released through dissolution of crystals caused by microorganisms, weathering and/or changes in soil pH. Perhaps calcium oxalate works like a "buffer," forming a biologically controlled system which "regulates" various aspects of the soil chemistry. Additional investigation of rodlets is also important and could be of fundamental value in understanding how the "calcium oxalate buffer system" operates. Fron an evolutionary point of view it would be extremely interesting to obtain sequence analyses for a series of rodlet proteins. Understanding the role of calcium oxalate, rodlets and litter will be of considerable importance in forest ecology. Naturally, future research should look toward a better understanding of calcium oxalate-forming litter organisms. Specific identification of the fungal species that are present in litter formed by different plants or different groups of plants also will be of use.

VI. ACKNOWLEDGMENTS

I thank the Department of Biology, The Center for Electron Microscopy and The Graduate School of The University of Texas at Arlington for their support; Linda E. Lopez, who was of inestimable help in both electron microscopy and in the preparation of this manuscript; Mary Alice Webb, Kenneth Whitney, Virginiae Blackmon and Mike Powell who shared a common interest in fungal calcium oxalate; and Mike Davis who assisted in the preparation of illustrations. I also thank my wife, Jean Arnott, for enduring many collection trips and for her amiable tolerance of the frenetic activity connected with the preparation of this article.

REFERENCES

Arnott, H. J., Calcium oxalate (weddellite) crystals in forest litter. Scanning Electron Microscopy, III, 1141, 1982a

Arnott, H. J., Three systems of biomineralization in plants with comments on the associated organic matrix, In *Biological Mineralization and Demineralization*, Nancollas, G.H., Ed, Springer-Verlag, New York, 192, 1982b.

Arnott, H. J., Rodlets associated with the cell wall and crystals of litter fungi, in Proc. 42nd Annual Meeting of the Elect. Microsc. Soc. Amer., Bailey, G. W., ed., The San Francisco Press, Inc., San Francisco, 322, 1984.

Arnott, H. J., and Webb, M. A., The structure and formation of calcium oxalate cyrstal deposits on the hyphae of a wood rot fungus, Scanning Electron Microscopy, IV, 1747, 1983.

Arnott, H. J., Rodlets associated with the cell wall and crystals of litter fungi. in Proc. 42nd Ann. Meet. of the Electron Micro. Soc. of Amer., Bailey, G. W., Ed., San Francisco Press, San Francisco, California, 1984, 322.

Arnott, H. J., and A. Fryar, Raphide-like fungal crystals from Arlington Texas compost. Scanning Electron Microscopy, IV, 1745, 1984.

Arnott, H. J., Lopez, L. E. and Webb, M.A., Electron beam damage in crystals of calcium oxalate produced in the leaves of *Vitis vulpina*. Texas Soc. Electron Microsocpy J. , 23, 39, 1992.

Arnott, H. J., and Pautard, F. G. E., Calcification in Plants, in *Biological Calcification: Cellular and Molecular Aspects*, H. Schraer, Ed., Appleton-Century-Crofts, New York, 1970, 375.

Arnott, H. J., and Webb, M. A., The structure and formation of calcium oxalate crystal deposits on the hyphae of a wood rot fungus. Scanning Electron Mircoscopy, IV, 1747, 1983.

Arnott, H. J., and Whitney, K. D. Calcium oxalate crystal formation in a species of Hypoderma (Basidiomycetes). in Proc. 42nd Ann. Meet of the Electron Micro. Soc. of Amer., Bailey, G. W., Ed., San Francisco Press, San Francisco, California, 1984, 320.

Bennett, A. R., Mycleial growth and oxalate production by five strains of *Cryphonectria parasitica* in selected culture media, Mycologia, 81, 554, 1989.

Bennett, A. R., and Hindal, D. F., Mycelium formation and calcium oxalate production by dsRNA-free virulent and dsRNA-containing hypovirulent strains of *Cryphonectria parasitica*, Mycologia, 821, 358, 1990.

Benny, G. L., and Khan, S. R., The Radiomycetaceae (Mucorales; Zygomycetes). Calcium oxalate crystals on the sporangiolar wall and aerial hyphae, Scanning Microscopy, 2, 1199, 1988.

Beever, R.E., Redgwell, R. J., and Dempsey, G. P., Purification and chemical characterization of the rodlet layer of *Neurospora crassa* conidiospores., J. Bacteriol., 140, 1063, 1979.

Cairney, J. W. G., Internal structure of mycelial cords of *Agaricus carminescens* from Heron Island, Great Barier Reef, Mycol. Res., 94, 117, 1991.

Cairney, J. W. G., and Clipson, N. J. W., Internal structure of rhizomorphs of *Techispora vaga*, Mycol. Res., 95, 764, 1991.

Cairney, J. W. G., Jennings, D. H., and Veltkamp, C J., A scanning electron microscope study of the internal structure of mature linear mycelial organs of four basidiomycete species, Can. J. Bot., 67, 2266, 1988.

Claverie-Martin, F., Diaz-Torres, M. R., and Geoghegan, M. J., Chemical composition and electron microscopy of the rodlet layer of *Aspergillus nidulans* conidiospores, Curr. Microbiol. 14, 221, 1986.

Cole, G. T., A correlation between rodlet orientation and conidiogenesis in Hyphomycetes, Can. J. Bot., 51, 745, 1973.

Cole, G. T., Sun, S.H., and Huppert, M., Isolation and ultrastructural examination of conidial wall components of *Coccidioides* and *Aspergillus*, Scanning Electron Microscopy, 1677, 1982.

Cromack, K. Jr., Sollins, P., Graustein, W. C., Speidel, K., Todd, A. W., Spycher, G., Li, C. Y., and Todd, R. L., Calcium oxalate accumulation and soil weathering in mats of the hypogeous fungus *Hysterangium crassum*, Soil Biol. Biochem 11, 463, 1979.

De Bary, A., *Comparative Morphology and Biology of the Fungi, Mycetozoa and Bacteria*, Clarendon Press, Oxford, 11.

Dempsey, G. P., and Beever, R. E., Electron microscopy of the rodlet layer of Neurospora crassa conidia, J. Bacteriol., 140, 1050, 1979.

Farley, M. L., Mabry, L., Mounz, L. A., and Diserens, H. W., Crystals occurring in pulmonary cytology specimens associated with Aspergillus infection., Acta cytol., 29, 737, 1985.

Foster, J. W., *Chemical Activities of Fungi*, Acad. Press, New York, 1949, 648 p.

Gardner, J., Hess, W. M., and Tripathi, R. K., Surface rodlets of *Tilletia indica* teliospores, J. Bacteriol., 154, 502, 1983.

Graustein, W. C., Cromack, K. Jr., and Sollins, P., Calcium oxalate: Its occurrence in soils and effect on nutrient and geochemical cycles, Sci. 198, 1252, 1977.

Hamlet, W. M., and Plowright, C. B., On the occurence of oxalic acid in fungi, Chem. News, 36, 93, 1877.

Hashimoto, T., Wu-Yuan, C. D., and Blumenthal, H. J., Isolation and characterization of the rodlet layer of *Trichophyton mentagrophytes* microconidial wall, J. Bacteriol., 127, 1543, 1976.

Havir, E. A., and Anagnostakis, S. L., Oxaloacetate acetylhydrolase activity in virulent and hypovirulent strains of *Endothia (Cryphonectria) parasitica*, Physio. Plant Path. 26, 1, 1985.

Hawker, L. E., Thomas, B., and Bekett, A., An electron microscope study of structure and germination of conidia of *Cunninghamella elegans* Lender., J. Gen. Microbiol. 60, 181, 1970.

Heckman, C. A., Pelok, S. D., Kimpel, S. A. and Wu, L., Scanning electron microscope studies on fruitbody primordium formation in *Agaricus bisporus*, Mycologia 81, 717, 1889.

Hein, I., Studies on the mycelium of *Psalliota campestris*, Amer. J. Bot. 17, 197, 1930.

Hess , W. M., Sassen, M. M. A., and Remsen, C.C., Surface characteristics of *Penicillium* conidia, Mycologia, 60, 290, 1968.

Hess, W. M., and Stocks, D. L., Surface characteristics of *Aspergillus* conidia, Mycologia, 61, 560, 1969.

Hintikka,V., Korhonen, K., and Naykki, O., Occurrence of calcium oxalate in relation to the activity of fungi in forest litter and humus., Karstenia 19, 58, 1979.

Hobot, J. A., and Gull, K., Changes in the organization of surface rodlets during germination of *Syncephalastrum racemosum* sporangioconidiospores, Protoplasma, 107, 339, 1981.

Hodgkinson, A., Oxalic acid metabolism in lower plants, in *Oxalic Acid in Biology and Medicine*, Academic Press, New York, 1977.

Hollós, L., Die Gasteromyceten Ungarns, Leipzig, O. Weigel, 1904.

Holt, S. C., and E. R. Leadbetter, Comparative ultrastructure of selected aerobic spore forming bacteria: a freeze-etching study, Bacteriol. Rev., 33, 346, 1969.

Horner, H. T., Tiffany, L. H., and Cody, A. M., Formation of calcium oxalate crystals associated with apothecia of the discomycete *Dasyscypha capitata*. Mycologia, 75, 423, 1983.

Horner, H. T., Tiffany, L. H., Cody, A. M., and Knaphus, G., Development of fungal calcium oxalate crystals associated with the basidiocarps of *Geastrum minus* Lycoperdales. Scanning Electon Microscopy, II, 789, 1985a.

Horner, H. T., Tiffany, L. H., and Cody, A. M., Calcium oxalate bipyramidal crystals on the basidiocarps of *Geastrum minus* Lycoperdales., Proc. Iowa Acad. Sci., 92, 70, 1985b.

Horner, H. T., Tiffany, L. H., and Knaphus, G., Oak-leaf-litter rhizomorphs from Iowa and Texas: Calcium oxalate producers, Mycologia, 87, 34, 1995.

Jackson, D. W., An SEM study of lichen pruina crystal morphology. Scanning Electron Microscopy, III, 279, 1981.

Jones, D., McHardy, W. J., and Wilson M. J., Ultrastructure and chemical composition of spines in Mucorales, Trans. Br. Mycol. Soc., 66, 153, 1976.

Keller, J., Les Cystides Crystalliferes des Aphyllophorales, Mycologia Helvetica 1, 277, 1985.

Krisai, I., and Mrazek, E., Calcium oxalate crystals in *Geastrum*, Pl. Syst. Evol., 154, 325, 1986.

Kurrein, F., Path, F. R. C., Green, G. H., and Rowles, S.S., Localized deposition of calcium oxalate around a pulmonary *Aspergillus niger* fungus ball, Am J. Clin Pathol., 64, 556, 1975.

Lapeyrie, F., Perrin, M., Pepin, R., and Bruchet, G., Extracellular formation of weddellite cultured in-vitro by *Paxillus involutus*: Significance of this production for ectomycorrhizal symbiosis. Can J. Bot., 62, 1116, 1984.

Lapeyrie, F., Picatto, C., Gerard, J., and Dexheimer, J, TEM study of intracellular and extracellular calcium oxalate accumulation by ectomycorrhizal fungi in pure culture or in association with *Eucalyptus* seedlings, Symbiosis, 9, 163, 1990.

Lohwag, H., Mykologische Studien. IX, Über die Fruchtkörpenentwicklung der Geastraceen, Beih. Bot. Centrabr., 52, 269, 1934.

Lohwag, H., Anatomie der Asco- und Basidiomyceten, in Handbuch der Pflanzenatomy, Linsbauer, K., Tischler, G., and Pascher, A., Eds, VI, Abt. II, 3c, Berlin, Gebr üder Borntraeger, 1941.

Malajczuk, N., and Cromack, K. Jr., Accumulation of calcium oxalate in the mantle of ecto-mycorrhizal roots of *Punus radiata* and *Eucalyptus marginata*. New Phyto., 93, 527, 1982.

McCarrol, R. D., and Thor, E., The role of oxalic acid in the pathogenesis of *Endothia parasitica*, In *Proceeding of the American Chestnut Symposium*, W. L. MacDonald, F. D. Ceck, J. Luckok and C. Smith, eds., West Virginia University Press, Morgantown 60, 1978.

Noyes, R. D., and Handcock, J. G., Role of oxalic acid in the Sclerotina wilt of sunflower. Physiological Plant Pathology, 18, 123, 1881.

Powell, M. D., and Arnott, H. J., Calcium oxalate crystal production in two members of the mucorales. Scanning Electron Microscopy, I, 183, 1985.

Punja, Z. K., Mycelial growth and pathogenesis by *Rhizoctonia carotae* on carrot. Can. J. Plant Path., 9, 24, 1987.

Punja, Z. K., Huang, J. S., and Jenkins, S. F., Relationship of mycelial growth and production of oxalic acid and cell wall degrading enzymes to virulence in *Sclerotium rolfsii*, Can. J. Pl. Pathol., 7, 109, 1985.

Rast, D., and Hollenstein, G. O., Architecture of the *Agaricus bisporus* spore wall, Can. J. Bot., 55, 2251, 1977.

Schoknecht, J. D., and Keller, H. W., Peridial composition of white fructifications in the Trichiales (*Perichena* and *Dianema*), Can. J. Bot., 55, 1807, 1977.

Smith, V. L., Punja, Z. K., and Jenkins, S. F., A histological study of infections of host tissues by *Sclerotium rolfsii*. Phytopath., 76, 755, 1986.

Snetselaar, K. M., and Whitney, K. M., Fungal calcium oxalate in mycorrhizae of *Monotropa uniflora*. Can J. Bot., 68, 533, 1990.

Sunhede, S., Geastraceae (Basidiomycotina) morphology, ecology, and systematics with speical emphasis on the North European Species, Synopsis Fungorum 1, Fungiflora, Oslo Norway, 1990.

Timdal, E., The delimitation of *Psora* (Ledideaceae) and related genera, with notes on some species. Nord. J. Bot., 4, 525, 1984.

Verrecchia, E. P., Dumont, J., and Verrecchia, K.E. Role of calcium oxalate biomineralization by fungi in the formation of calcretes: A case study from Nazareth, Israel, J. Sedimentary Petrology, 63, 1000, 1993.

Urbanus, J. F. L. M., van den Ende, H., and Koch, B., Calcium oxalate crystals in the wall of *Mucor mucedo*, Mycologia , 70, 829, 1978.

Wadsten, T., and Moberg, R., Calcium oxalate hydrates on the surface of lichens, Lichenologist, 17, 239, 1985.

Wessels, J. G. H., Kreger, D. R., Marchant, R., Regensburg, B. A., and DeBries, O. M. H., Chemical and morphological characterization of the hyphal wall surface of the basidiomycete *Schizophyllum commune*, Biochim. Biophys. Acta, 273, 346, 1972.

Whitney, K. D., and Arnott, H. J., Calcium oxalate crystals and basidiocarp dehiscence in *Geastrum saccatum* (Gasteromycetes).Mycologia 78, 649, 1986a.

Whitney, K. D., and Arnott, H. J., Morphologoy and development of calcium oxalate deposits in *Gilbertella persicaria* (Mucorales), Mycologia 78, 42, 1986b.

Whitney, K. D., and Arnott, H. J., Calcium oxalate crystal morphology and development in *Agaricus bisporus*. Mycologia, 79, 180, 1987.

Whitney, K. D., and Arnott, H. J., The effects of calcium on mycelial growth and calcium oxalate crystal formation in *Gilbertella persicaria* (Mucorales). Mycologia, 80, 707, 1988.

"Yucca, Yucca, Winnemucca." R. Cave, 1957.

Chapter 6

OXALATE BIOSYNTHESIS AND FUNCTION IN PLANTS AND FUNGI

Vincent R. Franceschi and Frank A. Loewus

I. OXALATES IN PLANTS AND FUNGI: FORMS AND DISTRIBUTION

Oxalates are common constituents of plants and are found in the majority of plant families [1, 2, 3, 4]. They occur as the free acid, soluble salts of potassium, sodium and magnesium and as the insoluble salt of calcium. The amount of oxalates in plants ranges from a few percent of dry weight to up to 80% of the total weight of the plant [5]. These compounds are usually accumulated within the vacuoles of plant cells, although crystalline calcium oxalate may form within cell walls of some higher plants [4, 6]. Since plant cells in general have a large vacuolar compartment, often from 75 to 90% of cell volume, there is the potential for massive accumulation of oxalates.

The distribution of oxalates within plant tissues and organs varies widely among species and can be found in most any part of the plant [4, 5]. For an individual species, the amount and cellular location of oxalates within a given organ appears to be under genetic regulation at some basal level, since there are very strong patterns set up within species or related taxa. The amount and distribution of oxalates within plant species has important implications to plant defense mechanisms and, with respect to crop and pasture plants, human and animal nutrition. This is an area of investigation which has not been extensively studied to date, but has drawn more attention recently and will be reviewed below.

In fungi, most of the oxalate appears to be secreted during growth of the mycelia, and the secreted oxalate may precipitate as crystals of calcium oxalate within or on the surface of cell walls. The morphology of the extracellular crystals varies from simple prismatic to complex druse-like aggregates [7]. Fungal cells can have extensive vacuoles, but little information is available on the amount of oxalate stored within the fungal cells. Intracellular calcium oxalate crystals have not been reported in fungi.

Important areas that need to be addressed with respect to oxalates in plants and fungi are the following: 1) biosynthetic pathways giving rise to oxalic acid, 2) function of oxalates, 3) biotic and abiotic factors involved in the regulation of oxalic acid synthesis, 4) developmental and molecular regulation of the spatial (cell location) and temporal accumulation of oxalates, 5) nutritional impact of oxalates in edible plant parts. Characterization of these features will be important to understanding the physiology of oxalate in plant systems and will be critical to designing chemical or genetic strategies aimed at improving plant nutritional status or defensive or physiological mechanisms related to oxalate formation.

0-8493-7673-4/95/$0.00+$.50

II. OXALATE BIOSYNTHESIS IN HIGHER PLANTS

The biosynthetic origin of oxalic acid intrigued plant scientists even in the last century [8]. but specifics remained largely unexplored until the advent of biochemical studies with putative radio- or stable-isotope labeled substrates. Attention then focused on three intermediary compounds of plant metabolism; oxaloacetate, glycolate and glyoxlate as well as cell constituents which generate these compounds. An oxaloacetate hydrolyase, EC 3.7.1.1, which cleaves oxaloacetate to acetate and oxalate, is found in spinach and red beets [9] but subsequent information regarding the role of this enzyme in oxalate production in plants has not appeared. A study of glycolate production via the photosynthetic carbon oxidation cycle with $^{18}O_2$ and $^{13}CO_2$ as stable isotopic markers failed to support the idea of direct oxidation of glycolate as a precursor of oxalate [10]. Glycolate oxidase, EC 1.1.3.15 [(S)-2-hydroxy acid oxidase], catalyzes an oxygen-dependent oxidation of glycolate to glyoxylate and H_2O_2 in plants [11]. Further oxidation of glyoxylate, either non-enzymatic or in the presence of catalase, leads to oxalate. Stepwise oxidation of glycolate to oxalate appears to be an attractive explanation for oxalate accumulation in plants in as much as glycolate is a major product of photorespiration [12]. Further, when glycolate oxidase inhibitors, 2-pyridylhydroxy-methanesulfonic acid or methyl 2-hydroxy-3-butynoic acid, are supplied to *Lemna minor* plants which have been exposed to [^{14}C]glycolic acid or [^{14}C]glyoxylic acid, incorporation of label into idioblastic crystals normally observed, is blocked, an indication that these two-carbon compounds can function as oxalic acid precursors [13]. In a subsequent study comparing developing crystal idioblasts to mesophyll cells of *L. minor* it was found that glycolate oxidase, glycine decarboxylase, and ribulose 1,5-bisphosphate carboxylase-oxygenase, enzymes necessary for photorespiratory glycolate synthesis and metabolism, are virtually absent from idioblasts, suggesting that idioblasts lack capacity for synthesis of oxalate from glycolate [14]. The fact that *L. minor* utilizes glycolate (and glyoxylate) for oxalate biosynthesis within crystal idioblasts which lack the machinery for this process indicates either a mechanism of oxalate transport from mesophyll cells or a biochemical pathway independent of glycolate (or glyoxylate) oxidation. It has been pointed out that oxalate transport faces physicochemical limitations to simultaneous movement of calcium and oxalate between source and sink cells, a process which would slow down crystal deposition. The alternative, a biochemical pathway independent of pre-formed glycolate, emerged when it was discovered that oxalic acid is a major product of ascorbate metabolism.

L-Ascorbic acid (L-*threono* 2-hexenono-1,4-lactone) is an ubiquitous constituent of plant cells, present predominately in the reduced form. Studies with ^{14}C-labeled precursors have clearly established a pathway of biosynthesis in which the carbon chain of glucose is conserved and its numerical sequence preserved; that is, C1 through C6 of D-glucose becomes C1 through C6 of L-ascorbic acid. This process stands in contrast to that found in L-ascorbic acid-synthesizing animals wherein the carbon chain sequence from D-glucose to L-ascorbic acid is inverted [15]. It should be mentioned that plants readily oxidize exogenously supplied L-galactono-1,4-lactone to L-ascorbic acid but this process fails to take into consideration the facts that L-galactono-1,4-lactone is not normally found in plants and that putative conversion of D-glucose or D-galactose to L-galactono-1,4-lactone involves a carbon chain inversion [16, 17].

In addition to its redox functions [18], ascorbic acid undergoes rapid catabolism in young or developing tissues; C2/C3 cleavage of the carbon chain of L-ascorbic acid is one aspect of this breakdown [19]. Since oxalic acid corresponds to C1+C2 of L-ascorbic acid, it also corresponds to C1+C2 of D-glucose, the hexose precursor of L-ascorbic acid. Chemical cleavage at the C2/C3 bond was first used to establish the constitution of L-ascorbic acid. The products were oxalic acid and L-threonic acid [20]. Subsequently, it was found that C2/C3 cleavage of L-ascorbic acid with alkaline H_2O_2 provided a convenient synthetic route to L-threonic acid [21]. More recently, mechanistic details regarding this cleavage have been provided [22].

A biosynthetic role for L-ascorbic acid in the production of oxalic acid emerged when it was noted that *Pelargonium crispum* apices which had been labeled with L-[1-^{14}C]ascorbic acid accumulated labeled oxalic acid whereas apices labeled with L-[6-^{14}C]ascorbic acid accumulated carboxyl-labeled L-tartaric acid, the oxidation product of L-threonic acid [23]. This biosynthetic C2/C3 cleavage of L-ascorbic acid to produce oxalic acid has since been observed in numerous oxalate-accumulating plant species [24, 25, 26]. When L-ascorbic acid was added to callus cultures of *Psychotria punctata*, idioblast formation was stimulated [27]. Similarly, *Lemna minor* growing in nutrient media containing L-[1-^{14}C]ascorbic acid incorporated ^{14}C into calcium oxalate crystals [28].

Although the mechanism of L-ascorbic acid-linked oxalic acid biosynthesis is unknown, one plausible scheme drawn from the studies of Hamilton and his colleagues deserves careful consideration [29]. Here, peroxygenation of L-ascorbic acid leads to a peroxydehydroascorbic acid intermediate which decomposes, either by hydrolysis or through formation of an oxalyl thioester, to yield oxalic acid and L-threono-1,4-lactone. Decomposition via an oxalyl thioester could conceivably involve glutathione as a thio-adduct of peroxydehydroascorbic acid. Glutathione readily forms an adduct with peroxygenated products [30] and is normally present in millimolar amounts in plant tissues [31]. It would be of great interest to determine the potential role of such an adduct or its oxidation product, the oxalyl thioester, in intracellular transport of oxalic acid. Al-Arab and Hamilton favor the mechanism involving an oxalyl thioester over that involving direct hydrolysis of peroxydehydroascorbic acid in as much as the former is about four orders of magnitude more stable than the corresponding H_2O adduct.

III. OXALATE BIOSYNTHESIS IN FUNGI

Oxalic acid and its calcium salt are common metabolic products of fungi but knowledge of the pathway(s) remain fragmentary despite research extending back to the previous century [8]. Oxaloacetase (oxaloacetate acetylhydrolyase, EC 3.7.1.1) which cleaves oxaloacetate to acetate and oxalate is present in many fungal species and this enzyme has been singled out as one possible biosynthetic source [32, 33, 34, 35]. Glyoxylate dehydrogenase (acylating), EC 1.2.1.17, has also been implicated in oxalate production in *Sclerotium rolfsii* [36]. A third process involving oxidation of glyoxylate by glyoxylate oxidase has been partially purified from a brown-rot fungus [37]. Most of these studies fail to link overall conversion of carbon from nutrient sources to oxalate production.

During a search for culturable, single-cell, L-ascorbic acid-producing organisms amenable to studies on L-ascorbic acid biosynthesis it was discovered that previous claims of the presence of L-ascorbic acid in yeasts and fungi were based on misleading analytical procedures which failed to distinguish L-ascorbic acid from a close analog [38, 39]. In the case of yeasts tested by Heick et al., the analog proved to be D-erythroascorbic acid (D-*glycero*-2-pentenono-1,4-lactone) [40, 41]. D-Erythroascorbic acid has also been found in *Candida* species [42], *Neurospora crassa* [43], *Sclerotinia sclerotiorum* [44] as well as *Sclerotium cepivorum* (Loewus and Suto, unpublished results) and many species of mushroom [45]. D-Erythroascorbic acid differs structurally from L-ascorbic acid only in the fact that it is a five carbon rather than a six carbon compound. Ring structures are identical which leaves the side-chain at C4 of D-erythroascorbic acid with a hydroxymethyl group as compared to the 1,2-dihydroxyethyl group on L-ascorbic acid. Clearly, peroxygenation of D-erythroascorbic acid will yield oxalic acid and D-glyceric acid just as similar treatment of L-ascorbic acid will yield oxalic acid and L-threonic acid. In fact, treatment of either compound with alkaline H_2O_2 produces a stoichiometric amount of oxalic acid. Moreover, D-erythroascorbic acid is readily oxidized by L-ascorbate oxidase, albeit at about 0.4 times the rate obtained with L-ascorbic acid (Loewus, unpublished results).

Murakawa et al. [42] reported that *Candida utilis* grown on D-arabinose accumulated substantially more erythroascorbic acid than when D-glucose was the carbon source. When D-arabinose was replaced by D-arabinono-1,4-lactone or the free acid, accumulation was even greater. A D-arabinono-1,4-lactone oxidase, catalyzing the final step in D-erythroascorbic acid biosynthesis was recently purified to apparent homogeniety by [46].

Loewus et al.[44] have broadened their study on D-erythroascorbic acid biosynthesis and metabolism to determine if this ascorbic acid analog is utilized by fungi as a precursor of oxalic acid. They chose oxalate-secreting *Sclerotinia sclerotiorum* (= *Whetzelinia sclerotiorum*) an agronomically important phytopathogen which initiates infection by secreting oxalic acid and cell wall-degrading enzymes at the infected site [47]. If D-[1-^{14}C]arabinose was supplied to *S. sclerotiorum*, ^{14}C was recovered in oxalic acid, erythroascorbic acid and an unknown erythroascorbic acid-like compound but not in D-glyceric acid. If D-[5-^{3}H]arabinose was used, erythroascorbic acid, the unknown erythroascorbic acid-like compound and D-glyceric acid were labeled. Oxalic acid appeared as the free acid and accumulated in mycelia, nutrient medium and exudate. Only limited information is available regarding D-arabinose as a nutrient carbon source for fungi. D-Xylose is also an efficient carbon source for erythroascorbic acid and oxalic acid formation in *S. sclerotiorum* (Loewus, unpublished observation) suggesting that pentose metabolism in *S. sclerotiorum* (and possibly other oxalate-secreting fungi) is directed toward production of D-arabinose or a related stereoisomeric intermediate precursor of D-erythroascorbic acid. Obviously, more work is needed before the role of ascorbic acid analogs as precursors of oxalic acid in fungi is firmly established.

IV. FUNCTION OF OXALATES IN PLANT SYSTEMS

There have been many functions attributed to oxalates in plants, including: ion balance, defensive properties, waste end product, structural component of tissues [4]. While the range of forms, amount and distribution within the plant of oxalates readily lends itself to hypothesis of a variety of potential functions, a review of the literature indicates that only the first two functions listed above are supported by experimental data. The last two functions do not merit extensive discussion. All plants are capable of producing a range of structural elements that have specifically evolved for that function while there is little evidence that insoluble oxalates are an efficient or even possible structural element of any relevance. With respect to oxalate as a waste end product, recent research shows quite clearly that oxalic acid is not produced passively as an end product of various metabolic pathways, but that its synthesis is regulated and oxalate can be metabolized if it is necessary to remove it from the system [48]. Pathogenic and saprophytic fungi, in particular, produce and excrete oxalic acid as part of a mechanism to invade or degrade host tissues [49, 50, 51], a topic which will be treated separately in this review.

A. OXALATE AND ION BALANCE IN PLANTS

There is ample evidence that oxalate can serve as a counter ion for regulation of charge balance or in regulation of calcium activity [52, 53, 54, 55]. It may play a direct role in the soluble phase of the vacuole, where it can charge balance for potassium, a major cationic component of plants, and to a lesser degree for sodium, magnesium and possibly positively charged amino acids, which are also stored in vacuoles. Secondly, oxalate can be used to regulate calcium activity through production of the insoluble salt, calcium oxalate. Both processes are important basic physiological processes in plants, but involve different mechanisms and will be treated separately in this discussion.

The vacuole of plant cells typically contains high amounts of solutes, both charged and uncharged. These solutes are important for generation of a positive internal hydrostatic pressure (turgor pressure) which is the physical force that allows plant cells to be rigid in the absence of any other true skeletal structure (exo- or endo-skeleton). The solutes used are dependent to some degree on the organ, but consist of compounds such as sugars, organic acids, amino acids, inorganic mineral elements (phosphates, sulfates, charged mineral elements such as magnesium, calcium, potassium, chloride, and sodium) and occasionally more complex organic solutes such as ureides, and monophenolic derivatives [56]. The total solute potential in the vacuole will produce an osmotic activity which will cause water to flow into the vacuole and lead to a positive hydrostatic pressure. This pressure is transmitted to the cytoplasm, and is balanced by an equal and opposite force from the elastic cell wall. In order to avoid a dangerously high charge imbalance and unfavorable transmembrane tonoplast potential, the ionic solutes will be balanced to a large degree by synthesis and transport of organic acids or by selective transport of inorganic ions such as potassium. Therefore, organic acid synthesis may be in response to accumulation of inorganic ions such as potassium and sodium. This has been shown to be the case with respect to oxalate synthesis in some plants which grow in soils high in these elements, and in particular in the halophyte plants such as *Atriplex* and *Halogeton* [reviewed in 5]. These plants accumulate large amounts of sodium, and also large amounts of soluble oxalate. This implicates a role of oxalate in charge balance in at least plants specialized for growth in high salt environments.

It is probable that oxalate serves a similar role, though to a lesser degree in many plants which grow under more hospitable soil salt conditions. The question remains however, why is oxalate utilized rather than other organic ions? While it is clear that other organic acids such as malate and citrate are also important, oxalate may be more beneficial in those plants where cations are accumulated to very high concentrations since in terms of carbon balance it provides the maximum charge balance capacity per carbon atom. It is also a more stable ion in the sense of entry into biochemical pathways, which are significant and multiple for malate and citrate but apparently very limited for oxalate (see below).

B. OXALATE AND CALCIUM REGULATION

Insoluble calcium oxalate is found in plants in the form of crystals of a range of morphologies and sizes. The most common types of crystals are styloids (large single needles), raphides (bundles of needles), druse (stellate conglomerates), prismatic (rhombohedral), and crystal sand (packets of angular microcrystals) [4]. While other morphologies occur, most are modifications of the above. In many instances, the crystals are formed in the vacuoles of cells that are structurally modified specifically for this process, called idioblasts [57]. In some plants, cells typical of the ground tissue of the organ become secondarily modified for calcium oxalate formation.

Physiological and biochemical data indicate calcium oxalate is produced as part of a mechanism for regulating calcium in many plant tissues and organs. Calcium enters the plant in the root zone and is transported to various parts of the plant primarily through the xylem of the vascular system. Transport in the transpiration stream of the xylem is driven by evaporation of water from the surface of the plant. As water is lost, dissolved nutrients in the xylem stream are accumulated in the apoplast and or symplast of cells. Herein lies a potential problem with respect to calcium and maintenance of normal metabolic function in the plant. Many plant species do not tightly regulate calcium uptake by the roots: the amount of calcium taken up is directly related to the level of calcium in the surrounding soil solution [58]. On the other hand, most plants do not have a mechanism for excretion of excess materials in the xylem stream. Thus, as water evaporates from the surface of the plant the dissolved minerals are left behind and are either used for continued growth of the plant, or may build up to very high levels. This is a particular problem for long lived organs which grow to a finite size. Calcium is perhaps the most difficult element to deal with since, as with animal cells, it is involved in signal transduction and biochemical regulation of a host of cellular processes [59, 60]. Calcium activity is carefully regulated to very critical levels in cells, and build up of high concentrations of soluble calcium within a plant tissue or organ can lead to severe physiological problems.

Structural and biochemical data indicate that cells specialized for calcium oxalate formation are primarily involved in bulk regulation of calcium activity in organs and tissues [61]. Calcium oxalate formation is not a simple precipitation phenomenon, but has been demonstrated to be a carefully regulated process involving coordination of cell growth and crystal growth, and production of specialized subcellular organelles [2, 4]. Formation of calcium oxalate and the specialized cells for this process, can be induced by increasing calcium in the growth medium [27, 61, 62, 63, 64]. Under conditions of calcium limitation, calcium oxalate crystals can be dissolved, presumably providing calcium for plant growth and development [65]. These observations indicate that calcium oxalate in many plants is produced in response to excess calcium and can be considered to be a high capacity mechanism for regulating calcium activity in tissues and organs.

A model of calcium oxalate formation in plants that takes into account structural, physiological and biochemical data in the published literature, can be described as follows. If a plant that is capable of forming calcium oxalate is growing in a soil with a relative abundance of soluble calcium, over time calcium will increase in the various organs as water is continuously transported to and evaporated from its surfaces, leaving the calcium behind. The calcium binding capacity of the cell wall is limited and eventually will be saturated, resulting in increased soluble calcium in the apoplast. The concentration difference (up to 1000 fold) between the apoplast (mM) and cytosol (sub-μM) and the inside negative potential across the plasma membrane produce a strong electrochemical potential gradient driving calcium into the cell. Under these conditions the low capacity mechanisms for regulating cytosolic calcium, such as pumping back into the apoplast and compartmentation in the various organelles, becomes saturated and cytosolic activity will increase. These are conditions which we hypothesize will induce the machinery necessary for calcium oxalate formation (transporters, vacuolar membranes, oxalate synthesis, etc.). The end result is the controlled formation of crystalline calcium oxalate. This salt is extremely insoluble, and is physiologically and osmotically inert. The cells that produce the calcium oxalate become high capacity sinks for calcium, sparing surrounding cells the effect of high apoplastic calcium and allowing them to continue with their normal functions, such as photosynthesis. As already mentioned, large amounts of calcium can be accumulated in this form, particularly in very long lived plants such as cacti [65].

A variation of this model is found in young developing tissues. In these tissues, accumulation of calcium in the apoplast can interfere with the normal process of cell expansion by cross-linking acidic residues of cell wall polymers. Another complication is that many of the young cells are still in a division phase, which is very sensitive to cytosolic calcium activity. These cells generally have little or no vacuolar compartment, and thus a reduced capacity for calcium sequestration. Under such conditions differentiation of localized strong calcium sinks in the form of calcium oxalate cells would alleviate the potential problem of calcium build up in the cell walls. This mechanism operates in leaves of plants such as *Lemna* and *Pistia* where there is a gradation of differentiation over a fairly long period giving rise to mature tissues at the tip of the leaf and differentiating tissues at the base [13].

C. OXALATE AND DEFENSE MECHANISMS

Soluble oxalate and calcium oxalate may have protective functions in some plants. High soluble oxalate content can inhibit the activity of sucking insects such as leafhoppers and aphids [66, 67]. Very little work has been done on the relations between plant oxalate content and insect resistance so it is difficult to determine if this is a general defense mechanism. It is also well established that ingestion of oxalate rich plants by livestock can lead to acute toxic effects and death [68], often due to renal failure. Grazing animals may be protected from the toxic effects of oxalates if they have acquired within their intestinal microflora the oxalate degrading anaerobic bacterium, *Oxalobacter formigenes* [69]. In addition, more subtle antinutritive effects may be induced by ingestion of plant oxalates, including reducing the availability of dietary calcium due to formation of insoluble calcium oxalate [reviewed in 5].

While the examples cited clearly demonstrate that plant oxalate can be toxic to animals, it is unlikely that plants accumulate soluble oxalate primarily as a protective compound. The toxicity associated with oxalates is probably, in most instances, a secondary effect of accumulation of oxalate for ion balance. For example, one of the highest accumulators of soluble oxalate is *Halogeton*, which is known to be responsible for poisoning of livestock [70]. This plant is a halophyte, capable of growing in high salt soils, and the oxalate content is likely related to charge balance for sodium taken up by the plant. This is supported by studies showing a direct relationship between oxalate content and sodium or potassium in the growth medium [71].

A stronger argument can be made for the function of calcium oxalate crystals as part of a plant defensive mechanism in some species. The clearest case can be made for specialized stinging hairs found in *Tragia* [72], a type of stinging nettle. The stinging hairs have a large conical cell, the lumen of which contains a sharply pointed calcium oxalate crystal with grooves along its sides. Upon contact with an animal (or human!) the thin walled tip of the cell ruptures, the skin is penetrated by the tip of the crystal and toxins within the cell, whose contents were under pressure, are channeled along the grooves and into the wound created by the crystal. In other plants, large raphide crystal cells are formed whose contents also appear to be under pressure [73], and upon being disturbed, a thin end wall is disrupted and the crystals are expelled. Studies have shown that such crystals can be directly linked to dermal irritation [74], particularly if the crystals exceed a certain length (approx. 100 μm). Toxins have been found associated with the crystals that are responsible for more severe reactions [75, 76] so it may be that the crystals provide a physical means for piercing the protective layer of the skin while the toxins are responsible for more severe reactions. In support of this is the observation that such crystals are often modified with grooves and barbs [77].

Examination of other plant species reveals anatomical features of the placement and/or size of calcium oxalate crystals which would seem by simple observation to present a formidable barrier to some grazing insects or animals. The ability of calcium oxalate crystals to deter feeding will, of course depend upon the relative size of the crystal and animal. The total amount of calcium oxalate present can also have an effect. Studies with peccaries grazing on prickly-pear cactus indicated that they preferentially fed on cactus morphs having lower calcium oxalate content [78]. The large quantities of calcium oxalate that can be found in some cacti undoubtedly result in textural qualities that can be detected by the peccaries (and other grazing mammals), leading to avoidance.

While calcium oxalate crystals may provide a degree of protection against grazing animals, this is probably a secondary function of calcium oxalate formation in an evolutionary sense. Formation of larger crystals or synthesizing toxins within the crystal forming cells can provide defense at the same time that the primary function of regulation of bulk calcium is occurring. It is important to note that while many plant species produce calcium oxalate crystals there is no strong *general* relationship between calcium oxalate crystals and plant toxicity [79].

V. FUNCTIONS OF OXALIC ACID IN FUNGI

Higher plants and fungi deal with oxalic acid accumulation in quite different ways. Generally speaking, higher plants possess intra- and intercellular mechanisms for sequestration, usually as soluble or insoluble salts, although exceptions are found in which the bulk of the acid is present in free form [80, 81]. Even then, secretion of oxalic acid to outer surfaces of unwounded plants is seldom evident. A rare case in which calcium oxalate crystals form extracellularly has been reported recently [82]. When in culture, fungi actively secrete oxalic acid into the medium. Whether calcium oxalate crystals found in association with fungal hyphae and fruiting bodies under natural conditions form following secretion or whether they develop within hyphae and then are expelled through the hyphal wall, is a matter still not fully resolved [7, 83].

Phytopathogenic fungi, notable examples being *Sclerotium rolfsii*, *S. cepivorum* and *Sclerotinia sclerotiorum*, elaborate oxalic acid, pectic enzymes and cellulase which act synergistically on the plant cell wall in the infection process [84, 85, 86, 87]. These agronomically important forms are probably representative of a broad spectrum of actinomycete phytopaths which disrupts the integrity of the plant in order to provide the fungus with an invasive foothold. Oxalic acid accumulates early in pathogenesis, lowering the pH at the infective site which enhances activity of lytic enzymes such as polygalacturonase, hemicellulases and cellulase. Oxalic acid also breaks down calcium pectate complexes which would otherwise resist enzymic attack, thus advancing host cell wall destruction and facilitating fungal colonization. Injection of oxalic acid or a cultural filtrate from oxalic acid-elaborating phytopathogenic species is often sufficient to induce disease symptoms [88]. Recently, five mutants deficient in production of oxalic acid which were recovered from UV-irradiated *S. sclerotiorum*, failed to colonize and incite necrotic lesions on bean pods unless an exogenous source or an inducer of oxalic acid was provided [50]. Pathogenicity appeared to be specifically associated with oxalic acid production and not with pectolytic or cellulolytic enzymes. Godoy et al. reported a close association between oxalic acid production and sclerotial development in that oxalic acid deficient mutants failed to form sclerotia. As these authors point out, impairment of oxalic acid biosynthesis in early stages of infection might provide a useful approach toward control of pathogenicity by organisms which elaborate this product. Use of *Arabidopsis thaliana* as a model for such studies has been demonstrated [89].

A biochemical role for oxalic acid in relation to lignin and cellulose biodegradation by processes associated with wood-rot is receiving increasingly more attention [37, 90, 91, 92, 93, 94, 95]. A detailed review which examines the possible role of oxalate metabolism in relation to the physiology of wood-rotting fungi has recently appeared [96]. The potential role for ascorbic acid and its analogs as precursors of oxalic acid which were described above and the fact that such analogs are present in basidiomycetes [45], suggests a need for further studies to seek out connections between such products of fungal metabolism and their roles as sources of oxalic acid as well as H_2O_2.

VI.   OXALATE DEGRADING ENZYMES IN PLANTS

Oxalate is generally considered to be a relatively stable end product in plants and there have been few investigations of its metabolism or degradation. Early studies on plant organic acids found that diurnal fluctuations of oxalate occurred in some species [97], and radiotracer experiments have shown that label from [^{14}C]oxalic acid can be incorporated into a number of organic compounds by plant tissues [98, 99]. Enzymes capable of oxalate "degradation" have been identified in higher plants and their role in the physiology of these organisms has only recently received much attention. In higher plants two enzymes, oxalate oxidase and oxalate decarboxylase, have been identified. A considerable amount of work has been done on oxalate oxidase, which produces CO_2 and H_2O_2, since it is an abundant protein in certain plant tissues [48] while oxalate decarboxylase may represent a minor or insignificant pathway in higher plants.

Dissolution of calcium oxalate has been seen in some plants under conditions where calcium may become limiting under normal growth [reviewed in 4] or where calcium deficiency is induced [61]. Oxalate degrading enzymes would be necessary to metabolize the oxalate released, which would otherwise lead to problems with ion balance and turgor regulation. Carbon released as CO_2 could be recovered by photosynthetic activity or the low level of PEP carboxylase found in the cytosol of most plant cells. This is supported by studies showing recovery of labeled carbon from oxalic acid into sugars, polysaccharides and lipids [13].

Recent work by Lane [48] suggests a very novel and important role for oxalate and oxalate oxidase in plant development. Lane has hypothesized that peroxide produced from oxalate oxidation can affect changes in the extensibility of the extracellular matrix (cell wall) of plants and thus influence cell expansion and growth. H_2O_2 may also be produced by oxalate oxidation in response to fungal attack, providing a strong oxidant as well as a potential signaling molecule to induce other defense mechanisms [48]. Oxalate would play central roles in development and defense in tissues where oxalate oxidase is expressed, extending the function of oxalate in plants to physiological processes never before considered.

The role of oxalic acid as a biosynthetic intermediate merits brief comment, since in a well documented case it has serious pharmacological effects. Certain legume species, notably grasspea (*Lathyrus sativus*) accumulate a compound called ß-N-oxalyl-L-α,β–diaminoproprionic acid (ODAP) in their seeds [100]. Oxalic acid is a substrate in the synthesis of ODAP. Seeds of grasspea are used as a protein source in some countries and can lead to a condition known as neurolathryism. This irreversible paralysis of the lower limbs is due to the antagonist effect of this non-protein amino acid with respect to glutamic acid function in neurotransmission.

VII. SUMMARY

A comparison of synthesis and function of oxalates in plants and fungi provides some insight into the importance of this compound which has so often been considered to be a useless endproduct of metabolism. In both organisms, oxalic acid synthesis is an active process occurring as a response to particular environmental cues. Plants utilize oxalate primarily for ion regulation and secondarily for defense. These functions are intracellular in nature. In contrast, fungi secrete oxalate into their immediate environment and the oxalate becomes part of the mechanism by which the fungi invade and degrade their plant host tissues. The unique chemical properties of oxalic acid are critical to these divergent functions, and could not easily be replaced by other organic acids. In both organisms, oxalic acid plays a direct and dynamic physiological role unlike other organic acids which are often intermediates of metabolism. The divergence of function of oxalic acid between plants and fungi is clearly a reflection of their different life histories.

Along with the divergence in function of oxalates in plants and fungi is a divergence in the possible chemical pathways of oxalic acid synthesis. However, while a number of pathways can give rise to oxalic acid in plants, most do not have the capacity for large scale synthesis without disruption of other basic biochemcial pathways of which the substrates are intermediates. A pathway using ascorbic acid as a precursor has been demonstrated to operate in plants and appears capable of large scale oxalic acid synthesis. Interestingly, oxalic acid synthesis in fungi may follow a related pathway. An ascorbic acid analog, erythroascorbic acid, can serve as a precursor of oxalic acid synthesis in fungi. The details of these pathways remain to be elucidated.

The oxalate content of plants represents a potential health hazard to both humans and livestock through its effect on renal function, among other things. Human suffering and monetary costs can be significant. Identification of the biochemical basis of oxalate formation in plants will help in the development of chemical or genetic strategies aimed at reducing the oxalate levels in food stuffs. Traditional plant breeding practices have shown that this potential does exist [5]. Similar strategies can be used to enhance plant defense against fungi, as indicated by recent work with the oxalate oxidase gene. Plants genetically engineered to express oxalate oxidase in the root cell walls have been found to be resistant to *Sclerotinia sclerotiorum*, a common and costly fungal pathogen of many crop plants [48]. Selective expression of the gene for this protein, or biosynthetic enzymes once identified, in other plant parts could reduce endogenous oxalate levels in edible plant organs. An increased understanding of oxalate synthesis, degradation and function in plants and fungi is worth pursuing since it can provide the tools necessary for improving various aspects of nutritional value and defense, as well as give plant biologists a true perspective of the importance of oxalates to basic plant function.

REFERENCES

1. Mcnair, J. B., The interrelation between substances in plants: essentail oils and resins, cyanogen and oxalate, *Amer. J. Bot.*, 19, 255, 1932.
2. Arnott, H. J. and Pautard, F. G. E., Calcification in plants, in *Biological Calcification; Cellular and Molecular Aspects*, Schraer, H., Ed., Appleton-Century-Crofts, New York, 1970, 375.
3. Gallaher, R. N.,The occurrence of calcium in plant tissue as crystals of calcium oxalate, *Comm. Soil Sci. Plant Anal.*, 6, 315, 1975.
4. Franceschi, V. R. and Horner, Jr., H. T., Calcium oxalate crystals in plants, *Bot. Rev.*, 46, 361, 1980.
5. Libert, B. and Franceschi, V. R., Oxalate in crop plants, *J. Agric. Food Chem.*, 35, 926, 1987.
6. Gambles, R. L. and Dengler, N. G., The leaf anatomy of hemlock, *Tsuga canadensis*, *Can. J. Bot.*,52, 1049, 1974.

7. Arnott, H. J. and Webb, M. A., The structure and formation of calcium oxalate crystal deposits on the hyphae of a wood rot fungus, *SEM*, IV, 1747, 1983.

8. Hodgkinson, A. Oxalic Acid in Biology and Medicine, Academic Press, London, 1977, 325 p.

9. Chang, C.C. and Beevers, H., Biogenesis of oxalate in plant tissues, *Plant Physiol.*, 43, 1821, 1968.

10. Raven, J.A., Griffiths, H., Glidewell, S.M., and Preston, T., The mechanism of oxalate biosynthesis in higher plants: investigations with the stable isotopes ^{18}O and ^{13}C, *Proc. R. Soc. Lond. B*, 216, 87, 1982.

11. Emes, M.J. and Erismann, K.H., Purification and properties of glycollate oxidase from *Lemna minor* L., *Int. J. Biochem.*, 16, 1373, 1984.

12. Tolbert, N.E., Photorespiration, in *The Biochemistry of Plants*, Vol. 2, Davies, D.D., Ed., Academic Press, New York, 1980, chap. 12.

13. Franceschi, V.R., Oxalic acid metabolism and calcium oxalate formation in *Lemna minor* L., Plant Cell Environ., 10, 397, 1987.

14. Li, X. and Franceschi, V.R., Distribution of peroxisomes and glycolate metabolism in relation to calcium oxalate formation in *Lemna minor* L., *Eur. J. Cell Biol.*, 51, 9, 1990.

15. Loewus, F.A. and Loewus, M.W., Biosynthesis and metabolism of L-ascorbic acid in plants, *CRC Crit. Rev. Plant Physiol.* 5,101, 1987.

16. Ôba, K., Fukui, M., Imai, Y., Iriyama, S. and Nogami, K., L-Galactono-1,4-lactone dehydrogenase: partial characterization, induction of activity and role in the synthesis of ascorbic acid in wounded white potato tuber tissue, *Plant Cell Physiol.*, 35, 473, 1994.

17. Ôba, K., Ishikawa, S., Nishikawa, M., Mizuno, H. and Yamamoto, T., Purification and properties of L-galactono-1,4-lactone dehydrogenase, a key enzyme for ascorbic acid biosynthesis, from sweet potato roots, *J. Biochem.*, 117, 120, 1995.

18. Foyer, C., Ascorbate, in *Antioxidants in Higher Plants*, Alscher, R.G. and Hess, J.L, Eds. CRC Press, Boca Raton, FL, 1993, p. 31.

19. Loewus, F.A., Ascorbic acid and its metabolic products, in *The Biochemistry of Plants*, Vol. 14, Preiss, J., Ed., Academic Press, New York, 1988, 85.

20. Herbert, R.W., Hirst, E.L., Percival, E.G.V., Reynolds, R.J. W., and Smith, F., The constitution of ascorbic acid, J. Chem. Soc., 1270, 1933.

21. Isbell, H.S. and Frush, H.L. Oxidation of L-ascorbic acid by hydrogen peroxide: perparation of L-threonic acid, *Carbohydr. Res.*, 72, 1854, 1979.

22. Kwon, M-B., Foote, C.S. and Khan, S.I., Photooxygenation of ascorbic acid derivatives and model compounds, *J. Am. Chem.Soc.*, 111, 1854, 1989.

23. Wagner, G. and Loewus, F.A., The biosynthesis of (+)-tartaric acid in *Pelargonium crispum, Plant Physiol.* 52, 784, 1974.

24. Yang, J. and Loewus, F.A., Metabolic conversion of L-ascorbic acid to oxalic acid in oxalate-accumulating plants, Plant Physiol., 56, 283, 1975.
25. Nuss, R.F. and Loewus, F.A., Further studies on oxalic acid biosynthesis in oxalate-accumulating plants, *Plant Plysiol.*, 61,590, 1978.
26. Wagner, G.J., Vacuolar deposition of ascorbate-derived oxalic acid in barley, *Plant Physiol.* , 198, 591, 1981.
27. Franceschi, V.R. and Horner, H.T., Jr., Use of *Psychotria punctata* callus in study of calcium oxalate crystal idioblast formation, Z. *Pflanzenphysiol.* 92, 61, 1979.
28. Franceschi, V.R., Oxalic acid metabolism and calcium oxalate formation in *Lemna minor* L., *Plant Cell Environ.*, 10, 397, 1987.
29. Al-Arab, M.M. and Hamilton, G.A., The hydrolysis of diethyl monothiooxalate and its reaction with nucleophiles, *Bioorg. Chem.*, 15, 81, 1987.
30. Gunshore, S, Brush, E.J. and Hamilton, G.A., Equilibrium constants for the formation of glyoxylate thiohemiacetals and kinetic constants for their oxidation by O_2 catalyzed by L-hydroxy acid oxidase, *Bioorg. Chem.* 13, 1, 1985.
31. Rennenberg, H. Glutathione metabolism and possible biological roles in higher plants, *Phytochemistry*, 21, 2771, 1982.
32. Maxwell, D.P., Oxalate formation in *Whetzelinia sclerotiorum* by oxaloacetate acetylhydrolyase, *Physiol. Plant Pathol.* 3, 279,1973.
33. Houck, D.R. and Inamine, E., Oxalic acid biosynthesis and oxaloacetate acetylhydrolyase activity in *Streptomyces cattleya*, *Arch. Biochem. Biophys.*, 259, 58, 1987.
34. Kubicek, C.P., Schreferl-Kunar, G., Wöhrer, W. and Röhr, M., Evidence for a cytoplasmic pathway of oxalate biosynthesis in *Aspergillus niger*, *Appl. Environ. Microbiol.* 54, 633, 1988.
35. Akamatsu, Y., Takahashi, M. and Shimada, M., Cell-free extraction of oxaloacetase from white-rot fungi, including *Coriolus vesicolor*, *Wood Res.*, 79, 1, 1993.
36. Maxwell, D.P. and Bateman D.F.,Oxalic acid biosynthesis by *Sclerotium rolfsii*, *Phytopathology*, 58, 1635, 1968.
37. Akamatsu, Y. and Shimada, M., Partial purification and characterization of glycolate oxidase from the brown-rot basidiomycete *Tyromyces palustris*, *Phytochemistry*, 17, 649, 1994.
38. Heick, H.M.C., Graff, G.L.A., Humpers, J.E.C., Occurence of ascorbic acid among the yeasts, *Can. J. Microbiol.*, 18, 597, 1972.
39. Rai, R.D. and Saxena, S., Effect of storage temperature on vitamin C content of mushrooms (*Agaricus bisporus*), *Current Sci.*, 57, 434, 1988.
40. Nick, J.A., Leung, C.T. and Loewus, F.A., Isolation and identification of erythroascorbic acid in *Saccharomyces cervisiae* and *Lypomyces starkeyi*, *Plant Sci.*, 46,181, 1986.

41. Kim, H.S., Seib, P.A. and Chung, O.K., Effect of D-erythroascorbic acid on wheat dough and its level in baker's yeast, *J. Food Sci.*, 58, 845, 1993.
42. Murakawa, S., Sano, S., Yamashita, H. and Takahashi, T., Biosynthesis of D-erythroascorbic acid by Candida, *Agric. Biol. Chem.*, 9,1799, 1977.
43. Dumbrava, V.A. and Pall, M.L., Control of nucleotide and erythroascorbic acid pools by cyclic AMP in *Neurospora crassa*, *Biochim. Biophys. Acta*, 926, 331, 1987.
44. Loewus, F.A., Saito, K., Suto, R.K. and Maring, E., Oxalic acid biosynthesis in the fungal phytopathogen *Sclerotinia sclerotiorum*, Plant Physiol. Suppl.. 105, 157, 1994.
45. Okamura, M., distribution of ascorbic acid analogs and associated glycosides in mushrooms, *J. Nutr. Sci. Vitaminol.* 40, 81, 1944.
46. Huh, W-K, Kim, A-T, Yang, K.S., Seok, Y-J., Hah, Y.C. and Kang, S-O., Characterization of D-arabinono-1,4-lactone oxidase from *Candida albicans* ATCC 10231, *Eur. J. Biochem.*, 225,1073, 1994.
47. Maxwell, D.P. and Lumsden, R.D., Oxalic acid production by *Sclerotinia sclerotiorum* in infected bean and in culture, Phytopathology, 60, 1395, 1970.
48. Lane, B. G., Oxalate, germin, and the extracellular matrix in higher plants, *FASEB J .*, 8, 294, 1994.
49. Rao, D. V. and Tewari, J. P., Production of oxalic acid by *Mycena citricolor*, causal agent of the American leaf spot of coffee, *Phytopathology*, 77, 780, 1987.
50. Godoy, G., Steadman, J. R., Dickman, M. B. and Dam, R., Use of mutants to demonstrate the role of oxalic acid in pathogenicity of *Sclerotinia sclerotiorum* on *Phaseolus vulgaris*, *Physiol. Molec. Plant Path.*, 37, 179, 1990.
51. Mouly, A., Rumeau, D., and Esquerre-Tugaye, M-T., Differential accumulation of hydroxyproline-rich glycoprotein transcripts in sunflower plants infected with *Sclerotinia sclerotiorum* or treated with oxalic acid, *Plant Sci.*, 85, 51, 1992.
52. Joy, K. W., Accumulation of oxalate in tissues of sugar-beet, and the effect of nitrogen supply, *Ann. Bot.*, 28, 689, 1964.
53. Smith, F. W., Potassium nutrition, ionic relations and oxalic acid accumulation in three cultivars of *Setaria sphacelata*, Aust. J. Agr. Res., 23, 969, 1972.
54. Austenfield, F. A., and Leder, U., Uber den Oxalathaushalt von *Salicornia europaea* L. unter dem Einflub variieter Erdalkalisalz-Gaben, Z. Pflanzenphysiol., 88, 403, 1978.
55. Kinzel, H., Calcium in the vacuoles and cell walls of plant tissue. Forms of deposition and their physiological significance, *Flora*, 182, 99, 1989.
56. Taiz, L., The plant vacuole, *J. exp. Biol.*, 172, 113, 1992.
57. Foster, A. S., Plant idioblasts: remarkable examples of cell specialization, *Protoplasma*, 46, 184, 1956.

58. Kirkby, E. A., and Pilbeam, D. J., Calcium as a plant nutrient, *Plant Cell Environ.*, 7, 397, 1984.
59. Marme, D., The role of calcium in the cellular regulation of plant metabolism, *Physiol. Veg.*, 23, 945, 1985.
60. Poovaiah, B. W., and Reddy, A. S. N., Calcium messenger system in plants. *CRC Critical Rev. Plant Sci.*, 6, 47, 1987.
61. Franceschi, V. R., Calcium oxalate formation is a rapid and reversible process in *Lemna minor* L., *Protoplasma*, 148, 130, 1989.
62. Zindler-Frank, E., On the formation of the pattern of crystal idioblasts in *Canavalia ensiformis* D. C. VII. Calcium and oxalate content of the leaves in dependence of calcium nutrition, *Z. Pflanzenphysiol.*, 77, 80, 1975.
63. Borchert, R., Calcium-induced patterns of calcium-oxalate crystals in isolated leaflets of *Gleditsia triacanthos* L. and *Albizia julibrissin* Durazz., *Planta*, 165, 301, 1985.
64. Borchert, R., Calcium acetate induces calcium uptake and formation of calcium-oxalate crystals in isolated leaflets of *Gleditsia triacanthos* L., *Planta*, 168, 571, 1986.
65. Cheavin, W. H. S., The crystals and cystoliths found in plant cells. Part I. Crystals, *Microscope*, 2, 155, 1938.
66. Yoshihara, T., Sogawa, K., Pathak, M. D., Juliano, B. O., Sakamura, S., Oxalic acid as a sucking inhibitor of the brown planthopper in rice (Delphacidae, Homoptera), *Entomol. Exp. Appl.*, 27, 149, 1980.
67. Massonie, G., Elevage d'un biotype de *Myzus persicae* Sulzer sur milieu synthetique. V.-influence des acides oxalique et gentisique sur la valeur alimentaire d'un milieu synthetique, *Ann. Nutr. Aliment.*, 34, 139, 1980.
68. Von Burg, R., Toxicology update, *J. Appl. Toxicol.*, 14, 233, 1994.
69. Allison, M. J., Dawson, K. A., Mayberry, W. R., and Foss, J. G., *Oxalobacter formigenes* gen. nov.: Oxalate-degrading anaerobes that inhabit the gastrointestinal tract, *Arch. Microbiol.*, 141, 1, 1985.
70. James, L. F., Locomotor disturbance of cattle grazing *Halogeton glomeratus*, *J. Amer. Vet. Med. Assoc.*, 156, 1310, 1970.
71. Williams, M. C., Effect of sodium and potassium salts on growth and oxalate content of *Halogeton*, *Plant Physiol.*, 35, 500, 1960.
72. Thurston, E. L., Morphology, fine structure and ontogeny of the stinging emergence of *Tragia ramosa* and *T. saxicola* (Euphorbiaceae), *Amer. J. Bot.*, 63, 710, 1976.
73. Rauber, A., Observations on the idioblasts of *Dieffenbachia*, *Clin. Toxicol.*, 23. 79, 1985.
74. Sakai, W. S., Sheronia, S. S., and Nagao, M. A., a study of raphide microstructure in relation to irritation, *SEM*, 2, 979, 1984.
75. Kuballa, B., Lugnier, A. A. J., and Anton, R., Study of *Dieffenbachia*-induced edema in mouse and rat hindpaw: respective role of oxalate needles and trypsin-like protease, *Toxicol. Appl. Pharmacol.*, 58, 444, 1981.
76. Schmidt, R. J., and Moult, S. P., The dermatic properties of black bryony (*Tamus communis* L.), *Contact Derm.*, 9, 390, 1983.

77. Sakai, W. S., Hanson, M., and Jones, R. C., Raphides with barbs and grooves in *Xanthosoma sagittifolium* (Araceae), *Science*, 178, 314, 1972.
78. Theimer, T. C., and Bateman, G. C., Patterns of prickly-pear herbivory by collard peccaries, *J. Wildl. Manage.*, 56, 234, 1992.
79. Doaigey, A. R., Occurrence, type, and location of calcium oxalate crystals in leaves and stems of 16 species of poisonous plants, *Amer. J. Bot.*, 78, 1608, 1991.
80. Crombie, W.M.L., oxalic acid metabolism in *Begonia semperflorens*, *J. Exper. Bot.*, 5, 173, 1954.
81. Sasaki, K., Studies on oxalic acid metabolism in *Begonia* plant, *Bot. Mag.* Tokyo, 76, 48, 1963.
82. Berg, R.H., A calcium oxalate-secreting tissue in branchlets of the Casuarinaceae, *Protoplasma*, 183, 29, 1994.
83. Snetselaar, K.M. and Whitney, K.D., Fungal calcium oxalate in mycorrhizae of *Monotroopa uniflora*, *Can. J. Bot.*, 68, 533, 1990.
84. Bateman, D.F. and Beer, S.V., Simultaneous production and synergistic action of oxalic acid and polygalacturonase during pathogenesis of *Sclerotium rolfsii*, *Phytopathology*, 55, 204, 1965.
85. Stone, H.E. and Armentrout, V.N., Production of oxalic acid by *Sclerotium. cepivorum* during infection of onion, *Mycologia*, 77, 526, 1985.
86. Maxwell, D.P. and Lumsden, R.D., Oxalic acid production by *Sclerotinia sclerotiorum* in infected bean and in culture, *Phytopathology*, 60, 1395, 1970.
87. Kritzman, G., Chet, I. and Henis, I., The role of oxalic acid in the pathogenic behavior of *Sclerotium rolfsii* Sacc., *Exper. Mycol.*, 1, 280, 1977.
88. Marciano, P., Di Lenna, P. and Magro, P., Oxalic acid, cell wall-degrading enzymes and pH in pathogenesis and their significance in the virulence of two *Sclerotinia sclerotiorum* isolates on sunflower, Physiol. Plant Pathol., 22, 339, 1983.
89. Dickman, M.B. and Mitra, A., *Arabidopsis thaliana* as a model for studying *Sclerotinia sclerotiorum* pathogenesis, *Physiol. Molec. Plant Pathol.*, 41, 255, 1992.
90. Akamatsu, Y., Takahashi, M. and Shimada, M., Influences of various factors on oxaloacetase activity of the brown-rot fungus *Tyromyces palustris*, *J. Japan Wood Res.. Soc.*, 39, 352, 1993.
91. Wariishi, H., Valli, K. and Gold, M.H., Manganese(II) oxidation by manganese peroxidase from the basidiomycete *Phanerochaete chrysosporium*, *J. Biol. Chem.*, 267, 23688, 1992.
92. Espejo, E. and Agosin, E., Production and degradation of oxalic acid by brown rot fungi, *Appl. Enviorn. Microbiol.* 57, 1980, 1991.
93. Goodwin, D.G., Barr, D.P., Aust, S.D. and Grover, T.A., The role of oxalate in lignin peroxidase-calalyzed reduction: protection from compound III accumulation, *Arch. Biochem. Biophys.* 315, 267, 1994.

94. Kuan, I-C. and Tien, M., Stimulation of Mn peroxidase activity: a possible role for oxalate in lignin biodegradation, *Proc. Natl. Acad. Sci. USA*, 90, 1242, 1993.

95. Dutton, M.V., Kathiara, M., Gallagher, I.M. and Evans, C.S., Purification and characterization of oxalate decarboxylase from *Coriolus versicolor*, *FEMB Microbiol. Lett.*, 116, 321, 1994.

96. Shimada, M., Ma, D-B., Akamatsu, Y. and Hattori, T., A proposed role of oxalic acid in wood decay systems of wood-rotting basidiomycetes, *FEMS Microbiol. Rev.*,, 13, 285, 1994.

97. Seal, S. N., and Sen, S. P., The photosynthetic production of oxalic acid in *Oxalis corniculata*, *Plant Cell Physiol.*, 11, 119, 1970.

98. Zbinovsky, V., and Burris, R. H., Metabolism of infiltrated organic acids by tobacco leaves, *Plant Physiol.*, 27, 240, 1952.

99. Calmes, J., and Piquemal, M., Variation saisonniere des cristaux d'oxalate de calcium des tissus de Vigne vierge, *Can. J. Bot.*, 55, 2075, 1977.

100. Smartt, J., Kaul, A., Araya, W. A., Rahman, M. M., and Kearney, J., Grasspea (*Lathyrus sativus* L.) as a potentially safe legume food crop, in *Expanding the Production and Use of Cool Season Food Legumes*, Muehlbauer, F. J., and Kaiser, W. J., Eds., Kluwer Academic Publishers, Dordrecht, The Netherlands, 1994, 144.

Chapter 7

OXALATE-DEGRADING BACTERIA

Milton J. Allison[1], Steven L. Daniel[2] and Nancy A. Cornick[1,3]

[1]Physiopathology Research Unit, National Animal Disease Center, USDA-ARS, Ames, IA 50010; [2]Lehrstuhl für Ökologische Mikrobiologie, BITÖK, 95440, Bayreuth, Germany; [3]Iowa State University, Ames, IA 50011

I. INTRODUCTION

Since oxalic acid is produced as an end product by a wide diversity of plants, animals and microbes, it should not be surprising that microbes able to attack oxalate are widely distributed in natural ecosystems. As is true with other natural products, microbes that degrade the products play important roles in the cycling of carbon. In the case of oxalate, its properties suggest that, were it not for microbial catabolism, its accumulation in large quantities might well be incompatible with life as we know it.

The ability to use oxalate as a carbon source is widespread among the fungi, and the presence or absence of this ability has been used in fungal classification schemes. Though fungi may play important roles in the oxalate loop of the carbon cycle, this review will be limited to considerations of oxalate-degrading bacteria.

From the viewpoint of a bacterium, oxalate would not seem to be a "first-choice foodstuff." The chemical nature of oxalate (a strong acid, the low solubility of many of its salts, its capacity as an enzyme inhibitor, etc.) would appear to limit microbial appetites for it. Furthermore, the high oxidation state of oxalate dictates that only a small amount of energy can be made available through further oxidations of the molecule. Yet, as we shall see, a wide variety of bacteria have filled the multitude of niches where oxalate is produced and is available as a substrate. The niches range from aerobic sites where oxygen is available and is used and may be required as a terminal electron acceptor to anaerobic sites where oxygen is virtually absent.   Some oxalate-degrading bacteria are "generalists" that are able to ferment many other substrates as well as oxalate, while others are "specialists" that depend upon oxalate for their very existence.

II. OXALATE-DEGRADING BACTERIA

A. ANAEROBIC ORGANISMS
Information about oxalate degradation by anaerobic microbes is mainly from studies of activities in the rumen and the mammalian hind gut, but some data have also been collected from terrestrial ecosystems.   While evidence for oxalate loss from such anaerobic habitats is not new, the organisms responsible for this activity have only recently been isolated.

1.   Oxalate Degradation in Animal Gastrointestinal Tracts:
During studies of the toxicity of oxalate in animal diets it became clear that animals did develop a tolerance to high oxalate diets. It was then hypothesized that this acquisition of tolerance might be due to increased oxalate degradation by intestinal microbes.   The significance of toxin degradation by gastrointestinal microbes is most readily appreciated when the degradative activity occurs in a forestomach fermentation site before the toxin has had a chance for absorption.   The reticulo-rumen is the best known of such forestomach fermentation sites, but a number of non-ruminants also use this digestive strategy and the survival value of such forestomach toxin degradations in an evolutionary context has been discussed.[1]

a.   Rumen Activity
The capacity of ruminal microbes to degrade oxalate was recognized with studies of oxalate degradation in rumen contents obtained from cattle[2,3] and from sheep.[4]   Results of these and of subsequent studies[5,6] showed that exposure of animals to increasing quantities of oxalate in diets led to increased rates of oxalate degradation by ruminal microbes and

provided support for the hypothesis that acquisition of tolerance was due to the increased oxalate degradation rates.

b.   Hind Gut Activity

Results of oxalate balance studies with swine,[7] and horses[8] suggested that appreciable amounts of ingested oxalate may be degraded by intestinal bacteria. Further evidence for oxalate degradation by intestinal microbes was obtained by Shirley and Schmidt-Nielsen who found that $^{14}CO_2$ was produced when ^{14}C-labeled oxalate was fed to pack rats, sand rats, and hamsters, but not when the labeled oxalate was injected intraperitoneally. With laboratory rats, however, little if any $^{14}CO_2$ was recovered after feeding the animals ^{14}C-labeled oxalate.[9] The importance of oxalate degradation by intestinal microbes in desert rodents that ingest high oxalate diets was discussed. It was then shown that microbial populations from the hind gut of swine, rabbits, guinea pigs, and a horse all degraded oxalate (Table 1) and, as had been noted with rumen populations, rates of degradation increased markedly when increased levels of dietary oxalate were supplied.[10] This was a strong indication that populations with increased proportions of oxalate-degrading bacteria were selected when increased amounts of oxalate were available. In agreement with the findings of Shirley and Schmidt-Nielsen,[9] however, rates of oxalate degradation by microbes in cecal contents from laboratory rats were negligible, or very low, and addition of oxalate to diets did not influence rates.[10] It was then found that although the ceca of wild rats were colonized by oxalate-degrading bacteria, this was not true for most laboratory rats.[11]

c.   Isolations of Intestinal Anaerobes

Initial attempts to isolate oxalate-degrading bacteria from oxalate-adapted rumen populations were unsuccessful.[5] Populations enriched with oxalate-degrading bacteria were then established *in vitro* in both batch[12] and continuous flow enrichment cultures[13] with oxalate as the principal carbon and energy source. The first isolations of anaerobic ruminal oxalate-degrading bacteria were made by Dawson from an enrichment culture of a population from the rumen of a sheep.[14] Subsequent studies with these isolates led to the conclusion that they were different from any previously described organism and to the establishment of the new genus and species,

Table 1.   Rates of Oxalate Degradation and Concentrations of Oxalate-degrading Bacteria from Gastrointestinal Samples

Host Animal	Site	Oxalate Degradation Rate		Concentration (log$_{10}$CFU/g)	Reference
		Normal Diet	High Oxalate Diet		
Sheep	Rumen	0.03[a]	0.3-0.8[a]	-	5
Sheep	Rumen	1.8[b]	26.0[b]	6.4[d]	18
Sheep	Rumen	1.6[b]	7.5[b]	8.3[d]	18
Sheep	Rumen	3.9[b]	19.0[b]	8.8[d]	18
Pig	Cecum	0-0.1[c]	4.0[c]	-	10
Rabbit	Cecum	0.5[c]	6.0[c]	-	10
Guinea pig	Cecum	0.7[c]	7.0[c]	-	10
Horse	Rectum	0.1[c]	0.8[c]	-	10
Wild rat	Cecum	2.5-20.6[b]	-	7.8	11
Lab rat[e]	Cecum	1.8-3.0[b]	1.8-2.6[b]	0	11
Lab rat[f]	Cecum	2.0[b]	23.1[b]	7.2[d]	11

[a]μmol degraded/ml ruminal contents per hour

[b]μmol degraded/g (dry weight) per hour

[c]μmol degraded/g (wet weight) per hour

[d]Estimates from high oxalate diet samples

[e]These rats were not colonized by *Oxalobacter*

[f]These rats were colonized by *Oxalobacter*

Oxalobacter formigenes.[15]   Isolations of *O. formigenes* have
since been made from cecal contents of a pig,[15] wild and
domestic rats,[11] and a guinea pig.[16] While information that is
now available suggests that bacteria similar to *O. formigenes*
are the most important oxalate-degrading anaerobic bacteria in
intestinal tracts of animals, the possible function of other
species in this activity cannot be excluded. There is mention of
the isolation of an oxalate-metabolizing *Clostridium* from donkey
dung, but no description of the organism was given. There is
also a very recent report indicating that oxalate could serve as
a substrate for growth of spore-forming, Gram-negative,
acetogenic bacterium that was isolated from the rumen of a
mature deer in France.[17]

Development of methods for detection of oxalate-degrading
bacteria based upon the appearance of zones of clearing of
calcium oxalate around colonies[15,18] has greatly facilitated
isolations and enumerations of oxalate-degrading bacteria.
Failures to detect *Oxalobacter* in previous studies of
gastrointestinal populations might be explained by their
requirements for: oxalate as growth substrate, for failure to
appreciate their requirements for strictly anaerobic growth
conditions, and the fact that they are outnumbered by other
bacteria. Most of the isolations and enumerations were
accomplished using modifications[19,20] of the Hungate roll tube
technique for anaerobiosis.

2. Oxalate Degradation in Human Gastrointestinal Tracts
a. Activity

In 1940, Barber and Gallimore[21] demonstrated that
microbes in human feces degraded oxalate during anaerobic *in
vitro* incubation. The incubation times were long and data
concerning rates *in vivo* were not collected. Hodgkinson[22]
estimated that gut bacteria degraded 70-100 mg oxalate per
day, but the source of data supporting the estimate is not clear.
Estimates of fecal oxalate degradation rates (based on
measurements of $^{14}CO_2$ production after addition of ^{14}C-oxalate
to fecal samples and incubation under conditions approximating
those expected *in vivo*) are given in Table 2. Extrapolations
from one set of these values gave an estimate that colonic
bacterial populations had the potential to degrade from 100 to
more than 1,000 mg of oxalate per 24 h.[23]

Table 2.   Oxalate-degrading Bacteria in Human Feces

Status	Oxalate Degradation Rate[a]	Concentration $(\log_{10} CFU/g)$[b]	#+/#Tested[c]	Reference
Normal	0.1-4.8	7.4	4/7	24
Normal	0.46	-	14/22	31
Normal	0.75	-	-	31
Normal	0.19-3.5	8.0	8/10	23
Normal	0.27-3.5	8.4	4/7	108
Normal	-	7.4	5/8	25
Low oxalate diet	0.15-0.57	-	-	24
Spinach diet	-	8.4	5/8	25
High oxalate diet	1.26	-	-	31
JIB[d]	0-0.006	-	-	24
Crohn's disease	0.11	-	1/9	31
Steatorrhea	0.08	-	1/4	31
Inflammatory bowel disease	0.05-0.44	5.4	2/10	23
Stone formers[e]	-	-	1/10	108,109
Control[f]	-	-	13/22	108,109

[a]μmol oxalate degraded/g (wet wt) of feces/h

[b]$\log_{10}$ cleared zones/g (wet wt) of feces (highest measurement reported)

[c]Number of individuals positive (fecal cultures with colonies showing cleared zones of Ca-oxalate)/number of individuals tested

[d]Samples from 8 patients that had undergone jejunoileal bypass surgery

[e]Patients with more than 4 stone episodes

[f]Patients without idiopathic oxalate urolithiasis

b.   Isolations

Although relatively few isolates of anaerobic oxalate-degrading bacteria from human intestinal specimens have been studied, the information available now indicates that the bacteria mainly responsible for oxalate degradation in humans are similar to *O. formigenes* strains that have been isolated from animals and from sediments.[24-26]   There is, however, also a report of the isolation from human feces of a *Eubacterium lentum* strain that was able to degrade oxalate.[27]   The latter strain was not dependent upon oxalate for growth, and so was not a specialist and differed in this regard from all known strains of *O. formigenes*. As the information now available is limited and most of it is recent, it should not be assumed that the only bacteria involved in intestinal oxalate degradation are those which have now been identified.

c.   Colonization Incidence in Humans

Concentrations of oxalate-degrading bacteria as high as 10^8 per g of feces have been estimated in several laboratories (Table 2).   Evidence is, however, accumulating which confirms the position that some persons who have been part of "normal subject populations" were not colonized by oxalate-degrading bacteria.   These individuals were either not colonized or concentrations of oxalate-degrading bacteria in fecal samples were so low that they were not detected by the cultural methods used.   Initially, it was presumed that this failure to detect oxalate-degrading bacteria was due to inadequacies of methods;[24] however, that explanation seems now to be in error since several other laboratories have reported that some of the persons in their "normal" group of subjects were not colonized. Information concerning the site of colonization was obtained by culture of fluid and brush samples taken at colonoscopy.[26] Oxalate-degrading bacteria that appeared to be similar to *O. formigenes* were detected in cecal and/or sigmoid brush samples from 22 of 24 subjects examined. It was suggested that longer incubation time for cultures may have increased the sensitivity of detection of oxalate-degrading bacteria. Concentrations of oxalate-degrading bacteria in cecal brush samples were estimated to be about 10-fold greater than in sigmoid brush samples.

d.   Significance of Colonization by O. formigenes

Doane et al.[25] compared urinary and fecal oxalate excretion patterns of a group of persons that were colonized by oxalate-degrading bacteria (n = 5) with those of a group that was not colonized (n = 3).   Means of measurements of urinary oxalate recovered were consistently higher from the group that was not colonized, but with the small sample size, statistical differences in this parameter were found during only one of the three test periods.   Oxalate recovery in feces was markedly greater (p < 0.01) from the non-colonized group during all test periods.

A number of gastrointestinal disorders have been associated with increased absorption of dietary oxalate and thus an increased risk for urolithiasis. Results indicate that the colon is an important site for the increased oxalate absorption and a multiplicity of etiologic factors have been implicated in the condition that is now known as enteric hyperoxaluria.[28,29] Loss of, or diminished activity, of oxalate-degrading bacteria in the colon has generally not been considered as a factor in enteric hyperoxaluria. It is our hypothesis, however, that loss of this activity is indeed important.[24,30] Support for this hypothesis is found in results from studies with groups of patients that suffer from, or are at increased risk for, enteric hyperoxaluria. Negligible or very low rates of oxalate degradation were found in fecal samples from patients that had undergone jejunoileal bypass surgery[24], patients with Crohn's disease or steatorrhea,[31] or with inflammatory bowel disease.[23]   Kleinschmidt and co-workers[32] reported that the concentrations of oxalate-degrading bacteria in feces from patients with highly active idiopathic calcium oxalate urolithiasis were 3 orders of magnitude less than concentrations in controls. They were able to detect oxalate- degrading bacteria in stool samples from only 1 of 10 patients that had suffered from more than 4 episodes of stone disease, while 13 of 22 patients in their control group were colonized by these bacteria (Table 2).

Reasons for loss of oxalate-degrading bacteria in these conditions are generally not known but diarrheas that are common with jejunoileal bypass might well preclude establishment of an active oxalate-degrading population.[24] The role of increased bile salt concentrations as inducers of enhanced permeability of the colon to oxalate has been considered important in enteric hyperoxaluria. The finding that growth of *O. formigenes* was inhibited by sodium deoxycholate[30] seems to be another reason to consider the possibility that loss of oxalate-degrading capacity may be important.

e. Animal Models

It has already been mentioned that tests with ^{14}C-oxalate indicated that some laboratory rats were not colonized by oxalate-degrading bacteria. When this situation was examined in more detail by Daniel and co-workers it was discovered that of 8 groups of rats that were tested (from 5 different suppliers), only one group was colonized by oxalate-degrading bacteria.[11] Establishment of a population of *Oxalobacter* in adult rats (that were not previously colonized) did not occur unless a relatively high level of sodium oxalate (3% or more) was added to the diet. Strains of *O. formigenes* that had been isolated from cecal contents of swine, guinea pigs, wild rats, laboratory rats and from human feces all colonized laboratory rats when high oxalate diets were fed. A strain isolated from the rumen of a sheep, however, failed to colonize these rats.[33] Since one group of rats was naturally colonized by *O. formigenes* and since this occurred without the benefit of a high oxalate diet, it seems reasonable to propose that in these rats *O. formigenes* populations became established at an early age.   Difficulties encountered during attempts to colonize adult rats suggest the operation of some type of competitive exclusion factor.

To evaluate the importance of oxalate degradation in the hind gut, experiments were conducted to compare the fate of dietary oxalate between laboratory rats that were colonized following inoculation with *O. formigenes* and rats that were not colonized. In colonized rats, rates of oxalate degradation and concentrations of *O. formigenes* in the cecum and colon were similar.   When ^{14}C-oxalate was given orally, colonized rats excreted less ^{14}C from labeled oxalate in feces and more as $^{14}CO_2$.   There was, however, not a significant effect of colonization on urinary oxalate excretion.[34] This suggests that colonization did not affect the amount of dietary oxalate absorbed. We suggest that the relatively high levels of calcium in the rat diets used for these experiments (0.93%) may have limited both oxalate absorption and oxalate metabolism by the bacteria so that effects of colonization were obscured.   This hypothesis was supported by comparisons of oxalate excretion patterns between rats fed diets with 0.02% and 1% calcium.[35]

Guinea pigs that were adapted to high oxalate diets developed populations of intestinal microbes that degraded oxalate at significantly increased rates and the animals absorbed less of the dose of ^{14}C-oxalate that was injected into the ceca. The protective effect of the increased oxalate degradation rates was lost when adapted guinea pigs were treated with oral antibiotics.[30]

f.   Properties of O. formigenes

Most information about properties of *O. formigenes* has come from studies with the type strain, strain OxB.[14] It is a Gram-negative, obligately anaerobic, rod-shaped bacterium with cells (0.4-0.6 x 1.2-2.5 μm) that are often curved and which occur singly, in pairs, or in short chains. Spores or flagella are not present and the G + C content of DNA is 48-51%. The requirement for oxalate for growth is absolute, and growth is proportional to oxalate supplied with about 1 g of cell (dry wt) synthesized per mole of oxalate metabolized. Acetate is the only other organic growth factor that has been found for the organism. Formate and carbon dioxide are the main products from oxalate catabolism, and they are produced in a nearly 1:1 ratio.[15]

g.   Taxonomy of and Diversity Within O. formigenes

The new genus and species *Oxalobacter formigenes* was established when phenotypic properties of strains indicated that no appropriate placement could be made into any existing genus.[15] Comparisons of 16S rRNA sequences of several *O. formigenes* strains with sequences in a large data bank also support the establishment of a new genus for this group.[36] Comparisons were made between strains of *O. formigenes* that included isolates from human feces, sheep rumen, pig cecum, guinea pig cecum and from lake sediments. Based on differences between patterns of cellular fatty acids, of cellular proteins and of nucleic acid fragments, the 21 strains studied could be placed into two main groups, but no obvious habitat-group relationship was observed.[37]

3.   Sediments

Anaerobic microbes appear to be important agents of oxalate degradation in sediments. Anaerobic oxalate degradation occurred in every sediment tested from a diversity of aquatic environments.[38] Production of $^{14}CO_2$ from ^{14}C-oxalate added to these sediments was markedly inhibited when oxygen was provided. Smith and co-workers isolated both rod-shaped and spiral-shaped anaerobic oxalate-degrading bacteria from fresh water reservoir sediments.[39] The rod-shaped isolates (strain Sox-4) were considered to belong to *O. formigenes* while the spiral-shaped isolates (strain Ox-8) appeared not to fit into any known taxon. Curved rods isolated from freshwater sediments by Dehning and Schink[40] were similar to strain Ox-8. Cells were, however, motile with 1-2 polar flagella and they were assigned to a new species *Oxalobacter vibrioformis*. These workers also isolated a Gram-positive, spore-forming motile rod which was designated as a new species, *Oxalophagus oxalicus (Clostridium oxalicum)*.[40,41]

All of the latter group of isolates appear to be physiologically similar to *O. formigenes* strains isolated from intestinal habitats in that all are specialists that require oxalate or oxamate as substrates and other substrates do not serve as energy sources. Formate is a major end product and acetate is assimilated, presumably for cell synthesis purposes.[39,40]

Postgate[42] used oxamate based enrichment cultures to isolate an anaerobic vibrio from a stream polluted with duck excrement. The isolate, now classified as *Desulfovibrio vulgaris* subsp. *oxamicus*[43] grew with lactate, pyruvate, choline or formate, as well as with oxamate or oxalate as energy sources. However, the stability of oxalate-degrading capabilities with this organism are questioned since neither oxalate or oxamate were degraded in tests with a strain of this organism which had been maintained in a stock culture collection with lactate as substrate.[40]

4. Thermophilic Habitats

Even though the *in situ* oxalate-degrading activities of microorganisms present in thermophilic habitats (e.g., hot springs) have yet to be examined, two thermophilic bacteria, *Clostridium thermoaceticum* and *Clostridium thermoautotrophicum*, have recently been shown to be capable of growth at the expense of oxalate.[44] These two obligate anaerobes possess several unique properties which distinguish them from the other anaerobic oxalate-degrading bacteria isolated to date (Table 3). One is the potential of these

Calcium Oxalate in Biological Systems

Table 3. General Properties of Anaerobic Oxalate-utilizing Bacteria

Organism	Source	Gram stain*	Spore former	Motile	T_{opt} (°C)	Nutritional requirement[b]	G + C (mol%)	Growth-supportive substrate	Reference
Clostridium thermoaceticum	Horse manure	+	+	-	60	Nicotinic acid	54	Oxalate, glyoxylate, pyruvate C_1-compounds (CO, HCOOH, H_2/CO_2, methanol, O-methyl groups), glucose, xylose	44, 47, 122, 123, 124
Clostridium thermoautotrophicum	Hot spring	+	+	+	60	Nicotinic acid	54	Oxalate, glyoxylate, pyruvate C_1-compounds (CO, HCOOH, H_2/CO_2, methanol, O-methyl groups), glucose, xylose	44, 47, 123, 124, 125
Desulfovibrio vulgaris subsp. *oxamicus*	Creek mud	-	-	+	30	Yeast extract sulfate[c]	63.1	Oxalate, oxamate, formate pyruvate, choline, lactate	42
Oxalobacter formigenes	Sheep rumen[d]	-	-	-	37	Acetate	49	Oxalate	14, 15
Oxalophagus oxalicus	Freshwater sediment	+	+	+	30	Acetate	36.3	Oxalate, oxamate	40, 41
Oxalobacter vibrioformis	Freshwater sediment	-	-	+	30	Acetate	51.6	Oxalate, oxamate	40
Strain Ox-8 sediment	Freshwater	-	-	-	25	Acetate	ND*	Oxalate	39

* All are rod-shaped organisms.
[b] The pH range for growth of these organisms is 6-7.
[c] Sulfate is not required for growth at the expense of pyruvate or choline.
[d] Strains of *O. formigenes* have been also isolated from human feces,[24] from the cecal contents of swine[15], rats[11], and a guinea pig[16], and from freshwater lake sediments.[39]
* ND, Not determined.

organisms to utilize a wide variety of substrates, as well as oxalate, for carbon and energy. This is in contrast to the other oxalate-degrading anaerobes (with the exception of *D. vulgaris* subsp. *oxamicus*) which can only utilize oxalate (or oxamate after deamination) as a growth-supportive substrate. Thus, *C. thermoaceticum* and *C. thermoautotrophicum* are considered "generalists", not "specialists", relative to substrate utilization. Another distinctive property of *C. thermoaceticum* and *C. thermoautotrophicum* is that both belong to a group of obligate anaerobes known as acetogenic bacteria.[44,45] This group of bacteria has the capacity to use an autotrophic process termed the acetyl coenzyme A (acetyl-CoA) or Wood pathway (i) for the reductive synthesis of acetate from CO_2, (ii) as a terminal electron-accepting, energy-conserving process, and (iii) for the synthesis of cell carbon from CO_2.[45-48] So far, the potential among acetogenic bacteria to utilize oxalate appears to be limited to *C. thermoaceticum* and *C. thermoautotrophicum* since none of the other species of acetogenic bacteria that have been tested (*Acetobacterium woodii*, *Acetogenium kivui*, *Clostridium aceticum*, *Clostridium formicoaceticum*, *Eubacterium limosum*, and *Peptostreptococcus productus*) were capable of oxalate-dependent growth.[44] Since acetogenic bacteria are present in most anaerobic environments including the gut and sediments[45-47], it is possible that this group of bacteria plays a role in the turnover of oxalate.

B.   AEROBIC ORGANISMS

1.   Terrestrial Ecosystems

Oxalate is commonly found in soils and is primarily derived from root exduates, breakdown products from plants, animal, and microbial tissues, and metabolites from bacteria and fungi.[49-51] Oxalate concentrations in soils range from 10^{-3} to 10^{-6} M with amounts being slightly higher in rhizosphere soils compared to non-rhizosphere soils.[49] The range in concentrations reflects seasonal cycles and the fact that oxalate in aerobic soils is constantly being synthesized and degraded;[51] the fate of oxalate in soils under anaerobic conditions is less obvious. By virtue of its capacity to chelate metals, oxalate is central to the solubilization and transport of soil metals and to the weathering of rock.[51-53] Through interactions with Al and Fe, oxalate plays a major role in plant nutrition by increasing the availability of P, K, Mg, and Ca in soils; furthermore, oxalate is also involved in the detoxification of excess Al and Ca by removing these cations from soil solutions.[49-52] Thus, aerobic or anaerobic microbial processes that affect oxalate-metal interactions or the availability of oxalate may ultimately influence the nutritional-toxicological status of terrestrial ecosystems. Unfortunately, while a considerable number of oxalate-degrading bacteria have been isolated from soils (Table 4);[54-56] information is limited regarding their interactions with metals, with oxalate or with other microbes or activities in soils.

a. Populations and Activities

Concentrations of aerobic oxalate-degrading microorganisms associated with Douglas fir mycorrhizal tissue and root sap, rhizomorphs of *Hysterangium crassum* (a species of fungus ectomycorrhizal with Douglas fir, and rhizosphere soil of Douglas fir) have been estimated.[54] The highest numbers of oxalate-degrading organisms were found in the rhizosphere soil from *H. crassum* (2.8×10^9 colonies per g of moist soil [35% of the total viable count]) and in the rhizosphere soil from Douglas-fir mycorrhizae (7.1×10^8 colonies per g of moist soil [33% of the total viable count]). With both soils, approximately one-third of the oxalate-degrading population were *Streptomyces*;[54] this group of microorganisms is considered to be the most active relative to the turnover of calcium oxalate in soils.[54,57] The increased concentrations of oxalate-degrading organisms associated with rhizosphere soils probably reflects the fact that mycorrhizal fungi are known producers of oxalate.[52] In this regard, in a recent study investigators found that within 21 days after oxalate was added to a field plot, microbial rates of oxalate degradation increased from 0.018 to 0.26 μmol CO_2 per g soil per 96 h and viable numbers of aerobic oxalate-degrading bacteria increased from approximately 0.1 to 1.57×10^7 organisms per g soil.[58] Interestingly, during the same time period, there was an initial increase (within the first 10 days) in extractable phosphate levels after the addition of oxalate but levels then decreased to near baseline values by day 21. Together, these results provide evidence to support the concept that oxalate additions to soils increase phosphate availability and that the oxalate-degrading activities of microorganisms can influence phosphorous cycling in soils.

Table 4. Aerobic Oxalate-utilizing Bacteria

Organism	Source	Reference
"Alcaligenes" eutrophus	Soil, sludge	63, 64
Bacillus oxalophilus	Soil	66
Carbophilus carboxidus	Soil	67
Methylobacterium extorquens[a]	Soil, earthworm excreta	68, 69, 70
Nocardia spp.	Soil	111
Oligotropha carboxidovorans	Soil, sewage wastewater	67
Proactinomyces citreus	Soil, water, sheep rumen	111
Pseudomonas carboxydohydrogena	Soil, mud	112
Streptomyces spp.	Soil, earthworm intestine	54, 59
Thiobacillus novellus	Soil	71
Vibrio oxalaticus	Soil	113
Xanthobacter spp.	Soil, mud, water	72
Actinomyces spp.	Earthworm intestine, mayfly, springtails, stonefly, mites	57
Mycobacterium lacticola	Earthworm intestine	114
"Pseudomonas oxalaticus"	Earthworm intestine	80
Uncertain Affiliation		
Pseudomonas sp. OD1	Soil	113
Pseudomonas sp. OX-53	Soil	86
Pseudomonas sp. KOx	Chicken dung	116
Pseudomonas sp. MOx	Chicken dung	116
Pseudomonas sp. YOx	Chicken dung	116
Alcaligenes sp. LOx	Chicken dung	116
Strain RO-16	Sheep rumen	116

[a] Based on recommendations from comparative studies,[69,70,119] the following oxalate-utilizing, pink-pigmented facultatively methylotrophic bacteria should probably be considered strains of *Methylobacterium extorquens*: *Pseudomonas* sp. M27 isolated from soil;[120] *Pseudomonas* sp. AM1 isolated as an aerial contaminant;[121] *Pseudomonas* sp. AM2 isolated as an aerial contaminant;[82] *Protaminobacter ruber*;[82] *Pseudomonas* sp. AM1 (variant 470 isolated from chicken dung;[117] and *Pseudomonas* sp. RJ, isolated from soil.[122]

b.   Isolations

As illustrated in Table 4, the ability to use oxalate appears to be a general trait among several genera of aerobic soil bacteria (e.g., *Streptomyces* and *Xanthobacter*). In fact, the ability to use oxalate has been used to classify species of the genus *Streptomyces*.[59-61] Given the number of oxalate-degrading bacteria isolated to date, it is beyond the scope of this review to deal with each organism individually; only selected organisms will be discussed in the following paragraphs. The reader is directed to several reviews[22,62] for an overview of oxalate-utilizing bacteria and to Section III (see below) on the physiology and biochemistry of oxalate-degrading bacteria.

"Alcaligenes" eutrophus is a diverse group of Gram-negative, strictly aerobic, rod-shaped bacteria that grow at the expense of a wide assortment of organic compounds, including oxalate.[63] *"A." eutrophus* strain H16 is a facultative chemolithoautotroph;[64] however, strains of *"A." eutrophus* that utilize oxalate but are not capable of H_2-dependent chemolithotrophic growth have been isolated, indicating H_2-dependent chemolithotrophy is no longer a stable phenotypic trait for *"A." eutrophus*.[63] In the 9th edition of Bergey's Manual of Systematic Bacteriology,[65] *"A." eutrophus* is listed as *species incertae sedis* because of phenotypic and phylogentic differences from *Alcaligenes faecalis* (the type strain of the genus) and will probably be assigned to a new genus in the future.

Bacillus oxalophilus is a mesophilic oxalate-degrading bacterium that was recently isolated from the rhizosphere of sorrel.[66] Cells of *B. oxalophilus* are motile, Gram-positive, sporeforming rods and utilize only oxalate as a source of carbon and energy. The latter characteristic as a specialist is unlike other known aerobic oxalate-degrading bacteria.

Carbophilus carboxidus, *Pseudomonas carboxydohydrogena*, and *Oligotropha carboxidovorans* are carboxidotrophic bacteria based on their ability to utilize CO as a sole source of carbon and energy for chemolithoautotrophic growth; these aerobic Gram-negative rods can also utilize chemoorganoheterotrophic substrates, including oxalate, for growth.[67]

Methylobacterium extorquens was the first oxalate-degrading bacteria to be described in the literature and was initially named *Bacillus extorquens* by Bassalik in 1913.[68] However, the taxonomic position of this organism and other pink-pigmented, facultative methylotrophs (PPFMs) was uncertain. Thus, they were subsequently assigned to several different genera (e.g., *Vibrio*, *Pseudomonas*, *Protomonas*) until it was proposed that PPFMs be grouped in the genus *Methylobacterium* and that *Protomonas extorquens* be reclassified as *M. extorquens*.[69,70] *M. extorquens* and related strains are able to use C_1 compounds (methanol and methylamine) as well as multi-carbon substrates such as oxalate (Table 4).[70]

Thiobacillus novellus is a strictly aerobic, facultatively chemolithoautotrophic bacterium, deriving energy for growth either from the oxidation of reduced sulfur compounds (i.e., thiosulfate or sulfite) or from the oxidation of oxalate and several C_1-compounds (formate, methanol, and formamide).[71] Interestingly, oxalate-grown cells of *T. novellus* do not oxidize thiosulfate whereas formate-, methanol-, and formamide-grown cells are capable of thiosulfate oxidation.[71]

Xanthobacter spp. are obligately aerobic, chemolithoautotrophic, hydrogen-oxidizing bacteria that are able to fix dinitrogen.[72] *Xanthobacter autotrophicum* and *Xanthobacter flavus* are two species that can utilize a number of organic acids (including oxalate), short-chain alcohols, certain sugars, and amino acids.[72]

Aerobic oxalate-degrading bacteria isolated from soils are also being used in biotechnological applications. Strains of oxalate-degrading bacteria have been recently used to protect *Arabidopsis thaliana* (a model host plant) from infection by *Sclerotinia sclerotiorum* (a oxalate-producing fungal pathogen).[55] Apparently, oxalic acid is involved in the infection process by *S. sclerotiorum*, and, when cultures of oxalate-degrading bacteria are applied to the leaves of *A. thaliana*, the infection is prevented by the oxalate-degrading bacteria. Future applications are to develop (through the introduction of bacterial genes involved in oxalate degradation) transgenic plants that are resistant to *S. sclerotiorum* infection.

2. Gastrointestinal Habitats

a. Earthworm intestine

Several species of aerobic oxalate-degrading bacteria have been isolated from earthworm intestines (Table 4). Of these,

"*Pseudomonas oxalaticus*" has been the most extensively studied relative to oxalate metabolism.[22,62,73-79] It was originally isolated in 1953 by Khambata and Bhat[80] but was invalidly named.   Jenni et al.[63] examined the taxonomic relationship between "*A.*" *eutrophus* and "*P. oxalaticus*" and found them to be very similar phenotypically, although "*P. oxalaticus*" is unable to grow at the expense of H_2/CO_2.   In addition, these two bacteria have a 50% DNA/DNA homology with each other, further supporting the concept that they are closely related.[63] "*P. oxalaticus*" is a strict aerobe (Gram-negative rod) that grows at the expense of variety of substrates (oxalate, formate, lactate, succinate, glyoxylate, glycollate, malonate, acetate, and glycerol).[76,80]   Formate is the only substrate that supports autotrophic growth of "*P. oxalaticus*" via the Calvin-Benson cycle.[76,77,81]

b. Rumen

Two aerobic oxalate-degrading bacteria have been isolated from the rumen, *Proactinomyces citreus* and strain RO-16 (Table 4). Given the anaerobic nature of the rumen and the fact that obligate anaerobes greatly outnumber the aerobic bacteria in rumen, the role that these aerobic oxalate-degrading bacteria may be playing, other than possible transient organisms, in such an environment is presently unclear. This may also apply to the *Pseudomonas* strains that have been isolated from chicken dung (Table 4).

III. OXALATE METABOLISM

A. CATABOLISM

Most bacteria metabolize oxalate by a three-step process: i) oxalate is activated to oxalyl-CoA; ii) oxalyl-CoA is decarboxylated to formyl-CoA; and iii) formate is oxidized to CO_2. Energy for growth of aerobic oxalate-degrading bacteria is derived from the oxidation of formate coupled to the reduction of NAD(P) or another electron carrier. The key enzymes of these reactions are oxalyl-CoA decarboxylase (EC 4.1.1.8), formyl-CoA transferase (EC 2.8.3.4) and formate dehydrogenase (EC 1.2.1.2, EC 1.2.2.1). Quayle et al. proposed that "*P. oxalaticus*"[74] and *M. extorquens*[82] activate oxalate to oxalyl-CoA by transferring CoA directly from formyl-

CoA or indirectly from succinyl-CoA to formyl-CoA. Oxalyl-CoA decarboxylase and formyl-CoA transferase are both cytoplasmic enzymes. "*P. oxalaticus*" transports oxalate into the cell using an inducible, active transport system.[83] Both soluble and membrane bound formate dehydrogenases (FDH) have been described. *T. novellus*[71] and *M. extorquens*[82] produce a soluble, NAD-dependent FDH. The enzyme from *C. thermoaceticum* requires NADP[84] and the FDH from *B. oxalophilus* reduced the artificial electron carrier, phenazinemethosulfate, but not NAD.[66] "*P. oxalaticus*"[74,85] and "*A.*" *eutrophus*[64] produce both soluble NAD-dependent and membrane bound, NAD-independent formate dehydrogenases. One strain of *Pseudomonas* (OX-53) oxidizes oxalate directly to CO_2 and H_2O_2 using the enzyme oxalate oxidase (EC 1.2.3.4).[86] This enzyme is generally found in plants and is a flavoprotein,[22] while the enzyme purified from *Pseudomonas* is a metalo-protein that contains Mn and Zn.[86]

In contrast to the majority of oxalate-degrading bacteria, *O. formigenes* produces formate as an end product of oxalate metabolism.[15] This makes energy generation linked to formate oxidation unlikely. The change in free energy associated with the decarboxylation of oxalate (-26.7 kJ/mole)[87] is less than the energy required to synthesize a mole of ATP, making substrate level phosphorylation linked to decarboxylation unlikely.[15]

Proteoliposomes of *O. formigenes* exchange both formate and oxalate.[88] This exchange does not require cations and is electrogenic. These observations led Maloney and co-workers[88] to conclude that membranes of *O. formigenes* contain an electrogenic oxalate^{2-}:formate^{1-} antiporter. They proposed a model linking the transport and decarboxylation of oxalate to energy conservation which is depicted in Figure 1. Oxalate is transported into the cell and activated to oxalyl-CoA by the transfer of CoA from formyl-CoA. Oxalyl-CoA is then decarboxylated to formyl-CoA plus CO_2 and the CO_2 diffuses out of the cell. The CoA is transferred from formyl-CoA to an incoming molecule of oxalate and formate exits the cell through the antiporter. A proton is consumed for each molecule of oxalyl-CoA that is decarboxylated resulting in a proton gradient which is used to drive ATP synthesis. This cycle of influx, decarboxylation and efflux constitutes an indirect proton pump. Measurements of membrane potential ($\Delta\psi$) and of pH gradients (ΔpH) across *O. formigenes* cell membranes are in agreement

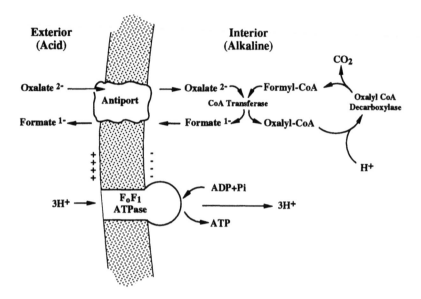

Figure 1. Proposed model for the synthesis of ATP from oxalate by *O. formigenes*.

with the proposed model.[89] These measurements also indicate that this organism has limited capacity to regulate internal pH and that $(\Delta\psi)$ is quantitatively a more important component of its proton motive force than is the ΔpH component.

The two key enzymes in this cycle, oxalyl-CoA decarboxylase and formyl-CoA transferase, have been purified from *O. formigenes*.[90,91] Both enzymes are cytoplasmic; only 0.4% of the transferase and 2.4% of the decarboxylase were contained in the membrane fraction.[92] Oxalyl-CoA decarboxylase makes up 10.5% of the soluble protein in the cell and requires thiamin pyrophosphate (TPP) as a cofactor.[91] Activity is also stimulated by Mg^{2+} ions. Formyl-CoA transferase makes up 0.2% of the soluble cell protein. Both oxalate and succinate act as acceptors for formyl-CoA transferase but acetate and malonate do not.[90] The model proposed by Maloney and co-workers[88] was substantiated by the isolation of the oxalate^{2-}:formate^{1-} antiporter of *O.*

formigenes.[93]   The antiporter (OxlT) is a single integral membrane protein that mediates the exchange of formate and oxalate without additional input of energy. OxlT catalyzes an extremely rapid turnover of oxalate, 1000/s, and constitutes 5-10% of the inner membrane protein of the cell. *O. vibrioformis* and *O.oxalicus (C. oxalicum)* also produce formate as an end product of oxalate metabolism.[40] It is not known how these organisms derive energy from oxalate but the mechanism could be similar to that of *O. formigenes*.

C. thermoaceticum converts oxalate to acetate and CO_2 in proportions described by the following equation:[44]

$$4 \, ^-OOC\text{-}COO^- + 4H_2O + H^+ \rightarrow CH_3COO^- + 6HCO_3^-$$
$$(\text{-}41.4 \text{ kJ/mol oxalate}^{[87]})$$

The acetogenic utilization of oxalate represents a new mechanism for the metabolism of oxalate by obligately anaerobic bacteria (Table 5). From these studies,[44] it was proposed that oxalyl- and formyl-CoA and formate were formed as intermediates during acetogenic oxalate catabolism. The formation of CoA level intermediates is based on the role of oxalyl-CoA decarboxylase and formyl-CoA transferase in oxalate degradation by *O. formigenes*.[15,90-92] If formate is formed as an intermediate in oxalate metabolism by *C. thermoaceticum*, it is likely oxidized via formate dehydrogenase to the level of CO_2 and reductant, the latter being utilized in the acetyl-CoA (Wood) pathway.[46-48]   Experiments to determine whether such a reaction sequence for oxalate-coupled acetogenesis occurs in *C. thermoaceticum* have recently been initiated. Enzymological studies indicate that oxalate degradation by *C. thermoaceticum* is dependent upon the presence of a utilizable electron acceptor (e.g., benzyl viologen) and that the addition of CoA to reaction mixtures does not influence oxalate-degrading activities.[94] This information suggests that a mechanism other than that described above is involved in the oxidation of oxalate by *C. thermoaceticum*. In this regard, it is interesting to note that several early studies to elucidate the acetyl-CoA pathway examined the possibility of oxalate as a potential intermediate in the formation of acetate from CO_2.[95,96]

The molar cell yields of oxalate-cultivated cells of *C. thermoaceticum* are greater than the reported molar cell yields of other oxalate-utilizing bacteria, including aerobic oxalate-

Table 5 Mechanisms for the Conservation of Oxalate-derived Energy by Oxalate-degrading Bacteria

Organism	Energy-conserving process coupled to oxalate catabolism	End product of oxalate catabolism	Molar cell yield (g/mol)	Reference
Anaerobes				
Clostridium thermoaceticum	Acetogenesis	Acetate, CO_2	4.9	44
Clostridium thermoautotrophicum	Acetogenesis	Acetate, CO_2	ND[a]	44
Desulfovibrio vulgaris subsp. *oxamicus*	Dissimilatory sulfate reduction	CO_2	ND	42
Oxalobacter formigenes	Electrogenic oxalate^{2-}: formate^{-1} antiport system	Formate, CO_2	1.1	15, 88
Oxalobacter vibrioformis	ND	Formate, CO_2	1.6	40
Oxalophagus oxalicus	ND	Formate, CO_2	1.8	40, 41
Strain Ox-8	ND	Formate, CO_2	0.07	39
Aerobes				
Pseudomonas oxalaticus	O_2 respiration	CO_2	3.8	126
Alcaligenes eutrophus	O_2 respiration	CO_2	3.9	64

[a] ND, Not determined.

degrading bacteria (Table 5). Lower molar cell yields by non-acetogenic anaerobes would be anticipated given the smaller change in free energy for oxalate metabolism. Another explanation for the low cell yields may be that, in contrast to C. *thermoaceticum* and aerobic oxalate-degrading bacteria, these anaerobes do not conserve oxalate-derived energy (as ATP) via substrate-level or electron-transport phosphorylation.[15,40] The energy derived from oxalate decarboxylation is therefore apparently conserved via an alternative mechanism(s). Certain decarboxylation reactions can be coupled to the generation of a Na^+-ion gradient via sodium-pumping decarboxylases.[97] Whether oxalate decarboxylation in C. *thermoaceticum* is coupled to trans-membrane, energy-conserving mechanisms has yet to be resolved. However, support for the concept that such reactions may be operational during oxalate catabolism in C. *thermoaceticum* is provided by the fact that, with the exception of glucose, oxalate supports higher cell yields per unit of reductant consumed than do other acetogenic substrates (e.g., H_2, CO, or methanol).[44]

Knowledge regarding the genetic control of oxalate degradation is limited. Lung et al.[98] cloned the oxalyl-CoA decarboxylase gene, *oxc*, from *O. formigenes* and expressed it in *E. coli*. Genomic clones containing the *oxc* gene were obtained by constructing a λ library using λ-GEM-11 *XhoI*.[99] Plaques were screened using a monoclonal antibody made against the purified enzyme. The *oxc* gene consists of a single open reading frame of 1,704 bp that encodes a 568-amino-acid protein.[98] Although oxalyl-CoA decarboxylase is only one of three proteins involved in the catabolism of oxalate by *O. formigenes*, *oxc* is probably not part of a polycistronic operon since both a promoter sequence and a rho-independent termination sequence were identified. Additional open reading frames were not found either upstream or downstream of the coding sequence.

B. BIOSYNTHESIS FROM OXALATE

Oxalate is a highly oxidized substrate and its assimilation into bacterial cells must involve reductive reactions. Figure 2 illustrates the two known pathways that are utilized by bacteria to reduce oxalate to 3P-glycerate, an important precursor metabolite. After oxalate is transported into the cell and activated to oxalyl-CoA, it is reduced to glyoxylate by the

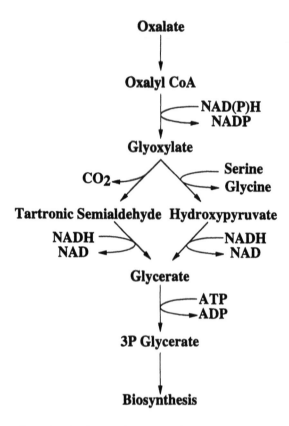

Figure 2. Anabolic pathways of oxalate metabolism. Left branch = glycerate pathway, right branch = serine pathway.

enzyme oxalyl-CoA reductase (EC 1.2.1.17). At this point the two pathways split. In the glycerate pathway (left branch), two glyoxylates are condensed by the enzyme glyoxylate carboligase (EC 4.1.1.47) to form the three-carbon compound tartronic semialdehyde plus CO_2. Tartronic semialdehyde is then reduced to glycerate by tartronic semialdehyde reductase (EC 1.1.1.60). In the serine pathway (right branch), a one-carbon group from serine is transferred to glyoxylate by the enzyme serine-glyoxylate aminotransferase (EC 2.6.1.45) to form the three-carbon compound hydroxypyruvate. Hydroxypyruvate is reduced to glycerate by hydroxypyruvate reductase (EC 1.1.1.81). In both pathways glycerate is phosphorylated to 3P-glycerate by glycerate kinase (EC 2.7.1.31). "*P. oxalaticus*",[74,100] "*A.*" *eutrophus*,[64] *T. novellus*,[71] and *O.*

formigenes[101] utilize the glycerate pathway through tartronic semialdehyde. *M. extorquens*[82] and *B. oxalophilus*[66] utilize the serine pathway through hydroxypyruvate. Zaitsev et al.[66] suggested that *B. oxalophilus* also assimilated oxalate via the glycerate pathway, but did not publish any data for the key enzymes of the pathway, glyoxylate carboligase and tartronic semialdehyde reductase.

O. *formigenes*,[15] O. *vibrioformis* and *O.oxalicus (C. oxalicum)*[40] all require small amounts (1 mM) of acetate to grow in defined medium. Since acetate does not support the growth of these bacteria, it is probably required for biosynthetic processes. When *O. formigenes* is grown in ^{14}C-acetate, all of the major cell fractions (protein, lipid, nucleic acid) are labeled.[101] Acetate is incorporated as a unit into glutamate, proline, arginine, and leucine and these four amino acids account for 60% of the labeled acetate that is incorporated into cell protein. The remaining acetate is split, and labels those amino acids that are derived from pyruvate and oxaloacetate. Approximately 8% of the total cell carbon of *O. formigenes* is derived from acetate, 55% is derived from oxalate and the remainder is probably derived from CO_2.

Several of the genes encoding biosynthetic reactions of the serine pathway have been cloned and partially sequenced from *M. extorquens*.[102] The genes encoding hydroxypyruvate reductase (*hprA*), malyl-CoA lyase (*mclA*), and PEP carboxylase are clustered on the chromosome.[103,104] The gene encoding serine-glyoxylate aminotransferase (*sgaA*) is 263 bp upstream from these genes and the gene that encodes 5,10 methylenetetrahydrofolate dehydrogenase is downstream from the cluster containing *hprA*.[102] Cells containing an insertion mutation of either *sgaA* or *hprA* lost the ability to grow on C-1 compounds but not on C-2 compounds. However, Chistoserdova and Lidstrom[102,103] did not report about growth of these insertion mutants on oxalate.

IV. POTENTIAL FOR MANIPULATION OF POPULATIONS OF INTESTINAL OXALATE-DEGRADING BACTERIA

A. Rumen Populations

Since oxalate is the essential substrate for *O. formigenes* growth, selections for increased proportions of *O. formigenes* in gastrointestinal populations and thus for increased rates of oxalate degradation are regulated by the supply of oxalate. This is thus an adaptation system where toxin degradation rates are regulated by the availability of the toxin itself. With ruminants this system seems well suited to provide protection from gradual changes in dietary oxalate. With abrupt, rather than gradual increases in dietary oxalate, however, acute toxicities occur[4,105] because time intervals of 4 or more days are needed for development of populations that degrade oxalate rapidly.[5,6] More rapid, and perhaps immediate, protection was provided by inoculation of the rumen of a sheep with a rumen population that had been previously adapted *in vitro* to high levels of oxalate.[6] It seems probable that inoculations with appropriate amounts of a pure culture of *O. formigenes* would also provide protection, but no tests of this have been reported.

B. Hind Gut Populations

All known isolates of *Oxalobacter* are obligate anaerobes and thus are probably not able to grow in the small intestine where conditions are much less anaerobic than in the large bowel. Results of a study in rats support this conclusion. The importance of colonic handling of dietary oxalate is discussed by Hatch (Chapter 3, this book). The effect of intestinal oxalate-degrading bacteria on oxalate absorption is not yet well defined but it is reasonable to believe that bacteria that are able to degrade and thus reduce the concentrations of free oxalate in the colon may indeed limit its absorption. This concept is supported by findings (Table 2) indicating that there was a decidedly lower incidence of intestinal colonization by oxalate-degrading bacteria in populations at increased risk for enteric hyperoxaluria and urolithiasis than in control populations.

If on-going investigations provide further support for a role for intestinal oxalate-degrading bacteria, the potential for manipulation of oxalate absorption by manipulation of populations of these bacteria becomes an important topic for investigation. Questions that may be posed include: What are the main factors regulating (limiting) populations of *Oxalobacter* in the gut? What is the diversity within the group of intestinal bacteria able to degrade oxalate? What interventions (inoculations) can be used to induce and/or enhance intestinal oxalate degradation? At what age do infants become colonized and what factors influence this? Do persons that are not colonized remain non-colonized? What are the effects of oral antibiotics or of diarrheal disease upon colonization? The need for specialized anaerobic methods for culture of the oxalate-degrading bacteria is perhaps the main factor that would limit research designed to answer the above questions. This obstacle may, however, be by-passed if modern probe methods can be developed to specifically and sensitively detect these bacteria.

ACKNOWLEDGMENTS

Current studies by S. L. Daniel are supported by funds from the Bundesministerium für Forschung und Technologie (0339476A0).

REFERENCES

1. **Janis, C.,** The evolutionary strategy of the equidae and the origins of rumen and cecal digestion, Evolution, 30, 757, 1976.
2. **Talapatra, S. K., Ray, S. C., and Sen, K. C.,** Calcium assimilation in ruminants on oxalate rich diet, J. Agric. Sci. Cambridge, 38, 163, 1947.
3. **Morris, M. P. and Garcia-Rivera, J.,** The destruction of oxalates by rumen contents of cows, J. Dairy Sci., 38, 1169, 1955.
4. **James, L. F. and Butcher, J. E.,** Halogeton poisoning of sheep: effect of high level oxalate intake, J. Anim. Sci., 35, 1233, 1972.
5. **Allison, M. J., Littledike, E. T., and James, L. F.,** Changes in ruminal oxalate degradation rates associated with adaptation to oxalate ingestion, J. Anim. Sci., 45, 1173, 1977.
6. **Allison, M. J. and Reddy, C. A.,** Adaptations of gastrointestinal bacteria in response to changes in dietary oxalate and nitrate, in Current Perspectives in Microbial

Ecology, Klug, M. J. and Reddy, C. A., Eds., Amer. Soc. Microbiol., Washington, DC, 1984, 248.

7. **Wilson, G. D. A. and Harvey, D. G.**, Studies on experimental oxaluria in pigs, *Brit. Vet. J.*, 133, 418, 1977.

8. **McKenzie, R. A., Balney, B. J., and Gartner, R. J. W.**, The effect of dietary oxalate on calcium, phosphorus and magnesium balances in horses, *J. Agric. Sci. Camb.*, 97, 69, 1981.

9. **Shirley, E. K. and Schmidt-Nielsen, K.**, Oxalate metabolism in the pack rat, sand rat, hamster and white rat, *J. Nutr.*, 91, 496, 1967.

10. **Allison, M. J. and Cook, H. M.**, Oxalate degradation by microbes of the large bowel of herbivores: the effect of dietary oxalate, *Science*, 212, 675, 1981.

11. **Daniel, S. L., Hartman, P. A., and Allison, M. J.**, Microbial degradation of oxalate in the gastrointestinal tracts of rats, *Appl. Environ. Microbiol.*, 53, 1793, 1987.

12. **Dawson, K. A., Allison, M. J., and Hartman, P. A.**, Characteristics of anaerobic oxalate-degrading enrichment cultures from the rumen, *Appl. Environ. Microbiol.*, 40, 840, 1980.

13. **Allison, M. J., Cook, H. M., and Dawson, K. A.**, Selection of oxalate-degrading rumen bacteria in continuous cultures, *J. Anim. Sci.*, 53, 810, 1981.

14. **Dawson, K. A., Allison, M. J., and Hartman, P. A.**, Isolation and some characteristics of anaerobic oxalate-degrading bacteria from the rumen, *Appl. Environ. Microbiol.*, 40, 833, 1980.

15. **Allison, M. J., Dawson, K. A., Mayberry, W. R., and Foss, J. G.**, *Oxalobacter formigenes* gen. nov., sp. nov.: oxalate-degrading anaerobes that inhabit the gastrointestinal tract, *Arch. Microbiol.*, 141, 1, 1985.

16. **Fisher, C. and Allison, M. J.**, Unpublished data, 1984.

17. **Rieu-Lesme, F., Fonty, G., and Dore, J.**, Isolation and characterization of a new hydrogen-utilizing bacterium from the rumen, *FEMS Microbiol. Lett.*, 125, 77, 1995.

18. **Daniel, S. L., Cook, H. M., Hartman, P. A., and Allison, M. J.**, Enumeration of anaerobic oxalate-degrading bacteria in the ruminal contents of sheep, *FEMS Microbiol. Ecol.*, 62, 329, 1989.

19. **Holdeman, L. V., Cato, E. P., and Moore, W. E. C.,** Anaerobe Laboratory Manual, Virginia Polytechnic Institute and State University, Blacksburg, 1977.
20. **Bryant, M. P.,** Commentary on the Hungate technique for culture of anaerobic bacteria, Am. J. Clin. Nutr., 25, 1324, 1972.
21. **Barber, H. H. and A Gallimore, E. J.,** The metabolism of oxalic acid in the animal body, Biochem. J., 34, 144, 1940.
22. **Hodgkinson, A.,** Oxalic Acid in Biology and Medicine, Academic Press, New York, 1977.
23. **Goldkind, L., Cave, D. R., Jaffin, B., Robinson, W., and Bliss, C. M.,** A new factor in enteric hyperoxaluria: *Oxalobacter formigenes*, Am. J. Gastro., 80, 860, 1985.
24. **Allison, M. J., Cook, H. M., Milne, D. B., Gallagher, S., and Clayman, R. V.,** Oxalate degradation by gastrointestinal bacteria from humans, J. Nutr., 116, 455, 1986.
25. **Doane, L. T., Liebman, M., and Caldwell, D. R.,** Microbial oxalate degradation: effects on oxalate and calcium balance in humans, Nutrition Research, 9, 957, 1989.
26. **Weaver, G. A., Krause, J. A., Allison, M. J., and Lindenbaum, J.,** Distribution of digoxin-reducing, oxalate-degrading, and total anaerobic bacteria in the human colon, Microb. Ecol. in Health and Dis., 5, 227, 1992.
27. **Ito, H., Kotake, T., Yamamoto, K., and Hara, T.,** Isolation and identification of oxalate decomposing bacteria from human faeces, Abstr. 5th European Urolithiasis Symposium, 58, 1994.
28. **Hofmann, A. F., Tacker, M. M., Fromm, H., Thomas, P. J., and Smith, L. H.,** Acquired hyperoxaluria and intestinal disease, Mayo Clin. Proc., 48, 35, 1973.
29. **Earnest, D. L.,** Enteric hyperoxaluria, in Advances in Internal Medicine, Vol. 25, Stollerman, G. H., Ed., Year Book Medical Publisher, St. Louis, 407, 1979.
30. **Argenzio, R. A., Liacos, J. A., and Allison, M. J.,** Intestinal oxalate-degrading bacteria reduce oxalate absorption and toxicity in guinea pigs, J. Nutr., 118, 787, 1988.
31. **Goldkind, L., Cave, D. R., Jaffin, B., Bliss, C. M., and Allison, M. J.,** Bacterial oxalate metabolism in the human colon: a possible factor in enteric hyperoxaluria, Gastroenterology, 90, 1431, 1986.

32. **Kleinschmidt, K., Mahlmann, A., and Hautmann, R.,** Oxalate-degrading bacteria in the gut: the key to the pathogenesis of calcium oxalate stone formation?, J. Urol., 147, 330A, 1992.

33. **Daniel, S. L., Hartman, P. A., and Allison, M. J.,** Intestinal colonization of laboratory rats with *Oxalobacter formigenes*, Appl. Environ. Microbiol., 53, 2767, 1987.

34. **Daniel, S. L., Hartman, P. A., and Allison, M. J.,** Intestinal colonization of laboratory rats by anaerobic oxalate-degrading bacteria: effects on the urinary and faecal excretion of dietary oxalate, Microbial Ecology in Health and Disease, 6, 277, 1993.

35. **Costello, J. and Allison, M. J.,** Unpublished data, 1993.

36. **Stahl, D.,** Unpublished data, 1994.

37. **Jensen, N. S. and Allison, M. J.,** Studies on the diversity among anaerobic oxalate-degrading bacteria now in the species *Oxalobacter formigenes*, Abstr. Ann. Mtg. Amer. Soc. Microbiol., I-29, 1994.

38. **Smith, R. L. and Oremland, R. S.,** Anaerobic oxalate degradation: widespread natural occurrence in aquatic sediments, Appl. Environ. Microbiol., 46, 106, 1983.

39. **Smith, R. L., Strohmaier, F. E., and Oremland, R. S.,** Isolation of anaerobic oxalate-degrading bacteria from freshwater lake sediments, Arch. Microbiol., 141, 8, 1985.

40. **Dehning, I. and Schink, B.,** Two new species of anaerobic oxalate-fermenting bacteria, *Oxalobacter vibrioformis* sp. nov. and *Clostridium oxalicum* sp. nov., from sediment samples, Arch. Microbiol., 153, 79, 1989.

41. **Collins, M. D., Lawson, P. A., Willems, A., Cordoba, J. J., Fernandez-Garayzabal, J., Garcia, P., Cai, J., Hippe, H., and Farrow, J. A. E.,** The phylogeny of the genus *Clostridium*: proposal of five new genera and eleven new species combinations, Int. J. Syst. Bacteriol., 44, 812, 1994.

42. **Postgate, J. R.,** A strain of *Desulfovibrio* able to use oxamate, Arch. Mikrobiol., 46, 287, 1963.

43. **Postgate, J. R.,** *Desulfovibrio*, in Bergey's Manual of Systematic Bacteriology, Vol. 1, 9th ed., Sneath, P. H. A., Ed., Williams and Wilkins, Baltimore, 1984, 666.

44. **Daniel, S. L. and Drake, H. L.**, Oxalate- and glyoxylate-dependent growth and acetogenesis by *Clostridium thermoaceticum*, Appl. Environ. Microbiol., 59, 3062, 1993.

45. **Drake, H. L.**, Acetogenesis, acetogenic bacteria, and the acetyl-CoA "Wood/Ljungdahl" pathway: past and current perspectives, in Acetogenesis, Drake, H. L., Ed., Chapman and Hall, New York, 1994, 3.

46. **Drake, H. L.**, Acetogenesis and acetogenic bacteria, in Encyclopedia of Microbiology, Vol. 1, Lederberg, J., Ed., Academic Press, San Diego, 1992, 1.

47. **Ljungdahl, L. G.**, The autotrophic pathway of acetate synthesis in acetogenic bacteria, Ann. Rev. Microbiol., 40, 415, 1986.

48. **Drake, H. L., Daniel, S. L., Matthies, C., and Küsel, K.**, Acetogenesis: reality in the laboratory, uncertainty elsewhere, in Acetogenesis, Drake, H. L., Ed., Chapman and Hall, New York, 1994, 273.

49. **Fox, T. R. and Comerford, N. B.**, Low-molecular-weight organic acids in selected forest soils of the southeastern USA., Soil Sci. Soc. Am. J., 54, 1139, 1990.

50. **Tani, M., Higashi, T., and Nagatsuka, S.**, Dynamics of low-molecular-weight aliphatic carboxylic acids (LACAs) in forest soils., Soil Sci. Plant Nutr., 39, 485, 1993.

51. **Stevenson, F. J.**, Organic acids in soil, in Soil Biochemistry, Mclaren, A. D. and Peterson, G. H., Eds., Marcel Dekker, New York, 1967, 119.

52. **Graustein, W. C., Cromack, K., Jr., and Sollins, P.**, Calcium oxalate: occurrence in soils and effects on nutrient and geochemical cycles, Science, 198, 1252, 1977.

53. **Johnston, C. G. and Vestal, J. R.**, Biogeochemistry of oxalate in the antarctic cryptoendolithic lichen-dominated community, Microb. Ecol., 25, 305, 1993.

54. **Knutson, D. M., Hutchins, A. S., and Cromack, K. J.**, The association of calcium oxalate-utilizing *Streptomyces* with conifer ectomycorrhizae, Antonie van Leeuwenhoek J. Microbiol. Serol., 46, 611, 1980.

55. **Dickman, M. B. and Mitra, A.**, *Arabidopsis thaliana* as a model for studying *Sclerotina sclerotiorum* pathogenesis, Physiol. Mol. Plant Pathol., 41, 255, 1992.

56. **Tamer, A. Ü. and Aragno, M.** Isolement, caractérisation et essai d'identification de bactéries capables d'utiliser l'oxalate comme seule source de carbone et d'énergie, Bull. Soc. Neuch. Sci. Nat., 103, 91, 1980.

57. **Cromack, K. J., Jr., Sollins, P., Todd, R. L., Fogel, R., Todd, A. W., Fender, W. M., Crossley, M. E., and Crossley, D. A., Jr.,** The role of oxalic acid and bicarbonate in calcium cycling by fungi and bacteria: some possible implications for soil animals, Ecol. Bull. (Stockholm), 25, 246, 1977.

58. **Morris, S. J. and Allen, M. F.,** Oxalate-metabolizing microorganisms in sagebrush steppe soil, Biol. Fertil. Soils, 18, 255, 1994.

59. **Khambata, S. R. and Bhat, J. V.,** Decomposition of oxalate by *Streptomyces*, Nature, 174, 696, 1954.

60. **Nitsch, B. and Kutzner, H. J.,** Decomposition of oxalic acid and other organic acids by *Streptomyces* as a taxonomic aid, Zeitschrift Allg. Mikrobiol., 9, 613, 1969.

61. **Robbel, L. and Kutzner, H. J.,** *Streptomyces* taxonomy: utilization of organic acids as an aid in the identification of species, Naturwissenschaften, 60, 351, 1973.

62. **Jakoby, W. B. and Bhat, J. V.,** Microbial metabolism of oxalic acid, Bacteriol. Rev., 22, 75, 1958.

63. **Jenni, B., Realini, L., Aragno, M., and Tamer, A. Ü.,** Taxonomy of non H_2-lithotrophic, oxalate-oxidizing bacteria related to *Alcaligenes eutrophus*, System. Appl. Microbiol., 10, 126, 1988.

64. **Friedrich, C. G., Bowien, B., and Friedrich, B.,** Formate and oxalate metabolism in *Alcaligenes eutrophus*, J. Gen. Microbiol., 115, 185, 1979.

65. **Kersters, K. and De Ley, J.,** *Alcaligenes*, in Bergey's Manual of Systematic Bacteriology, Vol. 1, 9th ed., Krieg, N. R. and Holt, J. G., Eds., Williams and Wilkins, Baltimore, 1984, 361.

66. **Zaitsev, G. M., Govorukhina, N. I., Laskovneva, O. V., and Trotsenko, Y. A.,** Properties of the new oxalotrophic bacterium *Bacillus oxalophilus*, Microbiol., 62, 378, 1993.

67. **Meyer, O., Stackebrandt, E., and Auling, G.,** Reclassification of ubiquinone Q-10 containing carboxidotrophic bacteria: transfer of *Pseudomonas carboxydovorans* OM5 to *Oligotropha*, gen. nov. as *Oligotropha carboxidovorans*, comb. nov., transfer of

Alcaligenes carboxydus DSM 1086 to *Carbophilus*, gen. nov. as *Carbophilus carboxidus*, comb. nov., transfer of *Pseudomonas compransoris*, DSM 1231 to *Zavarzinia*, gen. nov., as *Zavarzinia compransoris*, comb. nov. and amended descriptions of the new genera, System. Appl. Microbiol., 16, 390, 1993.

68. **Bassalik, K.**, Über die Verarbeitung der oxalsäure durch *Bacillus extorquens* n. sp., Jahrb. wiss Botan., 53, 255, 1913.

69. **Bousfield, I. J. and Green, P. N.**, Reclassification of bacteria of the genus *Protomonas* Urakami and Komagata 1984 in the genus *Methylobacterium* (Patt, Cole and Hanson) emend. Green and Bousfield 1983, Int. J. Syst. Bacteriol., 35, 209, 1985.

70. **Green, P. N. and Bousfield, I. J.**, A taxonomic study of some gram-negative facultatively methylotrophic bacteria, J. Gen. Microbiol., 128, 623, 1982.

71. **Chandra, T. S. and Shethna, Y. I.**, Oxalate, formate, formamide, and methanol metabolism in *Thiobacillus novellus*, J. Bacteriol., 131, 389, 1977.

72. **Jenni, B., Aragno, M., and Weigel, J. K. W.**, Numerical analysis and DNA-DNA hybridization studies on *Xanthobacter* and emendation of *Xanthobacter flavus*, System. Appl. Microbiol., 9, 247, 1987.

73. **Quayle, J. R.**, Carbon assimilation by *Pseudomonas oxalaticus* (OX1) 7. Decarboxylation of oxalyl-coenzyme A to formyl coenzyme A, Biochem. J., 89, 492, 1963.

74. **Quayle, J. R., Keech, D. B., and Taylor, G. A.**, Carbon assimilation by *Pseudomonas oxalaticus* (OX1) 4. Metabolism of oxalate in cell-free extracts of the organism grown on oxalate, Biochem. J., 78, 225, 1961.

75. **Blackmore, M. A. and Quayle, J. R.**, Choice between autotrophy and heterotrophy in *Pseudomonas oxalaticus*. Growth in mixed substrates, Biochem. J., 107, 705, 1968.

76. **Blackmore, M. A., Quayle, J. R., and Walker, I. O.**, Choice between autotrophy and heterotrophy in *Pseudomonas oxalaticus*. Utilization of oxalate by cells after adaptation from growth on formate to growth on oxalate, Biochem. J., 107, 699, 1968.

77. **Quayle, J. R. and Keech, D. B.**, Carbon assimilation by *Pseudomonas oxalaticus* (OX1). 1. Formate and carbon dioxide utilization during growth on formate, <u>Biochem. J.</u>, 72, 623, 1959.

78. **Quayle, J. R.**, Carbon assimilation by *Pseudomonas oxalaticus* (OX1). 6. Reactions of oxalyl-coenzyme A, <u>Biochem. J.</u>, 87, 368, 1963.

79. **Quayle, J. R. and Keech, D. B.**, Carboxydismutase activity in formate and oxalate grown *Pseudomonas oxalaticus* (strain OX1), <u>Biochem. Biophys. Acta</u>, 31, 587, 1959.

80. **Khambata, S. R. and Bhat, J. V.**, Studies on a new oxalate-decomposing bacterium, *Pseudomonas oxalaticus*, <u>J. Bacteriol.</u>, 66, 505, 1953.

81. **Quayle, J. R. and Keech, D. B.**, Carbon assimilation by *Pseudomonas oxalaticus* (OX1) 2. Formate and carbon dioxide utilization by cell-free extracts of the organism grown on formate, <u>Biochem. J.</u>, 72, 631, 1959.

82. **Blackmore, M. A. and Quayle, J. R.**, Microbial growth on oxalate by a route not involving glyoxylate carboligase, <u>Biochem. J.</u>, 118, 53, 1970.

83. **Dijkhuizen, L., Groen, L., Harder, W., and Konings, W. N.**, Active transport of oxalate by *Pseudomonas oxalaticus* OX1, <u>Arch. Microbiol.</u>, 115, 223, 1977.

84. **Li, L., Ljungdahl, L., and Wood, H. G.**, Properties of nicotinamide adenine dinucleotide phosphate-dependent formate dehydrogenase from *Clostridium thermoaceticum*, <u>J. Bacteriol.</u>, 92, 405, 1966.

85. **Dijkhuizen, L., Timmerman, J. W. C., and Harder, W.**, A pyridine nucleotide-independent membrane-bound formate dehydrogenase in *Pseudomonas oxalaticus* OX1, <u>FEMS Microbiol. Lett.</u>, 6, 53, 1979.

86. **Koyama, H.**, Purification and characterization of oxalate oxidase from *Pseudomonas* sp. OX-53, <u>Agric. Biol. Chem.</u>, 52, 743, 1988.

87. **Thauer, R. K., Jungermann, K., and Decker, K.**, Energy conservation in chemoautotrophic bacteria, <u>Bacteriol. Rev.</u>, 41, 100, 1977.

88. **Anantharam, V., Allison, M. J., and Maloney, P. C.**, Oxalate:formate exchange. The basis for energy coupling in *Oxalobacter*, <u>J. Biol. Chem.</u>, 264, 7244, 1989.

89. **Kuhner, C. H., Allison, M. J., and Hartman, P. A.,** Measurement of proton-motive force in the anaerobic oxalate-degrading bacterium *Oxalobacter formigenes*, Abstr. Ann. Mtg. Amer. Soc. Microbiol., Atlanta, GA, 108, 1987.

90. **Baetz, A. L. and Allison, M. J.,** Purification and characterization of formyl-coenzyme A transferase from *Oxalobacter formigenes*, J. Bacteriol., 172, 3537, 1990.

91. **Baetz, A. L. and Allison, M. J.,** Purification and characterization of oxalyl-coenzyme A decarboxylase from *Oxalobacter formigenes*, J. Bacteriol., 171, 2605, 1989.

92. **Baetz, A. L. and Allison, M. J.,** Localization of oxalyl-coenzyme A decarboxylase, and formyl-coenzyme A transferase in *Oxalobacter formigenes* cells, System. Appl. Microbiol., 15, 167, 1992.

93. **Ruan, Z. S., Anantharam, V., Crawford, I. T., Ambudkar, S. V., Rhee, S. Y., Allison, M. J., and Maloney, P. C.,** Identification, purification and reconstitution of OxlT, the oxalate:formate antiport protein of *Oxalobacter formigenes*, J. Biol. Chem., 267, 10537, 1992.

94. **Daniel, S. L., and Wagner, C.,** Enzymological studies on the mechanism of oxalate-dependent acetogenesis by *Clostridium thermoaceticum*. Abstr. Ann. Mtg. Amer. Soc. Microbiol., K-74, 1994.

95. **Ljungdahl, L. G., and Wood, H. G.,** Incorporation of C^{14} from carbon dioxide into sugar phosphates, carboxylic acids, and amino acids by *Clostridium thermoaceticum*., J. Bacteriol., 89, 1055, 1965.

96. **Ljungdahl, L. G.,** Fixation of carbon dioxide and formation of acetate by *Clostridium thermoaceticum*, Ph.D. Dissertation, Western Reserve University, 1964.

97. **Dimroth, P.,** Sodium ion transport decarboxylases and other aspects of sodium ion cycling in bacteria., Microbiol. Rev., 51, 320, 1987.

98. **Lung, H., Baetz, A. L., and Peck, A. B.,** Molecular cloning, DNA sequence, and gene expression of the oxalyl-CoA decarboxylase gene, *oxc*, from the bacterium *Oxalobacter formigenes*, J. Bacteriol., 176, 2468, 1994.

99. **Lung, H., Cornelius, J. G., and Peck, A. B.,** Cloning and expression of the oxalyl-CoA decarboxylase gene from the bacterium *Oxalobacter formigenes*: prospects for gene

therapy to control Ca-oxalate kidney stone formation, Amer. J. Kidney Dis., 17, 381, 1991.

100. **Quayle, J. R. and Keech, D. B.**, Formation of glycerate from oxalate by *Pseudomonas oxalaticus* OX1 grown on oxalate, Nature, 183, 1794, 1959.

101. **Cornick, N. A.**, Biosynthetic pathways in *Oxalobacter formigenes*, PhD Dissertation, Iowa State Univ, 1995.

102. **Chistoserdova, L. V. and Lidstrom, M. E.**, Genetics of the serine cycle in *Methylobacterium extorquens* AM1: Identification of *sgaA*, and *mtdA* and sequences of *sgaA*, *hprA*, and *mtdA*, J. Bacteriol., 176, 1957, 1994.

103. **Chistoserdova, L. V. and Lidstrom, M. E.**, Cloning, mutagenesis, and physiological effect of a hydroxypyruvate reductase gene from *Methylobacterium extorquens* AM1, J. Bacteriol., 174, 71, 1992.

104. **Arps, P. J., Fulton, G. F., Minnich, E. C., and Lidstrom, M. E.**, Genetics of serine pathway enzymes in *Methylobacterium extorquens* AM1: Phosphoenolpyruvate carboxylase and malyl coenzyme A lyase, J. Bacteriol., 175, 3776, 1993.

105. **James, L. F.**, Oxalate toxicosis, Clinical Toxicology, 5, 231, 1972.

106. **Stepanchuk, Y. B. and Shenderov, B. A.**, Role of intestinal microflora in oxalate metabolism, Zhurnal Mikrobiologii, Epidemiologii I Immunobiologii, 5, 58, 1992.

107. **Robinson, W. R., Cave, D. R., Bliss, C. M., Daniels, S., and Allison, M. J.**, Bacterial degradation of oxalate in human feces, Gastroenterology, 88, 1557, 1985.

108. **Kleinschmidt, K., Mahlmann, A., and Hautmann, R.**, Anaerobic oxalate-degrading bacteria in the gut decrease faecal and urinary oxalate concentrations in stone formers. Abstr. 5th European Urolithiasis Symposium, 1994, 69.

109. **Kleinschmidt, K.**, Personal communication, 1994.

110. **McClung, N. M.**, The utilization of carbon compounds by *Nocardia* species, J. Bacteriol., 68, 231, 1954.

111. **Müller, H.**, Oxalsäure als kohlenstoffquelle für mikroorganismen, Arch. Mikrobiol., 15, 137, 1950.

112. **Palleroni, N. J.**, Genus I. *Pseudomonas*, in Bergey's Manual of Systematic Bacteriology, Vol. 1, 9th ed., Sneath, P. H. A., Ed., Williams and Wilkins, Baltimore, 1984, 141.

113. **Bhat, J. V. and Barker, H. A.**, Studies on a new oxalate-decomposing bacterium, *Vibrio oxalaticus*, J. Bacteriol., 55, 359, 1948.

114. **Khambata, S. R. and Bhat, J. V.**, Decomposition of oxalate by *Mycobacterium lacticola* isolated from the intestine of earthworms, J. Bacteriol., 69, 227, 1955.

115. **Jayasuria, G. C. N.**, The isolation and characteristics of an oxalate-decomposing organism, J. Gen. Microbiol., 12, 419, 1955.

116. **Chandra, T. S. and Shethna, Y. I.**, Oxalate and formate metabolism in *Alcaligenes* and *Pseudomonas* species, Antonie van Leeuwenhoek J. Microbiol. Serol., 41, 465, 1975.

117. **O'Halloran, M. W.**, The effect of oxalate on bacteria isolated from the rumen, Aust. Soc. Anim. Prod. Proc., 4, 18, 1962.

118. **Stocks, P. K. and McCleskey, C. S.**, Identity of the pink-pigmented methanol-oxidizing bacteria as *Vibrio extorquens*, J. Bacteriol., 88, 1065, 1964.

119. **Anthony, C. and Zatman, L. J.**, The microbial oxidation of methanol. 1. Isolation and properties of *Pseudomonas* sp. M27, Biochem. J., 92, 609, 1964.

120. **Peel, D. and Quayle, J. R.**, Microbial growth on C_1 compounds. 1. Isolation and characterization of *Pseudomonas* AM1, Biochem. J., 81, 465, 1961.

121. **Metha, R. J.**, Studies on methanol-oxidizing bacteria. 1. Isolation and growth studies, Antonie van Leeuwenhoek J. Microbiol. Serol., 39, 295, 1973.

122. **Fontaine, F. E., Peterson, W. H., McCoy, E., Johnson, M. J., and Ritter, G. J.**, A new type of glucose fermentation by *C. thermoaceticum* n. sp., J. Bacteriol., 43, 701, 1942.

123. **Hippe, H., Andreesen, J. R., and Gottschalk, G.**, The genus *Clostridium*-nonmedical, in The Prokaryotes, Vol. 2, 2nd ed., Balows, A., Trüper, H. G., Dworkin, M., Harder, W., and Schleifer, K. H., Eds., Springer-Verlag, Berlin, 1992, 1800.

124. **Cato, E. P., George, W. L., and Finegold, S. M.**, Genus *Clostridium* Prazmowski 1880, in Bergey's Manual of Systematic Bacteriology, Vol. 2, 9th ed., Sneath, P. H. A., Ed., Williams and Wilkins, Baltimore, 1986, 1141.

125. **Weigel, J., Braun, M., and Gottschalk, G.,** *Clostridium thermoautotrophicum* species novum, a thermophile producing acetate from molecular hydrogen and carbon dioxide, <u>Curr. Microbiol.</u>, 5, 255, 1981.
126. **Dijkhuizen, L., Wiersma, M., and Harder, W.,** Energy production and growth of *Pseudomonas oxalaticus* OX1 on oxalate and formate, <u>Arch. Microbiol.</u>, 115, 229, 1977.

Chapter 8

EPIDEMIOLOGY OF CALCIUM OXALATE UROLITHIASIS IN MAN

D. S. Milliner, M.D.

I. INTRODUCTION

Urolithiasis is a common clinical disorder that is associated with considerable morbidity and a high recurrence rate. In the United States alone, urolithiasis accounts for approximately 200,000 hospitalizations per year.[1] The incidence of urolithiasis has been increasing steadily in industrialized regions of the world since early in the century. Calcium oxalate is by far the most common constituent of upper urinary tract calculi and may be important in endemic bladder calculi as well.

II. HISTORICAL CONSIDERATIONS

Urolithiasis has been known since antiquity. Urinary tract stones have been documented in a young Egyptian mummy who lived prior to 4800 B.C.[2] as well a in a mummified American Indian who lived more than 3000 years ago.[3] Urolithiasis was of sufficient clinical importance among the ancient Greeks to warrant special mention in the Oath of Hippocrites.[4]

Historically, bladder calculi appeared to predominate. Endemic stones of the bladder were particularly prevalent in children, with a peak age of occurrence of less than 10 years.[5,6] As recently as the late 19th and early 20th century, bladder stones were a major cause of morbidity among English school children.[5,7,8] Similar patterns of bladder stone disease were documented in India,[9] China,[10] Russia,[11] France,[12] Syria,[13] Israel,[14] and Turkey.[15] A common feature of these regions was a diet largely based on a single grain such as wheat, rice, or millet[7] and deficient in milk products.

Endemic bladder stones have conventionally been regarded as originating in the bladder and composed primarily of uric acid or ammonium urate. However, a number of reports have shown oxalate to be a frequent constituent of bladder stones[5,9,15-18] and nuclei of bladder stones resemble upper tract stones in composition.[5] Oxalate crystalluria was documented in 12 of 28 infant boys from a population with a high prevalence of bladder stones in children.[6] Autopsy studies of children in populations with a high prevalence of bladder stones show microliths in kidney tubules.[6] These observations suggest that vesicle calculi may, under some circumstances, originate in the upper tract with oxalate as a nucleus. Urates and ammonium urates may be deposited secondarily during residence in the bladder.

Endemic bladder stones in childhood continue to be a major problem in some areas of the world including Thailand,[6,19] India,[6,16] Turkey,[15] and Sudan[20] but have undergone a dramatic decline in Europe.[8,12] This decline has been attributed to a decrease in the cereal content of the diet in industrialized countries.[5,6,12]

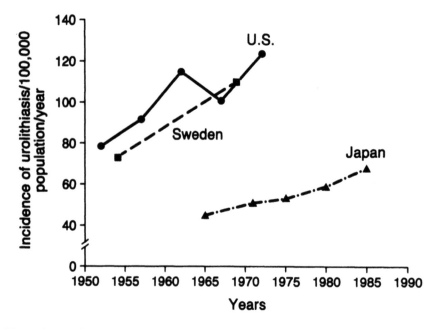

Figure 1: Population studies showing the increase in incidence of urolithiasis over time in men in the United States and Sweden and in men and women in Japan. Data adapted from Johnson et al.,[29] Norlin et al.,[24] and Yoshida and Okada.[22]

Concomitant with the decline in endemic bladder stones in children, there has been a striking increase in upper tract urolithiasis in adults. (Figure 1) This has particularly been true since World War II and has been attributed to an increase in dietary animal protein and possibly carbohydrates in increasingly affluent societies. The increase has been noted in Norway (more than two-fold),[12] Japan (three-fold),[21,22] Sudan,[20] Sweden,[4,23,24] Finland (two to three-fold),[25] Central Europe,[26,27] and the United States.[1,28,29]

Calcium oxalate accounts for the majority of upper tract calculi. In an analysis of 10,000 urinary calculi by crystallography and x-ray defraction,[30] 73 percent contained calcium oxalate with all other constituents being far less common: struvite 9 percent, calcium phosphate 8 percent, uric acid 7 percent, cystine 1 percent, and others 2 percent. Other recent studies have documented calcium oxalate to be responsible for 67 to 90 percent of urinary tract calculi in the United States,[29,31] Japan,[22] and Saudi Arabia.[32]

III. INCIDENCE AND PREVALENCE OF UROLITHIASIS

The predominance of calcium oxalate among upper tract stones allows a reasonable profile of incidence and prevalence despite the paucity of data specific to calcium oxalate urolithiasis among large, well studied populations. Accurate reports of the incidence or prevalence of urolithiasis are difficult to obtain due to the asymptomatic nature of many urinary tract calculi. Careful screening of populations including radiographic studies to detect asymptomatic calculi are, of necessity, limited with regard to the number of subjects evaluated. Few such studies have been published. McKay and colleagues screened 200 adult volunteers ages 16 to 80 years and found six subjects (3 percent) with confirmed renal calculi.[33] Scott et al.[34] studied 2,000 subjects randomly selected from the population at ages from early childhood to greater than 70 years. Subjects were evaluated by questionnaire, blood and urine testing, abdominal radiography, and, if the abdominal radiograph was abnormal, intravenous pyelography. Radiopaque stones were found in 3.83 percent of the population. Two studies in recent decades have reported prevalence rates of urolithiasis at postmortem examination of 0.7 to 2.5 percent.[33,35] However, postmortem assessments fail to reflect true prevalence rates since calculi may be formed and then lost from the urinary tract prior to death. Postmortem studies also select for a population of individuals affected by other diseases.

Hospital discharge data have been used to estimate the incidence and prevalence of urolithiasis. Hospitalization rates may differ appreciably by region and practice pattern and vary from 23 to 51 percent[29,36-38] of patients with symptomatic stone episodes. Thus, all studies based on hospital discharge data will underestimate the true incidence or prevalence of urolithiasis. From studies based on such data, incidence rates for urolithiasis for the population overall (from infancy through greater than 65 years of age) have ranged from 0.4 to 1.4 per thousand population per year in the United States[28,31,39] and 0.5 per thousand per year in England and Wales[40] to 7 per thousand per year in Australia.[41] The prevalence of urolithiasis was estimated from hospital discharge data by Schey, who in a U.S. population of all ages found a prevalence of 2.1 per thousand.[50]

Johnson et al.,[29] in a thorough population based study of patients seeking medical attention for symptomatic and documented stone episodes, found an overall incidence rate of 0.7 per thousand population per year with an age adjusted incidence of 0.4 per thousand per year in females and 11 per thousand per year in males. The lifetime prevalence rate in this study was estimated at 5 percent (50 per thousand) in women and 12 percent (120 per thousand) in men. Curhan[38] studied male health professionals, 40 to 75 years of age, by a detailed questionnaire. There were 45,289 men without a history of stone disease followed for six years. The incidence rate was 3 per thousand per year. The prevalence based on those with urolithiasis at the start of the study was 8 percent (80 per thousand). Despite the bias of selected populations, these two studies provide some of the most reliable information available to date.

The results of studies based on population surveys are summarized by geographic region in Tables I and II. Because of changes in incidence and prevalence of urolithiasis since the turn of the century and in view of the considerable changes in diagnostic methods and intervention techniques over that time period, only those studies that include evaluations of populations after 1970 are tabulated. Since prevalence of urolithiasis is influenced by the age of the population, the age range is provided for each study. The largest and best studied populations are confined to the United States, Europe, and Japan. Information regarding other geographic regions is much more limited. Incidence rates (Table 1) estimated from population surveys in the U.S., Europe, and Japan range from 0.7 to 5.4 per thousand population per year. The incidence was much higher (22 per thousand per year) in a single study from Turkey.[42] The prevalence of urolithiasis (Table II) ranges from 3 to 68 per thousand population in the United States and Europe, to 148 to 280 per thousand in the Middle East, and 4 percent to 54 per thousand in Thailand and Japan. Questionnaires and national surveys are subject to the bias of self reporting of conditions, usually without verification. However, such studies can be performed in very large populations and thus provide useful information.

Repeated stone episodes contribute to the morbidity and economic burden of urolithiasis. Combined data from six large retrospective studies confirm recurrence rates of idiopathic calcium urolithiasis of 14 percent at one year, 35 percent at five years, and 52 percent at ten years.[4,29,43-47]

IV. BIOLOGIC FACTORS INFLUENTIAL IN THE EPIDEMIOLOGY OF UROLITHIASIS

A. AGE AND SEX

Symptomatic urolithiasis in industrialized countries is largely a disorder of males in the third to fifth decade of life (Figure 2). The lifetime prevalence of urolithiasis is estimated to be from 5.9 to 12.5 percent in men[29,31,48] and 3.7 to 5 percent in women.[29,48,49] The male to female ratio is generally quoted as 2:1 to 3:1,[1,4,39,49-51] although occasional studies have shown nearly equal representation of males and females among stone formers.[52,53] In population studies that include radiographic assessments to detect asymptomatic stones, male to female ratios range from 0.8:1 to 1.17:1[23,34] and in autopsy studies the male to female ratio ranges from 0.5:1 to 4:1.[23,33] One author,[54] in a postmortem study, found equal numbers of renal stones in males and females but more ureteral stones in males. These data suggest that adult males are more likely to have symptomatic stones but that the actual incidence of stone formation may be similar in men and women. The disproportionate representation of males among stone formers is largely a feature of mid adulthood. Male to female ratios reach a peak of as high as 4.85:1 at ages 40 to 60 years but decline at both ends of the age spectrum, approaching 1:1 in childhood as well as in individuals over the age of 70 years.[39,31,50,51,55]

TABLE I
Annual Incidence of Urolithiasis per 1,000 Population

Region	Ages of Subjects	Males	Females	Overall	% Calcium Oxalate	Ref
USA						
California	0 to > 65 yrs	1.8	0.6	1.2	81%	31
Rochester, MN	0 to > 70 yrs	1.1	0.4	0.7	67%	29
All U.S.	40 to 75 yrs	3.0	--	--	--	38
Europe						
Great Britain	10 to > 75 yrs	0.5	0.3	0.8	--	37
Germany	> 18 y.o.	6.7	4.2	5.4	--	53
Sweden	35 to 63 yrs	6.2	2.0	--	--	4
Middle East						
Turkey	15 to 70 yrs	--	--	22	--	42
Asia						
Japan	0 to 90 yrs	--	--	0.9	79%	22

TABLE II
Prevalence of Urolithiasis per 1,000 Population

Region	Ages of Subjects	Males	Females	Overall	% Calcium Oxalate	Ref
USA						
Rochester, MN	0 to > 70 yrs	120	50	--	67%	29
All U.S.	40 to 75 yrs	80	--	--	--	38
California	0 to > 65 yrs	32	21	26	90%/60%	31
All U.S.	> 30 y.o.	75	41	--	--	48
Europe						
Spain	10 to > 65 yrs	52	34	43	--	140
Germany	15 to 65 yrs	69	31	68	--	52
Germany	> 18 y.o.	40	40	40	67%	53
England	18 to > 65 yrs	38	--	--	--	145
Sweden	35 to 63 yrs	89	32	52	--	4
Middle East						
Lebanon	Not specified	290	310	280	--	103
Turkey	15 to 70 yrs	--	--	148	--	42
Asia						
Thailand	< 50 yrs	--	--	4	--	146
Japan	0 to >90 yrs	--	--	54	79.4%	22

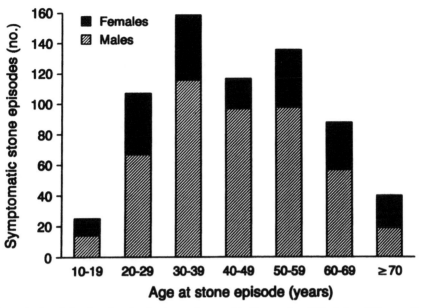

Figure 2:   Population based study showing an increased incidence of urolithiasis in middle age, the predominance of males among stone formers in midlife, and similar representation of males and females among stone formers in childhood and old age.   Data adapted from Johnson et al.[29]

The factors that predispose males in midlife to symptomatic urolithiasis are speculative but have included purported higher dietary purine/protein intake when compared with women and the very young or older men.   Some studies comparing urinary constituents in healthy men and women show higher calcium and oxalate excretion and lower citrate excretion in men.[56]   Hesse and colleagues[57] found no difference in urinary excretion of oxalate and no difference in calcium oxalate supersaturation when healthy men were compared with healthy women.   However, among calcium oxalate stone formers, the men had a higher urinary excretion rate of oxalate and higher calcium oxalate supersaturation than the women.[57]   In women, calcium and citrate excretion vary with the phase of the menstrual cycle suggesting a role for estrogen and progesterone.   An increase in urinary calcium excretion has been observed in women after menopause[58] and after ovariectomy.[59]   Women have been found to have a higher urine pH than men, perhaps related to progesterone.[55]   Crystal matrix protein, the most potent inhibitor of calcium oxalate crystal aggregation in urine, when derived from the urine of women possesses a stronger binding infinity for calcium oxalate than crystal matrix protein from the urine of men[60] and may play a protective role in women.

The urolithiasis of industrialized societies occurs predominantly in mid adulthood with a much lower incidence in childhood and in the elderly[22,29,31,35,50] (Figure 2).   Urolithiasis at the extremes of age warrants comment, nonetheless. Children and adolescents in industrialized countries, like adults, have a large percentage of upper tract calculi (87 to 94.4 percent).[61-63]   Calcium oxalate is the most frequent constituent of stones in children, though in several series, it accounted for only 33 to 56 percent of stones analyzed.[61-63,64]   Although idiopathic calcium oxalate urolithiasis often found, children are more likely than adults to have stones related to stasis or infection secondary to congenital abnormalities of the urinary tract.   Boys and girls are more equally represented than in the adult population with male to female ratios ranging from 1.05 to 1.4 to 1.[61-65]

Nephrocalcinosis and calcium containing stones have been recognized with increasing frequency in premature infants. In a recent prospective study, nephrocalcinosis developed in 67 percent of premature infants with birth weights of less than 1200 grams.[66] Other prospective and retrospective studies have identified nephrocalcinosis in 27 to 65 percent of premature infants.[67-69] Analysis of material deposited in the kidneys confirms both calcium oxalates and calcium phosphates in the limited reports available to date.[70] Risk factors for the development of nephrocalcinosis appear to include very low birth weight, bronchopulmonary dysplasia requiring long-term diuretic treatment, osteopenia of prematurity, hypercalciuria, and a family history of urolithiasis.[66,71] Recently, very low birth weight infants receiving parenteral nutrition have been observed to have significantly elevated urinary oxalate concentrations[72] and an increase in calcium oxalate supersaturation.[73] As larger numbers of very low birth weight infants survive the neonatal period, nephrocalcinosis and urolithiasis in infants are an increasing concern.

In geriatric populations, the incidence of urinary tract stones is much less than in early to mid adulthood.[22,29,31,50,55,74] The most common type of stone in the elderly is a calcium containing stone of the upper urinary tract, although the proportion of calcium containing stones is less than in younger adults.[75,76] In one study, pure calcium oxalate stones accounted for 40.2 percent of upper tract stones in men and 45 percent of those in women. Mixed calcium oxalate and calcium phosphate stones accounted for another 34.8 percent and 25 percent, respectively.[55] A decline in urinary calcium excretion has been observed with advancing age[77] perhaps due to a decline in intestinal calcium absorption.[78,79] Kohri et al. studied calcium oxalate nucleation rates and calcium oxalate crystal growth rates in both middle aged and elderly control subjects and stone formers.[55] Calcium oxalate nucleation rates were lower in both the elderly controls and elderly stone formers compared with middle aged subjects, although the difference did not reach statistical significance, perhaps due the small numbers of subjects evaluated. A higher proportion of uric acid stones and infected stones relative to calcium oxalate in the elderly may be related in part to an increased frequency of bladder outlet obstruction in older males and an increased frequency of urinary tract infections in geriatric individuals of both sexes.

B. RACE

There has been general agreement that blacks have a significantly lower prevalence of urolithiasis than whites.[31,64,80-83] It has been suggested that environmental factors, rather than genetic predisposition, could explain the difference. However, studies of the chemical composition of the urine in black and white normal subjects consuming similar amounts of dietary calcium and oxalate showed black subjects to have a lower calcium excretion and higher excretion of glycosaminoglycans.[84] Urinary inhibition of calcium oxalate crystal formation was higher in blacks of both sexes compared with whites.[84] The observation of significantly less nephrocalcinosis in black compared with white premature infants during hospitalization in a neonatal intensive care unit also argues in favor of genetic factors.[66]

The incidence of upper tract calculi among Chinese residents of San Francisco was found in one study to be twice that among Caucasians of the same city.[85] The authors attributed this finding to a high purine content of the diets of the Chinese residents. A more contemporary population study[48] of 1,185,124 individuals demonstrated a reduced prevalence of urolithiasis among Asian men and women when compared with whites. Relative to whites, the adjusted odds of developing urinary tract stones for men and women were, respectively: blacks 0.40 and 0.61, Asians 0.56 and 0.54, and Hispanics 0.66 and 0.85. The age adjusted prevalence of renal lithiasis derived from hospital diagnoses in North Carolina was 2.48 per thousand population in whites compared with 0.60 in nonwhites.[50] Factors responsible for these differences remain to be elucidated.

C. HEREDITY

Individuals who form urinary tract stones are more likely than non stone formers to have a family history of urolithiasis.[4,82,86-89] Among stone patients with frequent recurrences, the likelihood of a positive family history is even higher. In one study, 15 percent of control subjects, 29 percent of stone formers, and 63 percent of stone formers with frequent recurrences were found to have a family history of urolithiasis.[90] In another study of calcium oxalate stone formers, the frequency of renal calculi was 15.1 percent in fathers, 11.3 percent in mothers, 20.5 percent in brothers, and 8.2 percent in sisters compared with a control group who had 4.1 percent, 4.0 percent, 7.0 percent, and 4.1 percent, respectively, of family members with urolithiasis.[87]

First degree relatives of stone formers are more likely than distant relatives to form stones. Since first degree relatives also frequently share the same environment and similar diets, there has been debate as to whether the increased incidence of urolithiasis reflects genetic or environmental and nutritional factors. In two studies[91,92] an oral calcium loading test was used to evaluate stone formers, their spouses, and their first degree relatives. In both studies, urine calcium excretion was greater among stone formers and in their first degree relatives than in their spouses or unrelated control subjects. Calcium oxalate crystalluria was also found to be significantly greater in stone formers and in their first degree relatives than in spouses or controls.[91] These data, as well as observations of higher likelihood of nephrocalcinosis in premature infants who have a family history of stone disease compared with those who do not,[66] all support the contention that genetic factors play an important role in the familial predisposition to calcium oxalate stone disease. The data obtained in a study of calcium oxalate stone formers by Resnick and colleagues[87] was not consistent with monogenic inheritance but was compatible with hypothesis that the tendency to form calcium oxalate renal stones is regulated by a polygenic system.

Idiopathic hypercalciuria found in many calcium oxalate stone formers has well described hereditary features.[88,93-96] In some families, inheritance patterns of idiopathic hypercalciuria suggest autosomal dominant transmission.[97] Along these lines, the report of an abnormal red blood cell calcium pump in patients with idiopathic hypercalciuria[98] and the report of an inherited cellular defect in oxalate transport in patients with recurrent idiopathic urolithiasis[99] are of interest. These studies in patients with "idiopathic" calcium oxalate urolithiasis should not be confused with the variety of known hereditary syndromes that may lead to calcium urolithiasis such as primary hyperoxaluria, renal tubular acidosis, familial hyperparathyroidism, or familial renal tubular disorders associated with nephrocalcinosis and renal stones.[100]

V. ENVIRONMENTAL FACTORS INFLUENTIAL IN THE EPIDEMIOLOGY OF UROLITHIASIS

A. GEOGRAPHY AND CLIMATE
Regional differences in the prevalence of urolithiasis are well documented. Regions of higher stone incidence ("stone belts") have been recognized in the southeast portion of the United States,[48,101] Wales and the southern regions of England,[40] areas of South America,[102] and areas of the Middle East[42,103,104] and Asia.[104] A large population based study in the United States[48] showed a statistically significant correlation between latitude and the prevalence of urolithiasis in both men and women. Residents of the southeast United States were nearly twice as likely to have been diagnosed as having stones as residents of the northwest. Unfortunately, geographic factors are inextricably linked to other environmental variables such as climate, composition of the diet, mineral content of drinking water, and sometimes ethnic and/or religious practices unique to certain regions. Among these factors, there is convincing evidence that climate plays a role, with urolithiasis being much more prevalent in hot dry climates[105] and rare in the wet tropics.[102] Some have observed seasonal variations with a higher incidence of symptomatic stones in the summer and fall.[81,106-109] There has been speculation that the increased incidence of urolithiasis in the summer months occurs due to lower urine volumes and increased urine calcium excretion resulting from greater sunlight exposure and increased activation of vitamin D.[104] In temperate climates, an increase in daily urinary excretion of calcium and oxalate and an increase in calcium oxalate supersaturation in the summer months have been documented.[110] Juuti and colleagues[111] noted a seasonal variation in urinary excretion of calcium, oxalate, magnesium, and phosphorus when subjects were on unrestricted diets but not while subjects were on defined mineral diets, suggesting that seasonal variations in calcium and oxalate excretion (and perhaps the incidence of urolithiasis) are due to seasonal dietary changes rather than differences in sun exposure.

B. DIET
There is a large amount of epidemiologic evidence that diets low in animal protein and phosphorus and high in cereals favor the formation of endemic bladder stones, particularly in children.[5,7,12,15,16] By contrast, the high animal protein, high acid ash diets of more affluent societies appear to predispose to the formation of upper tract, largely calcium oxalate stones.[12,21,74,112-114] Robertson and colleagues observed a relationship between affluence (as determined by social class, annual salary, and weekly food expenditure) and the rate of hospitalizations for urinary stone disease not only in the United Kingdom[115] but throughout the world[116] (Figure 3). A strikingly similar correlation was found between animal protein intake and hospi-

talization rates for urolithiasis[116] (Figure 3) suggesting that increased affluence correlates with greater animal protein consumption.

Stone formers have been found to have a higher dietary animal protein intake than control subjects in some studies[112,113] but not in others.[117] When the diet of normal subjects (55 g of animal protein) was supplemented with an additional 34 g of animal protein, urine calcium excretion increased by 23 percent, oxalate by 24 percent, and uric acid by 48 percent.[112] The overall effect was an increase in the calculated relative probability of stone formation by 250 percent.[112] Normal volunteers studied on a high (92 g), medium (42 g), and then low (1 g) animal protein diets showed a progressive decrease in urinary calcium, oxalate, and uric acid with decreasing protein intake. Calcium excretion decreased to approximately 40 percent of the value observed on the high protein diet. This occurred despite the fact that calcium and phosphorus intakes were kept constant throughout the study.[112] Iguchi and colleagues[21] found that the dietary content of protein correlated more closely with urine calcium excretion than any other variable including dietary calcium intake. Other investigators[117] demonstrated a linear relationship between dietary protein intake and urinary calcium excretion and found that the slope of this relationship was much greater for patients with recurrent nephrolithiasis than for control subjects.

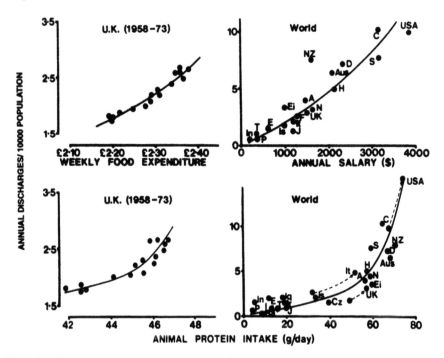

Figure 3: The annual hospital discharge rate for urinary stone disease correlates with annual salary, weekly food expenditure, and animal protein intake both in the United Kingdom and throughout the world.[116]

Codes: In = India; T = Thailand; P = Pakistan; E = Egypt; Is = Israel; J = Japan; It = Italy; F = Finland; Ei = Eire; A = Austria; UK = United Kingdom; N = Norway; Aus = Australia; NZ = New Zealand; H = Holland; D = Denmark; S = Sweden; C = Canada; and USA. Reproduced with permission.[116]

A diet high in animal protein favors acid ash on chemical analysis due to a high content of sulfur containing amino acids. A vegetarian diet, by contrast, forms an alkaline ash. The higher acid content may affect absorption and excretion of calcium.[118] Meat protein also contains more amino acids that are partially metabolized to oxalate including tryptophan, tyrosine, phenylalanine, and hydroxyproline.[114] In addition to increasing urinary excretion of calcium, oxalate, and uric acid, a high animal protein diet also reduces urine pH and urinary excretion of citrate, all of which are risk factors for calcium oxalate stone formation.[114] In keeping with these observations, vegetarians appear to have a lower prevalence of upper tract urinary stones than populations consuming animal protein.[113]

Among populations with a generally high consumption of a dietary constituent (for example, animal protein), a larger proportion of the population will be at risk of altered urinary composition that favors stone formation. The hypothesis has been advanced that individuals with calcium urolithiasis are sensitive to certain dietary constituents. It follows that at any given dietary intake, some individuals will be particularly susceptible to stone formation. This hypothesis helps to explain why populations show differences based on affluence, whereas individuals in a given socioeconomic class may consume similar diets but still differ with regard to the risk of stone formation.[119]

An increase in dietary calcium intake in healthy subjects can be expected to result in an increase in urinary calcium excretion equal to about 8 percent of the amount by which the dietary calcium is increased.[120] Contrary to what might be expected, however, population studies show that a high dietary calcium intake *decreases* the risk of symptomatic upper tract urolithiasis.[74] In a study by Curhan,[74] the incidence of symptomatic kidney stones during follow-up was lower by nearly 50 percent among the men with the highest energy adjusted calcium intake (mean 1326 mg/day) than among those whose intake was lowest (516 mg/day). These observations could be explained by increased absorption of oxalate in the gastrointestinal tract in individuals receiving less dietary calcium. Less calcium available to bind to oxalate in the intestinal tract may result in more oxalate available for absorption. Bataille[121] studied the effect of dietary calcium restriction without oxalate restriction in patients with a history of idiopathic calcium stones. Urinary calcium decreased and the urinary excretion of oxalate increased in both stone formers and controls. Similar results have been reported by others.[122,123] Since small changes in urinary oxalate concentration have a relatively greater effect on the supersaturation of calcium oxalate than similar changes in urine calcium concentration,[124] even a modest increase in urine oxalate may adversely affect calcium oxalate supersaturation and override any benefit of reduced urine calcium. An increase in urine oxalate has not been seen when dietary oxalate restriction accompanies reductions in dietary calcium.[122]

The role of dietary oxalate in idiopathic calcium oxalate stone formation appears to be less than that of animal protein or calcium. In healthy individuals, endogenous biosynthesis of oxalate accounts for 80 to 90 percent of the oxalate in the urine.[125,126] There is evidence to suggest that some hyperoxaluric stone formers have an even greater contribution of endogenous synthesis of oxalate to urinary oxalate.[127] Only 2 to 14 percent of ingested oxalate is absorbed in healthy subjects.[126] Absorption appears to be somewhat higher at 16 to 20 percent in certain subgroups of idiopathic stone formers.[126,128] In particular, there is evidence that stone formers with hyperabsorption of calcium may also absorb more oxalate.[128] Some have observed the relationship between urinary oxalate excretion and dietary oxalate intake to be relatively flat up to an intake of 2 mmol/day. Only at intakes exceeding this does urine oxalate increase proportionately to intake.[114] In addition, even though certain foods may be high in oxalate, the bioavailability of the oxalate is often low. The degree of oxalate binding to other dietary constituents appears important in the amount of oxalate that is absorbed from any given diet. In this regard, the ratio of oxalate to calcium in the intestinal lumen appears to play a dominant role.[129] Particularly disadvantageous is a diet that consists of relatively high oxalate in the face of a relatively low calcium intake. Such a circumstance has been described by Robertson in residents of the Arabian Peninsula. Among this population, mild hyperoxaluria occurs in more than 65 percent of the population, and the incidence of stones is very high.[114] Approximately half of the dietary oxalate in man is decomposed by bacteria in the large intestine.[130] Variations in intestinal flora might also play a role in the amount of dietary oxalate absorbed.

Other dietary constituents appear to influence the likelihood of stone formation. Dietary potassium intake has been observed to be inversely proportional to the risk of kidney stone formation,[74] possibly due to potassium-induced natriuresis and volume contraction as well as the obligatory increase in intake of bicarbonate and phosphate that accompanies the potassium. Both of these factors serve to decrease urinary calcium excretion.[120] Carbohydrate ingestion results in an increase in urinary calcium excretion and also leads to a reduction in urinary flow rate, exaggerating the increase in urinary calcium concentration.[131] These effects were found to be much more prominent in calcium stone formers and their relatives than in normal subjects.[132] In addition, postprandial hyperoxaluria has been reported in response to carbohydrate ingestion in both control subjects and recurrent idiopathic stone formers.[130-132]

High dietary sodium intake increases urinary calcium excretion in both healthy controls and individuals with recurrent calcium stone formation.[133,134] The calciuric effect is exaggerated in stone formers when compared with control subjects.[119] In hypercalciuric individuals, dietary sodium restriction can result in a significant decrease in urinary calcium excretion.[134] Despite these observations, most epidemiologic studies show no differences in dietary sodium intake or in urinary sodium excretion in idiopathic stone formers when compared with controls.[74,117,118] Magnesium in the urine inhibits nucleation and growth of calcium phosphate and calcium oxalate crystals[129] and dietary magnesium deficiency could potentially predispose to calcium stone formation. However, the role of urinary magnesium concentration in the absence of dietary magnesium deficiency has not been well studied in man. In one large population study,[74] there was no association between the risk of symptomatic kidney stones and dietary intake of sodium, magnesium, or carbohydrates.

A diet rich in fiber may inhibit intestinal calcium absorption but may also facilitate absorption of oxalate. Little is known of the relationship between dietary fat and idiopathic calcium stone formation. However, in a group of hypercalciuric recurrent stone formers, urinary oxalate and calcium were both reduced by ingestion of fish oil.[135] It has been observed that urolithiasis is very unusual in Eskimos, whose diet is rich in eicosapentanoic acid.[135] The epidemiologic significance of these dietary factors with regard to the risk of idiopathic calcium oxalate stone formation is largely unknown.

The role of fluid intake with regard to idiopathic stone formation is clear. With a urine volume of less than 1 liter daily, the risk of nucleation of constituents leading to calcium stones rises dramatically.[82,119] In a large population study, a higher fluid intake was associated with a reduced risk of idiopathic stone formation. Individuals with a fluid intake of greater than 2,500 mL daily had an age adjusted relative risk of stone formation of 0.52 compared with those who had a fluid intake of less than 1,300 mL per day.[74]

A number of studies have been concerned with calcium and magnesium concentration of drinking water, expressed as water hardness. Contrary to initial expectations, most studies have shown a negative association between water hardness and prevalence of urolithiasis[136-140] whereas other studies have shown no correlation between the hardness of drinking water and the prevalence of stone disease.[141,142]

C. OCCUPATIONS

Certain occupations appear to be associated with a greater likelihood of developing urolithiasis. Occupational exposure to cadmium (for example, in coppersmiths) is associated with an increased propensity to form urinary tract stones.[143] Direct renal tubular toxicity associated with increased urine calcium excretion appears to be the cause. In some circumstances (for example, air force pilots) a higher incidence of asymptomatic calculi in pilots as compared with nonflight personnel appears to be related to decreased fluid intake in pilots before and during flights.[144] Sedentary occupations seem associated with a higher likelihood of renal stones.[4,22,44,82] Unfortunately, the large number of potential confounding variables makes interpretation of such data difficult.

VI. CONCLUSION

Insoluble calcium oxalate in the urinary tract poses significant problems for individuals of all ages. Efforts to better understand the epidemiologic factors important in calcium oxalate stone formation continue to be necessary in order to define more effective preventative and treatment strategies.

ACKNOWLEDGEMENTS

The author is grateful to Dr. David M. Wilson for his helpful review of the manuscript and to Ms. Monica Poncelet for secretarial assistance.

REFERENCES

1.   **DeVita, M. V. and Zabetakis, P. M.,** Laboratory investigation of renal stone disease, *Clin. Lab. Med.,* 13, 225, 1993.

2.   **Shattock, S.,** On the microscopic structure of uric acid calculi, *Proc. R. Soc. Med.* 4, 110, 1911.

3.   **Williams, G. D.,** An ancient bladder stone, *JAMA,* 87, 941, 1926.

4.   **Ljunghall, S.,** Renal stone disease. Studies of epidemiology and calcium metabolism, *Scand. J. Urol. Nephrol.,* (supplement 41) 6, 1977.

5.   **Andersen, D. A.,** The nutritional significance of primary bladder stones, *Br. J. Urol.,* 34, 160, 1962.

6.   **Valyasevi, A. and Van Reen, R.,** Pediatric bladder stone disease; current status of research, *J. Pediatr.,* 72, 546, 1968.

7.   **Ashworth, M.,** Endemic bladder stones, *Br. Med. J.,* 301, 826, 1990.

8.   **Thomas, J. M. R.,** Vesical calculus in Norfolk, *Brit. J. Urol.,* 21, 20, 1949.

9.   **McCarrison, R.,** The causation of stone in India, *Br. Med. J.,* I, 1009, 1931.

10.   **Thomson, J. O.,** Urinary calculus at the Canton Hospital, Canton, China, *Surg. Gynecol. Obstet.,* 32, 44, 1921.

11.   **Assendelft, E.,** Berichtüber 630 stationär behandelte steinkränke, *Arch. Klin. Chir.,* 60, 669, 1900.

12.   **Andersen, D. A.,** A survey of the incidence of urolithiasis in Norway from 1853 to 1960, *J. Oslo City Hosp.,* 16, 101, 1966.

13.   **Brown, R. K. and Brown, E. L.,** Urinary stones. A study of their etiology in small children in Syria, *Surgery,* 9, 415, 1941.

14.   **Levy, D. and Falk, W.,** Urinary calculus disease among Israeli immigrants and Arab children, *J. Pediatr.,* 51, 404, 1957.

15.   **Eckstein, H. B.,** Endemic urinary lithiasis in Turkish children, *Arch. Dis. Child.,* 36, 137, 1961.

16.   **Singh, P. P., Singh, L. B. K., Prasad, S. N., and Singh, M. G.,** Urolithiasis in Manipur (north eastern region of India). Incidence and chemical composition of stones, *Am. J. Clin. Nutr.,* 31, 1519, 1978.

17.   **Sutor, D. J., Wooley, S. E., and Illingworth, J. J.,** A geographical and historical survey of the composition of urinary stones, *Br. J. Urol.,* 46, 393, 1974.

18.   **Gershoff, S. N., Prein, E. L., and Chandrapanand, A.,** Urinary stones in Thailand, *J. Urol.,* 90, 285, 1963.

19.   **Passmore, R.,** Observations on the epidemiology of stone in the bladder in Thailand, *Lancet,* I, 638, 1953.

20.   **Kambal, A., Wahab, S. M. A., and Khattab, A. H.,** The pattern of urolithiasis in the Sudan, *Br. J. Urol.,* 50, 376, 1978.

21.   **Iguchi, M., Kataoka, K., Kohri, K., Yachiku, S., and Kurita, T.,** Nutritional risk factors in calcium stone disease in Japan, *Urol. Int.,* 39, 32, 1984.

22.   **Yoshida, O. and Okada, Y.,** Epidemiology of urolithiasis in Japan: a chronological and geographical study, *Urol. Int.,* 45, 104, 1990.

23.   **Ljunghall, S.,** Incidence of upper urinary tract stones, *Miner. Electrolyte Metab.,* 13, 220, 1987.

24.   Norlin, A., Lindell, B., Granberg, P., and Lindvall, N., Urolithiasis: A study of its frequency, *Scand. J. Urol. Nephrol.*, 10, 150, 1976.

25.   Sallinen, A., Some aspects of urolithiasis in Finland, *Acta Chir. Scand.*, 118, 479, 1959.

26.   Grossmann, W., The current urinary stone wave in central Europe, *Brit. J. Urol.*, 10, 46, 1938.

27.   Hesse, A., Bach, D., and Vahlensieck, W., Epidemiologic studies in urolithiasis in West Germany, in *Urinary Calculus. Proceedings of the International Urinary Stone Conference*, Brokis, J. G. and Finlayson, B., Eds., PSG Publishing Company, Inc., Littleton, MA, 1981, 25.

28.   Sierakowski, R., Finlayson, B., Landes, R. R., Finlayson, C. D., and Sierakowski, N., The frequency of urolithiasis in hospital discharge diagnoses in the United States, *Invest. Urol.*, 15, 438, 1978.

29.   Johnson, C. M., Wilson, D. M., O'Fallon, W. M., Malek, R. S., and Kurland, L. T., Renal stone epidemiology: a 25-year study in Rochester, Minnesota, *Kidney Int.*, 16, 624, 1979.

30.   Herring, L. C., Observation on the analysis of ten thousand urinary calculi, *J. Urol.*, 88, 545, 1962.

31.   Hiatt, R. A., Dales, L. G., Friedman, G. D., and Hunkeler, E. M., Frequency of urolithiasis in a prepaid medical care program, *Am. J. Epidemiol.*, 115, 255, 1982.

32.   Abomelha, M. S., Al-Khader, A. A., and Arnold, J., Urolithiasis in Saudi Arabia, *Urology*, 35, 31, 1990.

33.   McKay, I., Sinclair, J., Scott, R., and Duncan, J. G., Radiologic and postmortem survey of abdominal lesions, *Urology*, 4, 274, 1974.

34.   Scott, R., Freeland, R., Mowat, W., Gardiner, M., Hawthorne, V., Marshall, R. M., and Ives, J. G. J., The prevalence of calcified upper urinary tract stone disease in a random population - Cumbernauld health survey, *Br. J. Urol.*, 49, 589, 1977.

35.   Larsen, J. F. and Phillip, J., Studies on the incidence of urolithiasis, *Urol. Int.*, 13, 53, 1962.

36.   Ljunghall, S., and Hedstrand, H., Epidemiology of renal stones in a middle-aged male population, *Acta Med. Scand.*, 197, 439, 1975.

37.   Currie, W. J. C. and Turmer, P., The frequency of renal stones within Great Britain in a gouty and non-gouty population, *Br. J. Urol.*, 51, 337, 1979.

38.   Curhan, G. C., Rimm, E. B., Willett, W. C., and Stampfer, M. J., Regional variation in nephrolithiasis incidence and prevalence among United States men, *J. Urol.*, 151, 838, 1994.

39.   Graves, E. J., 1983 Summary: National hospital discharge survey, *NCHS Advance Data*, 101, 1, 1984.

40.   Barker, D. J. P. and Donnan, S. P. B., Regional variations in the incidence of upper urinary tract stones in England and Wales, *Brit. Med. J.*, 1, 67, 1978.

41.   Bennett, R. C. and Hughes, E. S. R., Urinary calculi and ulcerative colitis, *Br. Med. J.*, 2, 494, 1972.

42.   Akinci, M., Esen, T., and Tellaloglu, S., Urinary stone disease in Turkey: An updated epidemiological study, *Eur. Urol.*, 20, 200, 1991.

43.   **Uribarri J., Oh, M. S., and Carroll, H. J.**, The first kidney stone, *Ann. Int. Med.*, 111, 1006, 1989.
44.   **Blacklock, N. J.**, The pattern of urolithiasis in the Royal Navy, in *Renal Stone Research Symposium*, Hodgkinson, A., Nordin, B. E., eds, Churchill, London, 1969.
45.   **Williams, R. E.**, Long-term survey of 538 patients with upper urinary tract stone, *Br. J. Urol.*, 35, 416, 1963.
46.   **Marshall, V., White, R. H., Chaput De Saintonge, M., Tresidder, G. C., and Blandy, J. P.**, The natural history of renal and ureteric calculi, *Br. J. Urol*, 47, 117, 1975.
47.   **Sutherland, J. W., Parks, J. H., and Coe, F. L.**, Recurrence after a single renal stone in a community practice, *Miner. Electrolyte Metab.*, 11, 267, 1985.
48.   **Soucie, J. M., Thun, M. J., Coates, R. J., McClellan, W., and Austin, H.**, Demographic and geographic variability of kidney stones in the United States, *Kidney Int.*, 46, 893, 1994.
49.   **Hiatt, R. A. and Friedman, G. D.**, The frequency of kidney and urinary tract diseases in a defined population, *Kidney Int.*, 22, 63, 1982.
50.   **Schey, H. M., Corbett, W. T., and Resnick, M. I.**, Prevalence rate of renal stone disease in Forsyth county, North Carolina during 1977, *J. Urol.*, 122, 288, 1979.
51.   **Wardlaw, H. S. H.**, Observations on the incidence and composition of urinary calculi, *Med. J. Aust.*, 1, 180, 1952.
52.   **Tschöpe, W., Ritz, E., Haslbeck, M., Mehnert, H., and Wesch, H.**, Prevalence and incidence of renal stone disease in a German population sample, *Klin. Wochenschr.*, 59, 411, 1981.
53.   **Vahlensieck, E. W., Bach, D., Hesse, A., and Strenge, A.**, Epidemiology, pathogenesis and diagnosis of calcium oxalate urolithiasis, *Intern. Urol. Nephrol.*, 14, 333, 1982.
54.   **Hughes, J., Coppridge, W. M., Roberts, L. C., and Mann, V. I.**, Oxalate urinary tract stones, *JAMA*, 172, 774, 1960.
55.   **Kohri, K., Ishikawa, Y., Katoh, Y., Kataoka, K., Iguchi, M., Yachiku, S., Kurita, T.**, Epidemiology of urolithiasis in the elderly, *Int. Urol. and Nephrol.*, 23, 413, 1991.
56.   **Sarada, B. and Satyanarayana, U.**, Urinary composition in men and women and the risk of urolithiasis, *Clin. Biochem.*, 24, 487, 1991.
57.   **Hesse, A., Classen, A., Klocke, K., and Vahlensieck, W.**, The significance of the sexual dependency of lithogenic and inhibitory substances in urine, in, *Urolithiasis and Related Clinical Research*, Schwille, P. O., Smith, L. H., Robertson, W. G., and Vahlensieck, W., Eds, Plenum Press, New York and London, 1985, 25.
58.   **Young, M. M., Durh, M. B., and Nordin, B. E. C.**, Effects of natural and artificial menopause on plasma and urinary calcium and phosphorus, *Lancet*, 2, 118, 1967.
59.   **Gallacher, J. C., Young, N. M., and Nordin, B. E. C.**, Effect of artificial menopause on plasma and urine calcium and phosphate, *Clin. Endocrinol.*, 1, 57, 1972.

60. **Doyle, I. R., Marshall, V. R., and Ryall, R. L.**, Crystal matrix protein - sorting the sexes, in *Urolithiasis 2*, Ryall, R. L., Bais, R., Marshall, V. R., Rofe, A. M., Smith, L. H., and Walker, V. R., Eds, Plenum Press, New York and London, 1994, 289.

61. **Milliner, D. S. and Murphy, M. E.**, Urolithiasis in pediatric patients, *Mayo Clin. Proc.*, 68, 241, 1993.

62. **Cheah, W. K., King, P. A., and Tan, H. L.**, A review of pediatric cases of urinary tract calculi, *J. Pediatr. Surg.*, 29, 701, 1994.

63. **Gearhart, J. P., Herzberg, G. Z., and Jeffs, R. D.**, Childhood urolithiasis: experiences and advances, *Pediatrics*, 87, 445, 1991.

64. **Polinsky, M. S., Kaiser, B. A., and Baluarte, H. J.**, Urolithiasis in childhood, *Pediatr. Clin. North Am.*, 34, 683, 1987.

65. **Nimkin, K., Lebowitz, R. L., Share, J. C., and Teele, R. L.**, Urolithiasis in a children's hospital: 1985-1990, *Urol. Radiol.*, 14, 139, 1992.

66. **Karlowicz, M. G., Katz, M. E., Adelman, R. D., and Solhaug, M. J.**, Nephrocalcinosis in very low birth weight neonates: family history of kidney stones and ethnicity as independent risk factors, *J. Pediatr.*, 122, 635, 1993.

67. **Robinson, C. M. and Cox, M. A.**, The incidence of renal calcifications in low birth weight (LBW) infants on Lasix for bronchopulmonary dysplasia (BPD), *Pediatr. Res.*, 20, 359a, 1986.

68. **Jacinto, J. S., Houchang, D., Modanlou, M. D., Crade, M., Strauss, A. A., and Bosu, S. K.**, Renal calcifications: incidence in very low birth weight infants, *Pediatrics*, 81, 31, 1988.

69. **Short, A. and Cooke, R. W. I.**, The incidence of renal calcification in preterm infants, *Arch. Dis. Child.*, 66, 412, 1991.

70. **Hufnagel, K. G., Shadid, N. K., Penn, D., Cacciarelli, A., Williams, P.**, Renal calcifications: a complication of long-term furosemide therapy in preterm infants, *Pediatrics*, 70, 360, 1982.

71. **Adams, N. D. and Rowe, J. C.**, Nephrocalcinosis, *Clin. Perinat.*, 19, 179, 1992.

72. **Campfield, T. and Braden, G.**, Urinary oxalate excretion by very low birth weight infants receiving parenteral nutrition, *Pediatrics*, 84, 860, 1989.

73. **Hoppe, B., Hesse, A., Neuhaus, T., Fanconi, S., Forster, I., Blau, N., and Leumann, E.**, Urinary saturation and nephrocalcinosis in preterm infants: effect of parenteral nutrition, *Arch. Dis. Child.*, 69, 299, 1993.

74. **Curhan, G. C., Willett, W. C., Rimm, E. B., and Stampfer, M. J.**, A prospective study of dietary calcium and other nutrients and the risk of symptomatic kidney stones, *N. Engl. J. Med.*, 328, 833, 1993.

75. **Coe, F. L, Keck, J., and Norton, E. R.**, The natural history of calcium urolithiasis, *JAMA*, 238, 1519, 1977.

76. **Elliott, J. S.**, Calcium oxalate urinary calculi, *Clin. Chem. Aspects Med.*, 62, 36, 1983.

77. **Balusu, L., Hodgkinson, A., Nordin, B. E., Peacock, M.**, Urinary excretion of calcium and creatinine in relation to age and body weight in normal subjects and patients with renal calculus, *Clin. Sci.*, 38, 601, 1970.

78. **Bullamore, J. R., Wilkinson, R., Gallagher, J. C., Nordin, B. E. C.,** Effect of age on calcium absorption, *Lancet,* 2, 535, 1970.
79. **Nordin, B. E., Wilkinson, R., Marshall, D. H., Gallagher, J. C., William, A, and Peacock, M.,** Calcium absorption in the elderly, *Calcif. Tissue Reg.,* 21, 442, 1976.
80. **Reaser, E. F.,** Racial incidence of urolithiasis, *J. Urol.,* 34, 148, 1935.
81. **Milbert, A. H. and Gersh, I.,** Urolithiasis in the soldier, *J. Urol.,* 53, 440, 1945.
82. **Finlayson, B.,** Renal lithiasis in review, *Urol. Clin. North Am.,* 1, 181, 1974.
83. **Boyce, W. H., Garvey, F. K., and Strawcutter, H. E.,** Incidence of urinary calculi among patients in general hospitals, 1948 to 1952, *JAMA,* 161, 1437, 1956.
84. **Meyers, A. M., Whalley, N., Zakolski, W. J., and Shar, T.,** Chemical composition of the urine in the normal black and white population, in *Urolithiasis II,* Ryall, R., Bais, R., Marshall, V. R., Rofe, A. M., Smith, L. H., Walker, V. R., Eds, Plenum Press, New York and London, 1994, 422.
85. **Fay, R.,** Calculus disease of upper urinary tract in San Francisco Chinese, *Urol,* 18, 123, 1981.
86. **McGeown, M. G.,** Heredity in renal stone disease, *Clin. Sci.,* 19, 465, 1960.
87. **Resnick, M., Pridgen, D. B., and Goodman, H. O.,** Genetic predisposition to formation of calcium oxalate renal calculi, *N. Engl. J. Med.,* 278, 1313, 1968.
88. **Coe, F. L., Parks, J. H., and Moore, E. S.,** Familial idiopathic hypercalciuria, *N. Engl. J. Med.,* 300, 337, 1979.
89. **Melick, R. A. and Henneman, P. H.,** Clinical and laboratory studies of 207 consecutive patients in a kidney-stone clinic, *N. Engl. J. Med.,* 259, 307, 1958.
90. **Ljunghall, S.,** Family history of renal stones in a population study of stone-formers and healthy subjects, *Br. J. Urol.,* 51, 249, 1979.
91. **Marya, R. K., Dadoo, R. C., and Sharma, N. K.,** Genetic predisposition to renal stone disease in the first-degree relatives of stone-formers, *Urol. Int.,* 36, 245, 1981.
92. **Kaul, P., Sidhu, H., Vaidyanathan, S., Thind, S. K., and Nath, R.,** Study of urinary calcium excretion after oral calcium load in stone formers, their spouses and first-degree blood relatives, *Urol. Int.,* 52, 93, 1994.
93. **Pak, C. Y. C., McGuire, J., Peterson, R., Britton, F., and Harrod, M. J.,** Familial absorptive hypercalciuria in a large kindred, *J. Urol.,* 126, 717, 1981.
94. **Weinberger, A., Schechter, J., Pinkhas, J., Sperling, O.,** Hereditary hypercalciuric urolithiasis, *Br. J. Urol.,* 53, 285, 1981.
95. **Aladjem, M., Modan, M., Lusky, A., Georgi, R., Orda, S., Eshkol, A., Lotan, D., and Boichis, H.,** Idiopathic hypercalciuria: a familial generalized renal hyperexcretory state, *Kidney Int.,* 24, 549, 1983.
96. **Harangi, F. and Mehes, K.,** Family investigations in idiopathic hypercalciuria, *Eur. J. Pediatr.,* 152, 64, 1993.

97.   **Mehes, K. and Szelid, Zs.**, Autosomal dominant inheritance of hypercalciuria, *Eur. J. Pediatr.*, 133, 239, 1980.
98.   **Bianchi, G., Vezzoli, G., Cusi, D., Cova, T., Elli, A., Soldati, L., Tripodi, G., Surian, M., Ottaviano, E., Rigatti, P., and Ortolani, S.**, Abnormal red-cell calcium pump in patients with idiopathic hypercalciuria, *N. Engl. J. Med.*, 319, 897, 1988.
99.   **Baggio, B., Gambaro, G., Marchini, F., Cicerello, E., Tenconi, R., Clementi, M., and Borsatti, A.**, An inheritance anomaly of red-cell oxalate transport in "primary" calcium nephrolithiasis correctable with diuretics, *N. Engl. J. Med.*, 314, 599, 1986.
100.   **Pook, M. A., Wrong, O., Wooding, C., Norden, A. G. W., Feest, T. G., and Thakker, R. V.**, Dent's disease, a renal Fanconi syndrome with nephrocalcinosis and kidney stones, is associated with a microdeletion involving DXS255 and maps to Xp11.22, *Hum. Mol. Genet.*, 2, 2129, 1993.
101.   **Thun, M. J. and Schober, S.**, Urolithiasis in Tennessee: an occupational window into a regional problem, *Am. J. Public Health*, 81, 587, 1991.
102.   **Davalos, A.**, The rarity of stones in the urinary tract in the wet tropics, *J. Urol.*, 54, 182, 1945.
103.   **Gottlieb, D. and Dolev, E.**, High prevalence of urinary stones in Marj-Ayoun, Lebanon, *Isr. J. Med. Sci.*, 20, 158, 1984.
104.   **Kotinis-Zambakas, S. J.**, Climatic characteristics of the high urinary tract calculi areas all over the world, *Geographia Medica*, 22, 25, 1990.
105.   **Frank, M., Atsmon, A., Sugar, P., and De Vries, A.**, Epidemiological investigation of urolithiasis in the hot arid southern region of Israel, *Urol. Int.*, 15, 65, 1963.
106.   **Elliott, J. P., Gordon, J. O., Evans, J. W., and Platt, L.**, A stone season: a 10-year retrospective study of 768 surgical stone cases with respect to seasonal variation, *J. Urol.*, 114, 574, 1975.
107.   **Prince, C. L., Scardino, P. L., and Wolan, C. T.**, The effect of temperature and humidity and dehydration on the formation of renal calculi, *J. Urol.*, 75, 209, 1956.
108.   **Black, J. M.**, Oxaluria in British troops in India, *Br. Med. J.*, I, 590, 1945.
109.   **Pierce, L. and Bloom, B.**, Observations on urolithiasis among American troops in a desert area, *J. Urol.*, 54, 466, 1945.
110.   **Robertson, W. G., Peacock, M., Marshall, R. W., Speed, R., and Nordin, B. E. C.**, Seasonal variations in the composition of urine in relation to calcium stone formation, *Clin. Sci. Mol. Med.*, 49, 597, 1975.
111.   **Juuti, M., Heinonen, O. P., and Alhava, E. M.**, Seasonal variation in urinary excretion of calcium, oxalate, magnesium and phosphate on free and standard mineral diet in men with urolithiasis, *Scand. J. Urol. Nephrol.*, 15, 137, 1981.
112.   **Robertson, W. G., Peacock, M., Heyburn, P. J., Hanes, F. A., Rutherford, A., Clementson, E., Swaminathan, R., and Clark, P. B.**, Should recurrent calcium oxalate stone formers become vegetarians, *Brit. J. Urol.*, 51, 427, 1979.
113.   **Robertson, W. G., Peacock, M., and Marshall, D. H.**, Prevalence of urinary stone disease in vegetarians, *Eur. Urol.*, 8, 334, 1982.

114. **Robertson, W. G.,** Epidemiology of urinary stone disease, *Urol. Res.,* 18, S3, 1990.

115. **Robertson, W. G. and Peacock, M.,** The pattern of urinary stone disease in Leeds and in the United Kingdom in relation to animal protein intake during the period 1960-1980, *Urol. Int.,* 37, 394, 1982.

116. **Robertson, W. G., Peacock, M., Heyburn, P. J., Hanes, F. A., and Swaminathan, R.,** The risk of calcium stone formation in relation to affluence and dietary animal protein, in *Urinary Calculus,* Brockis, J. G. and Finlayson, B., Eds., PSG Publishing Company, Inc., Littleton, Massachussetts, 1981, 3.

117. **Wasserstein, A. G., Stolley, P. D., Joper, K. A., Goldfarb, S., and Agus, Z. S.,** Case control study of risk factors for idiopathic calcium nephrolithiasis, *Miner. Electrolyte Metab.,* 13, 85, 1987.

118. **Lemann, J., Gray, R. W., Maierhofer, W. S., and Cheung, H. S.,** The importance of renal net acid excretion as a determinant of fasting urinary calcium excretion, *Kidney Int.,* 29, 743, 1986.

119. **Goldfarb, S.,** Dietary factors in the pathogenesis and prophylaxis of calcium nephrolithiasis, *Kidney Int.,* 34, 544, 1988.

120. **Lemann, J.,** Composition of the diet and calcium kidney stones, *N. Engl. J. Med.,* 328, 880, 1993.

121. **Bataille, P., Charransol, G., Gregoire, I., Daigre, J. L., Coevoet, B., Makdassi, R., Pruna, A., Locquet, P., Sueur, J. P., and Fournier, A.,** Effect of calcium restriction on renal excretion of oxalate and the probability of stones in the various pathophysiological groups with calcium stones, *J. Urol.,* 130, 218, 1983.

122. **Marshall, R. W., Cochran, M., and Hodgkinson, A.,** Relationships between calcium and oxalic acid intake in the diet and their excretion in the urine of normal and renal stone forming subjects, *Clin. Sci.,* 43, 91, 1972.

123. **Jaeger, P., Portmann, L., Jacquet, A. F., and Burckhart, P.,** Influence of the calcium content of the diet on the incidence of mild hyperoxaluria in idiopathic renal stone formers, *Am. J. Nephrol.,* 5, 40, 1985.

124. **Robertson, W. G. and Peacock, M.,** The cause of idiopathic calcium stone disease: hypercalciuria or hyperoxaluria, *Nephron,* 26, 105, 1980.

125. **Hodgkinson, A.,** *Oxalic Acid in Biology and Medicine,* Academic Press, London, New York, and San Francisco, 1977, 3.

126. **Lindsjö, M.,** Oxalate metabolism in renal stone disease with special reference to calcium metabolism and intestinal absorption, *Scand. J. Urol. Nephrol.,* suppl 119, 1, 1989.

127. **Hatch, M.,** Oxalate status in stone formers. Two distinct hyperoxaluric entities, *Urol. Res.,* 21, 55, 1993.

128. **Marangella, M., Fruttero, B., Bruno, M., and Linari, F.,** Hyperoxaluria in idiopathic calcium stone disease: further evidence of intestinal hyperabsorption of oxalate, *Clin. Sci.,* 63, 381, 1982.

129. **Schwille, P. O. and Herrmann, U.,** Environmental factors in the pathophysiology of recurrent idiopathic calcium urolithiasis (RCU), with emphasis on nutrition, *Urol. Res.,* 20, 72, 1992.

130. **Allison, M. J., Cook, H. M., Milne, D. B., Gallagher, S., and Clayman, R. V.,** Oxalate degradation by gastrointestinal bacteria from humans, *J. Nutr.,* 116, 455, 1986.

131.   Lemann, J., Piering, W. F., and Lennon, E. J., Possible role of carbo-hydrate-induced calciuria in calcium oxalate kidney-stone formation, *N. Engl. J. Med.*, 280, 232, 1969.

132.   Nguyen, N. U., Dumoulin, G., Wolf, J. P., Bourderont, D., and Berthelay, S., Urinary calcium and oxalate excretion in response to oral glucose load in man, *Horm. Metab. Res.*, 18, 869, 1986.

133.   Sabto, J., Powell, M. J., Breidahl, M. J., and Gurr, F. W., Influence of urinary sodium on calcium excretion in normal individuals, *Med. J. Aust.*, 140, 354, 1984.

134.   Muldowney, F. P., Freaney, R., and Maloney, M. F., Importance of dietary sodium in the hypercalciuria syndrome, *Kidney Int.*, 22, 292, 1982.

135.   Buck, A. C., Davies, R. L., and Harrison, T., The protective role of eicosapentanoic acid in the pathogenesis of urolithiasis, *J. Urol.*, 146, 188, 1991.

136.   Shuster, J., Finlayson, B., Scheaffer, R., Sierakowski, R., Zoltek, J., and Dzegede, S., Water hardness and urinary stone disease, *J. Urol.*, 128, 422, 1982.

137.   Churchill, D., Bryant, D., Fodor, G., and Gault, M. H., Drinking water hardness and urolithiasis, *Ann. Int. Med.*, 88, 513, 1978.

138.   Sierakowski, R., Hemp, G., and Finlayson, B., Water hardness and the incidence of urinary calculi, unpublished data.

139.   Frank, M., De Vries, A., Atsmon, A., Lazebnik, J., Kochwa, S., Epide-miological investigation of urolithiasis in Israel, *J. Urol.*, 81, 497, 1959.

140.   Ramirez, C. T., Morales, E. F., Gomez, A. Z., Alcaraz, L. G., and Samper, S., Del Rio: An epidemiological study of renal lithiasis in gypsies and others in Spain, *J. Urol.*, 131, 853, 1984.

141.   Donaldson, D., Pryce, J. D., Rose, G. A., and Tovey, J. E., Tap water calcium and its relationship to renal calculi and 24 h urinary calcium output in Great Britain, *Urol. Res.*, 7, 273, 1979.

142.   Ljunghall, S., Regional variations in the incidence of urinary stones, *Brit. Med. J.*, 1, 439, 1978.

143.   Scott, R. F., Cunningham, C., McLelland, A., Fill, A. S., Finch, O. P., and McKellan, N., The importance of cadmium on a factor in calcified upper urinary tract stone disease - a prospective 7 year study, *Br. J. Urol.*, 54, 584, 1982.

144.   Kohler, F. P., Asymptomatic upper urinary tract lithiasis in flying person-nel, *J. Urol.*, 86, 370, 1961.

145.   Robertson, W. G., Peacock, M., Baker, M., Marshall, D. H., Pearlman, B., Speed, R., Sergeant, V., and Smith, A., Studies on the prevalence and epidemiology of urinary stone disease in men in Leeds, *Br. J. Urol.*, 55, 595, 1983.

146.   Sriboonlue, P., Prasongwatana, K., Chata, K., and Tungsanga, K., Prevalence of upper urinary tract stone disease in a rural community of north-eastern Thailand, *Br. J. Urol.*, 69, 240, 1992.d

Chapter 9

ENZYMOLOGY AND MOLECULAR GENETICS OF
PRIMARY HYPEROXALURIA TYPE 1.
CONSEQUENCES FOR CLINICAL MANAGEMENT

Christopher J. Danpure[1] and Gillian Rumsby[2]

Departments of [1]Biology and [2]Chemical Pathology,
University College London, London, UK.

I. ENZYMOLOGY & MOLECULAR GENETICS

A. PH1 & AGT DEFICIENCY
1. Clinical Description of PH1[1-3]

Primary hyperoxaluria type 1 (PH1, McKusick 259900) is an inherited disorder of glyoxylate metabolism, characterised biochemically by the overproduction of oxalate and glycolate, and clinically by recurrent calcium oxalate (CaOx) stone formation in the urinary tract (urolithiasis) and/or diffuse deposition of CaOx throughout the renal parenchyma (nephrocalcinosis). In most patients, PH1 follows a progressive course that leads to chronic renal failure in mid-to-late childhood or early adulthood. However, in a minority of patients, disease follows a more rapid aggressive course, presenting in the neonate with metabolic acidosis, nephrocalcinosis and renal failure. "Acute neonatal" PH1 frequently results in death by the age of 1 year. Following the onset of renal failure, the effects of increased oxalate synthesis are compounded by decreased clearance, leading to the deposition of CaOx crystals throughout the body (systemic oxalosis).

2. AGT Deficiency & Its Metabolic Consequences

PH1 is caused by a deficiency of the liver-specific enzyme alanine:glyoxylate aminotransferase (AGT, E.C. 2.6.1.44), which catalyses the transamination of glyoxylate to glycine using pyridoxal-5-phosphate as a cofactor.[4] In the absence of AGT, glyoxylate is instead either oxidised to oxalate within the peroxisomes (catalysed by glycolate oxidase) or diffuses across the peroxisomal membrane into the cytosol, where it is oxidised to oxalate (catalysed by lactate dehydrogenase) and reduced to glycolate (catalysed by glyoxylate reductase and possibly lactate dehydrogenase) (Figure 1). Secretion of oxalate and glycolate by the liver into the blood and thence clearance by the kidneys leads to the biochemical hallmarks of PH1, namely hyperoxaluria and hyperglycolic aciduria.[1,5] All of the

0-8493-7673-4/95/$0.00+$.50
© 1995 by CRC Press

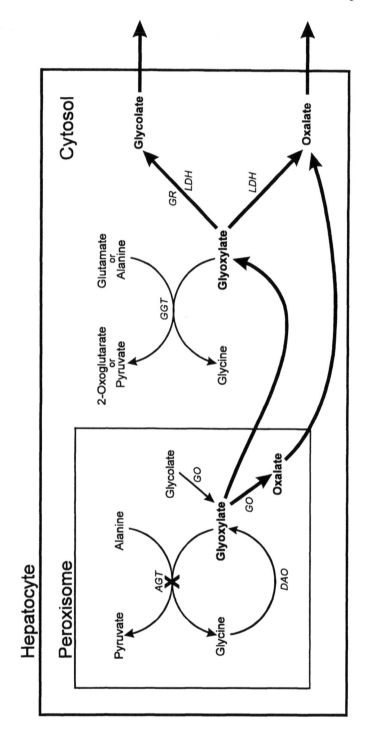

Figure 1. Pathways involved in the metabolism of glyoxylate in the human hepatocyte. GO, glycolate oxidase; DAO, D-amino acid oxidase; AGT, alanine:glyoxylate aminotransferase; LDH, lactate dehydrogenase; GR, glyoxylate reductase; GGT, glutamate:glyoxylate aminotransferase. Heavy arrows indicate the consequences of the metabolic block (X) in PH1.

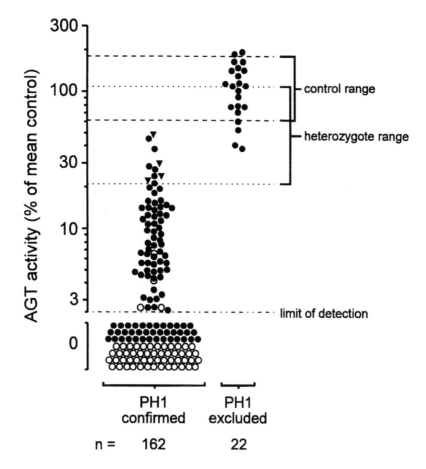

Figure 2. AGT enzymic analysis of 184 liver samples for the diagnosis of PH1. Control[7] and heterozygote[8] ranges are taken from the literature. Diagnosis of PH1 was confirmed in 162 cases and excluded in 22 cases. Solid symbols, CRM+ samples; open symbols, CRM- samples; circles, unambiguous confirmation or exclusion of PH1; triangles, diagnosis of PH1 probable due to other family and biochemical data, but not absolutely certain due to lack of any subcellular data.

pathophysiological characteristics of PH1 appear to be due to the increased oxalate synthesis and the low solubility of CaOx.

B. ENZYMIC PHENOTYPES

1. AGT Expression

PH1 is extremely heterogeneous at the enzymic level (Figure 2).[6-8] Based on the analysis of AGT in 162 PH1 liver biopsies (116 in the laboratory of CJD at the MRC Clinical Research Centre, Harrow, UK, between 1985 and 1992, and 46 in the laboratory of GR at the University College London

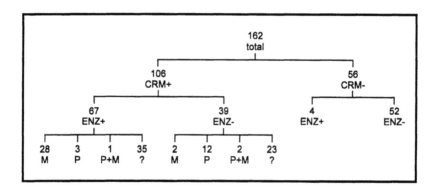

Figure 3. Distribution of enzymic phenotypes in 162 confirmed PH1 patients. CRM and ENZ are defined in the text; M, mainly mitochondrial AGT; P, exclusively peroxisomal AGT; P+M, both peroxisomal and mitochondrial AGT (that in the peroxisomes being aggregated into cores); ?, subcellular distribution of AGT unknown.

Hospitals, London, UK, between 1993 and 1994), three main enzymic phenotypes have been identified (Figure 3):- 1) absence of both immunoreactive AGT protein and AGT catalytic activity (CRM-/ENZ-) (32.1% of patients), 2) presence of immunoreactive AGT protein, but absence of AGT catalytic activity (CRM+/ENZ-) (24.1% of patients), and 3) presence of both immunoreactive AGT protein and AGT catalytic activity (CRM+/ENZ+) (41.4% of patients). In the latter group of patients, the level of hepatic AGT activity can vary from only just above the lower limit of detection (2.5% of the mean control value) up to nearly 50% (Figure 2). Some CRM+/ENZ+ patients cannot be distinguished from asymptomatic obligate carriers on the basis of enzyme activity alone. Although there is an approximate relationship between hepatic AGT activity and clinical severity, it is not clear enough to be prognostically useful.[6]

2. Subcellular Distribution of AGT

AGT is localised exclusively in the peroxisomes of liver parenchymal cells in most normal individuals.[9] However, in some people a small but significant amount (5-10%), is also found in the mitochondria.[10]

In most CRM+/ENZ- PH1 patients, the immunoreactive but catalytically defunct AGT protein is localised totally within the peroxisomes.[9] However, in most CRM+/ENZ+ patients, AGT appears to be selectively mistargeted, so that about 90% is localised in the mitochondria and only about 10% in the peroxisomes.[11] Such mistargeting of an enzyme from one intracellular organelle to another is without parallel in human genetic disease. The peroxisome is the major site of glyoxylate synthesis in human liver (see Figure 1) and, consequently, AGT is unable to carry out properly its metabolic function of glyoxylate detoxification when located at another intracellular site (i.e. the mitochondria).

A small proportion of PH1 patients (three so far identified) express a variation of the mistargeting phenotype. These CRM+/ENZ± patients have low, but detectable, levels of immunoreactive AGT protein but little or no AGT catalytic activity. In these individuals, AGT is approximately equally divided between peroxisomes and mitochondria, that in the peroxisomes being aggregated into unusual core-like structures.[12]

C. GENOTYPES

1. The Normal AGT Gene (*AGXT*)

The gene encoding AGT (i.e. *AGXT*) has been mapped to the tip of the long arm of chromosome 2 at 2q36-q37.[13] *AGXT* is composed of eleven exons ranging in size from 65 bp to 407 bp and is contained within approximately 10 kb of genomic DNA (see Figure 4). The mature mRNA is approximately 1.6 kb in length and translates into a 392 amino acid protein with a molecular mass of about 43 kD.[14]

A number of polymorphic variations have been identified in *AGXT* (see Figure 4). The less common "minor *AGXT* allele" differs from the more common "major *AGXT* allele" in at least three positions, two of which (i.e. C154T and A1142G point base substitutions) lead to amino acid replacements (i.e. Pro11Leu and Ile340Met, respectively).[10] The third difference is the presence of a 74 bp duplication in intron 1 in the minor allele.[15] The minor allele has a frequency of up to 20% in Caucasian populations, but only about 2% in Japanese populations.[16] Ile340Met appears to be without structural or functional consequence. However, Pro11Leu has a marked effect on the properties of AGT, in so far as it enables the N-terminus to fold into an amphiphilic α-helix with properties of a mitochondrial targeting sequence.[10,17] Whereas normal individuals homozygous for the major *AGXT* allele target their AGT exclusively to the peroxisomes, normal individuals homozygous for the minor allele target 5-10% of their AGT immunoreactive protein to the mitochondria (the remainder being peroxisomal).

An additional normally-occurring variation in *AGXT*, which was originally identified as a *Taq*I restriction-fragment-length polymorphism (RFLP),[18] was subsequently shown to be due to a variable number tandem repeat (VNTR) in intron 4.[16] The repeating unit is a 29/32 bp sequence which shares a degree of homology with the IR3 repeat sequence from Epstein Barr virus. Three different alleles have been identified so far in Caucasians consisting of 12 (type III), 17 (type II) and ~ 38 (type I) repeats with frequencies of 33%, 7% and 60% respectively. In contrast, a study of this polymorphism in the Japanese showed a quite different distribution with the type II allele as the most common form occurring in 45% of cases and the existence of a fourth allele containing ~32 repeats.[16] The minor *AGXT* allele (see above) was always found on a background of the type I intron 4 polymorphism in Caucasians, but this relationship did not extend to the Japanese.

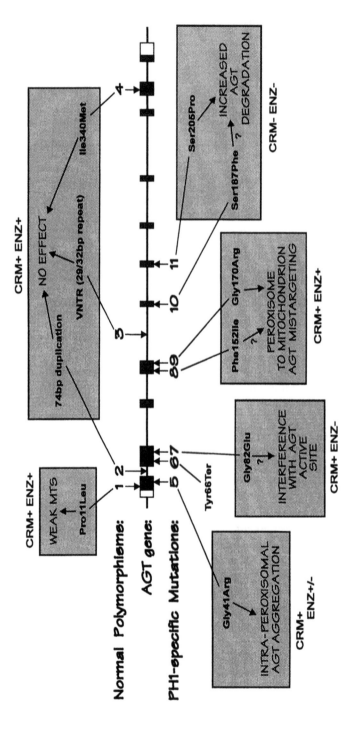

Figure 4. Normal polymorphisms and PH1-specific mutations in the AGT gene together with their associated enzymic phenotypes. Blocks represent the eleven exons of the AGT gene, the shaded areas being the coding regions. 1-4, the polymorphisms; 5-11, the mutations described in the text; ?, possible, but unconfirmed, mechanisms of action; CRM and ENZ are defined in the text.

Table 1
Mutations and Polymorphisms in the AGT Gene

PH1-specific mutations	Location	Amino acid substitution	Method of detection	Ref
G243→A	exon 1	Gly41→Arg	PCR and *Msp*I digestion	12
C320→G	exon 2	Tyr66→Ter	PCR and *Rsa*I digestion	15
G367→A	exon 2	Gly82→Glu	PCR and *Ava*I digestion	19
T576→A	exon 4	Phe152→Ile	PCR and *Mbo*I digestion	12
G630→A	exon 4	Gly170→Arg	PCR* and *Msp*I digestion	10
C682→T	exon 5	Ser187→Phe	PCR and *Bsm*I digestion	20
T735→C	exon 6	Ser205→Pro	PCR and *Sma*I digestion	21

Normal polymorphisms	Location	Amino acid substitution	Method of detection	
C154→T	exon 1	Pro11→Leu	PCR and *Sty*I digestion	10
A1142→G	exon 10	Ile340→Met	PCR and *Ava*II digestion	10
74bp duplication	intron 1	....	PCR	15
29/32bp VNTR	intron 4	....	*Pst*I digest + Southern blot	16

Data taken from reference 8; * special primer required.

2. Mutations in *AGXT*

Seven mutations have been described in the *AGXT* gene to date (Table 1). Six are expressed in mRNAs, while the seventh appears to be present on a non-expressed null allele. The mutations can best be described in the context of the enzymic phenotypes to which they lead (Figure 4).

a. Peroxisome-to-Mitochondrion Mislocalisation

Analysis of the *AGXT* gene from a CRM+/ENZ+ patient with the mistargeting phenotype identified the presence of a G630A point mutation which specifies a Gly170Arg amino acid replacement. Of the eight patients studied with the same enzymic phenotype, two were homozygous and six were heterozygous for this mutation.[10] In three of the heterozygotes, in whom RNA was available for analysis, only the mutant allele was expressed. Subsequent analysis of nearly 30 CRM+/ENZ+ patients with mitochondrial AGT has shown the presence, in all but two cases, of at least one allele containing the G630A mutation.[8] In all cases this mutation is present on the minor *AGXT* allele. The Gly170Arg mutation lies within a highly conserved internal region of 58 amino acids, and appears to act in concert with the

Pro11Leu polymorphism to achieve the peroxisome-to-mitochondrion mistargeting of AGT by an as yet unidentified mechanism.

Two CRM+/ENZ± patients have been identified with low levels of immunoreactive AGT protein, most of which is located in the mitochondria, whose genotype is somewhat different. These patients do not possess the G630A mutation, but instead they have a T576A mutation, again on the background of the minor *AGXT* alelle. This mutation, which is present on only one allele in each patient, specifies a Phe152Ile amino acid replacement and appears to have a similar effect to the Gly170Arg.[12]

b. Partial Mislocalisation and Intraperoxisomal Aggregation

Analysis of three CRM+/ENZ± patients with mitochondrial AGT and AGT aggregated into peroxisomal cores demonstrated complex genotypes.[12] All three patients were homozygous for the minor *AGXT* allele but compound heterozygotes for PH1-specific mutations, two patients possessing G243A + T576A (Gly41Arg + Phe152Ile) mutations and one possessing G243A + G630A (Gly41Arg + Gly170Arg) mutations. The mitochondrial mislocalisation of AGT could be accounted for by the presence of the Gly170Arg or Phe152Ile mutations in conjunction with the Pro11Leu polymorphism. On the other hand, the intraperoxisomal aggregation of AGT appeared to be the result of the Gly41Arg mutation.[12] The mechanism of action of the latter is currently unknown.

c. Normal Localisation but Loss of Catalytic Activity

Three unrelated CRM+/ENZ- patients were found to be homozygous for G367A mutation which would lead to a Gly82Glu amino acid replacement.[19] A number of other patients with the same enzymic phenotype were shown not to possess this mutation. Although the mechanism of action of the Gly82Glu substitution is not known, it might disrupt the secondary structure of the enzyme and interfere with the functioning of the active site.

d. Absence of Catalytic Activity and Immunoreactive Protein

A large proportion of PH1 patients have the overtly homogeneous CRM-/ENZ- phenotype. Such presentations could, however, arise from a variety of different types of mutation, including failure of transcription, nonsense mutations, unstable mRNA, or unstable protein. Of the two patients with this phenotype studied, both were found to have normal levels of mRNA but only trace amounts of immunoreactive AGT protein.[20,21] In one patient, cDNA analysis revealed the presence of a C682T transition on one allele which would lead to a Ser187Phe amino acid substitution.[20] The other allele was not expressed. A Japanese CRM-/ENZ- patient was shown to be homozygous for a T735C mutation which causes a Ser205Phe substitution.[21] These two mutations occur close together in the linear protein sequence and both change conserved amino acids. From the biochemical data, these mutations could be expected to play a role in decreasing protein stability, a finding that has recently been confirmed in the Japanese patient where the proteolytic degradation of the mutant AGT is accelerated.[22] A number of other CRM-/ENZ- patients have been screened for both of these mutations but no other examples have been found.[8]

e. Miscellaneous

A premature termination codon (Tyr66Ter) caused by a C320G mutation has been described in the non-expressed allele from a patient with the peroxisome-to-mitochondrion targeting defect.[15] The functional significance, if any, of this observation, which has so far only been found in one patient, is unclear.

II. CONSEQUENCES FOR CLINICAL MANAGEMENT

A. DIAGNOSIS

1. Clinical and Biochemical Diagnosis

Conventionally, diagnosis of PH1 is made following the clinical observation of renal CaOx deposition and its consequences, and elevated oxalate and glycolate excretion.[2,3] Secondary causes of renal CaOx crystalisation should be excluded. Although concomitant hyperoxaluria and hyperglycolic aciduria is indicative of PH1, the relative increases in oxalate and glycolate excretion vary considerably. Some PH1 patients (defined by AGT deficiency) do not have hyperglycolic aciduria.[6] Urinary oxalate and glycolate excretion becomes less reliable as a diagnostic indicator of PH1 as the patient approaches renal failure.

2. Enzymic Diagnosis

Definitive diagnosis of PH1 can be made, independently of renal function, by AGT analysis of liver biopsies.[23] Various methods have been developed to enable enzymic diagnosis to be achieved on as little as 2mg of liver tissue.[7,24-27] Unfortunately, another enzyme present in the liver, glutamate:glyoxylate aminotransferase (GGT), can also use alanine and glyoxylate as substrates.[28] Therefore, GGT has to be assayed along side AGT and a correction made for the crossover.[7] Immunoreactive AGT protein is much more stable than AGT catalytic activity. Therefore, as a general rule, diagnosis should be confirmed by measuring the amount of immunoreactive AGT protein by immunoblotting.[29]

A noted above, patients with AGT activities greater than 15-20% of the mean normal level cannot be distinguished from carriers on the basis of enzyme activity alone (see Figure 2). In these cases, it is essential that the intracellular distribution of immunoreactive AGT protein is ascertained by immunoelectron microscopy.

3. Molecular Genetic Diagnosis

Although it has not proved to be widely useful, it is possible to diagnose PH1 in some patients by mutational analysis. This approach is applicable possibly in only 5-10% of patients whose disease is caused by homozygosity or compound heterozygosity of previously identified mutations. For example, individuals who were shown to be homozygous for G630A or G367A (see above) could have been diagnosed in this manner, as could those who are compound heterozygotes for G630A + G243A or T576A + G243A. The main benefit of this method over AGT assays on liver biopsies is that it can be carried out on more easily accessible tissues, such as whole blood. All of the currently known mutations can be detected easily by PCR (polymerase chain reaction) and restriction digestion (see Table 1).

B. PRENATAL DIAGNOSIS
1. Biochemical Prenatal Diagnosis

Attempts to diagnose PH1 prenatally by the assay of oxalate or glycolate in amniotic fluid have not been successful.[30-32] The reason for this is unclear, but might be due to the metabolic defect not being manifested in the fetal liver, or due to the aberrant glyoxylate metabolites being cleared by the placenta more efficiently than by the fetal kidneys.

2. Enzymic Prenatal Diagnosis

Prenatal diagnosis can be performed by the measurement of AGT enzyme activity in fetal liver biopsies.[33-36] This procedure however requires a microassay, careful preservation of fetal tissue and, due to difficulties with the sampling procedure, cannot be carried out until at least 16 weeks gestation. Enzyme activity in fetal liver is lower than in postnatal liver, possibly due to the immaturity of hepatic enzyme systems. In addition, AGT activity varies with gestational age,[34] so that it is important to have a separate reference range for fetal tissue. To our knowledge, eight prenatal diagnoses have been carried out so far by hepatic AGT analysis, most having occurred well into the second trimester (20-24 weeks). Although delayed diagnosis is a major disadvantage of the technique, hepatic AGT assay coupled with immunoblotting and immunoelectron microscopy can identify all known PH1 enzymic phenotypes with no previous family studies.

3. Molecular Genetic Prenatal Diagnosis

An alternative approach utilises DNA analysis of chorionic villus biopsy at 10-12 weeks gestation or amniocytes at 16 weeks. Two methods of DNA analysis are possible employing either mutation detection or the indirect approach of linkage analysis.

a. Mutational Analysis

As no single common mutation causes all cases of PH1, mutation analysis is not feasible in many cases. There is however one mutation, G630A, which

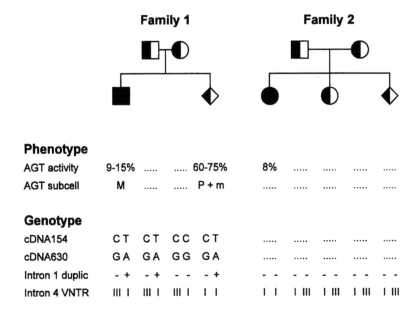

Figure 5. Approaches to the prenatal diagnosis of PH1. Data shows prenatal diagnosis in 2 families by enzyme, mutation and linkage analysis. In family 1, enzyme analysis implied that the fetus was either normal or a carrier, while mutational analysis (C154T + G630A) implied that the fetus was either affected or a carrier. The situation was resolved by intron 4 linkage analysis which showed the fetus to be a carrier. In family 2, enzyme and mutation analysis was not available. In this family, intron 1 linkage analysis was uninformative, but the intron 4 polymorphism showed that the fetus was a carrier. AGT activity is expressed as a percentage of the mean control level; M, mainly mitochondrial; P+m, mainly peroxisomal but with some mitochondrial. Genotypes are described in the text.

is found in a high proportion of patients with PH1 and which can be detected relatively easily by PCR amplification of DNA (see Table 1) and, therefore, can be included as part of a genetic screen.

b. Linkage Analysis

Linkage analysis, on the other hand, is more generally applicable than mutational analysis, as it requires no prior knowledge of the disease-causing mutation in a particular family. However, the procedure does require a family study with DNA available from the affected family member and parents. In some cases the affected child may have died. However, providing a sample of tissue has been stored either frozen or in the form of paraffin embedded material, it is possible to obtain sufficient material for analysis by PCR. Linkage analysis does require the identification of linkage markers either within or close to the disease gene. The intron 1 and intron 4 polymorphisms described above are both useful markers and have been used for prenatal diagnosis in four cases to date.[37,38] Examples of the various possible approaches to the prenatal diagnosis of PH1 are shown in Figure 5.

C. TREATMENT
1. Symptomatic treatment
Conventional management of PH1 attempts to delay the onset of renal failure by the restriction of dietary oxalate intake, the maintenance of high urine flow to minimise urinary CaOx supersaturation, and the administration of inhibitors of urinary CaOx crystallisation.[2,3,39] The eventual terminal failure of the kidney due to overwhelming CaOx deposition is usually followed by haemodialysis and renal transplantation. The only conventional treatment that actually addresses the basic cause of the disease is the administration of pharmacological doses of pyridoxine. Although pyridoxal phosphate is an essential cofactor of AGT, its mechanism of action in the minority of PH1 patients who are pyridoxine-responsive is not clearly understood.

2. Enzyme replacement therapy
Liver transplantation is now generally considered to be the definitive treatment in the majority of PH1 patients who are unresponsive to pyridoxine.[40-42] As AGT is expressed almost entirely within the liver in a normal individual, liver transplantation is almost ideal as a form of enzyme replacement therapy (ERT). Not only does it provide almost all of the body's requirement for AGT, but also it re-introduces the AGT in the correct organ (liver), cell (hepatocytes) and intracellular compartment (peroxisomes). In increasing numbers of patients, liver transplantation is able to provide an effective "cure", with the resolution of both biochemical and clinical symptoms. World-wide, well over 60 liver transplantations have been carried out for PH1, most being combined with kidney transplantations. In such cases, the liver corrects the biochemical defect while the kidney corrects the pathophysiological defect. Due to the slow resolubilisation of the systemic CaOx stores laid down during the years of compromised renal function, complete biochemical correction, as manifested by normalisation of urinary oxalate excretion, might take many years.[43,44] On the other hand, because glycolate does not form an insoluble calcium salt, urinary glycolate excretion is usually normalised very rapidly.[5]

The specific combination of enzymic and metabolic characteristics that are possibly unique to PH1 (Table 2) preclude the use of certain ERT strategies.[45] Auxiliary or partial liver transplants are not appropriate therapeutic options as they would not be expected to lead to significant decreases in hepatic oxalate production (Figure 6).[45,46] The AGT activity in the newly-introduced normal liver cells can in no way exert any influence on oxalate production in AGT-deficient liver cells remaining *in situ*. In some strategies, total hepatic oxalate production could even increase. Only complete orthotopic liver transplantation is likely to achieve the desired effect of normalising hepatic oxalate synthesis (Figure 6).

Table 2

**Enzymic & Metabolic Properties of PH1 that Determine the Effective
Strategies for Enzyme Replacement Therapy & Gene Therapy**

1. Metabolic abnormality is one of increased synthesis (of oxalate),
    not decreased degradation.

2. Pathologically accumulated material (oxalate) is not further metabolisable.

3. AGT-catalysed glyoxylate detoxification occurs mainly in hepatocyte peroxisomes.

4   Proteins cannot be taken up from the extracellular fluid and
    translocated to the peroxisomes.

3. Gene therapy

Although liver transplantation can be considered as a form of gene therapy as well as a form of ERT,[46] it is far from ideal, not least because it involves the replacement of thousands of perfectly normal genes and their products, just to replace the one that is abnormal. A more sensible approach to gene therapy would be to replace only the abnormal gene, leaving all the other normal ones intact.

As outlined previously,[46] there are a number of characteristics of PH1 that make it a suitable candidate for gene therapy. For example, at least at the level of genetic malfunction, if not at the level of clinical pathology, PH1 is a liver-specific disease. The liver represents an excellent target organ for gene therapy and a considerable amount of current research concerns the development of suitable vectors and clinical protocols aimed at maximising the expression of transgenes in hepatocytes. However, the same unique combination of enzymic and metabolic characteristics of PH1 that influence the strategies for ERT (Table 2) must also be taken into consideration when designing strategies for gene therapy.

Asymptomatic PH1 heterozygotes can have total liver AGT activities as low as 20% of the mean normal level[8] and, therefore, gene therapy should aim at this overall level of expression as a minimum. However, although only the liver needs to be targeted, it is important that the great majority of the hepatocytes (say >75%) take up and express the transgene (Figure 6). If only 10% of hepatocytes could be made to take up the transgene, the level of hepatic oxalate synthesis would hardly change, even if each transfected cell expressed 10 times the normal level of AGT (i.e. the total liver activity of AGT would be normal). The 90% of hepatocytes that fail to take up or express the AGT transgene will continue to synthesise oxalate at the same rate as before. In the present state of the art, whether *in vivo* or *ex vivo* approaches were to be used for the gene therapy of PH1, using retroviral or adenoviral vectors, treatment would need to be repeated many times in order

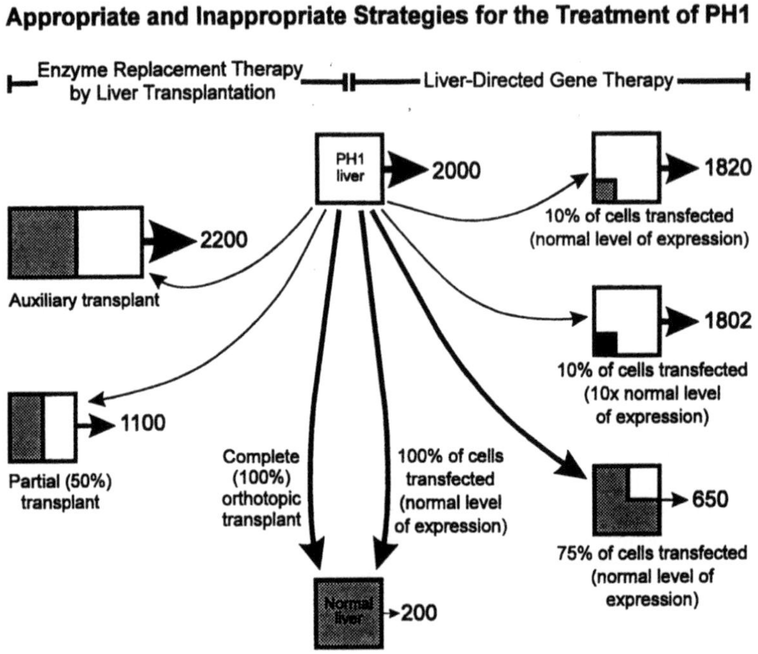

Appropriate and Inappropriate Strategies for the Treatment of PH1

Figure 6. Strategies for the treatment of PH1 by enzyme replacement therapy (left section) and gene therapy (right section). Numbers indicate an approximation of the daily rate of oxalate synthesis by the liver in μmol (based on typical daily excretion rates and assuming most of the urinary oxalate is derived from endogenous synthesis by the liver). Open boxes represent uncorrected PH1 liver cells; shaded boxes represent normal or completely corrected liver cells; black box represents liver cells over-expressing AGT. Favoured treatment strategies are denoted by the bold extended arrows.

to achieve the levels of AGT expression necessary to significantly decrease hepatic oxalate synthesis.

Despite the anticipated problems, gene therapy for PH1 is a distinct possibility within the next decade, the potential benefits being expected to outweigh the potential difficulties. Just as symptomatic approaches to the treatment of PH1, such as kidney transplantation, have slowly given way to more "rational" treatments aimed at tackling the basic cause of the disease, such as liver transplantation, so liver transplantation itself will give way to the more selective rational treatment of gene therapy.

REFERENCES

1. **Danpure, C.J. and Purdue, P.E.** Primary hyperoxaluria, in *The Metabolic and Molecular Bases of Inherited Disease*, 7th ed., Vol. 2, Scriver, C.R., Beaudet, A.L., Sly, W.S., and Valle, D., Eds., McGraw-Hill, New York 1995, chap. 75.

2. **Barratt, T.M. and Danpure, C.J.** Hyperoxaluria, in *Pediatric Nephrology*, 3rd ed., Holliday, M.A., Barratt, T.M., and Avner, E.D., Eds., Williams & Wilkins, Baltimore, 1994, chap. 29B.

3. **Danpure, C.J. and Smith, L.H.** The primary hyperoxalurias, in *Kidney stones: medical and surgical management*, Coe, F.L., Favus, M.J., Pak, C.Y., Parks, J., and Preminger, G., Eds., in press.

4. **Danpure, C.J. and Jennings, P.R.** Peroxisomal alanine:glyoxylate aminotransferase deficiency in primary hyperoxaluria type I. *FEBS Lett.* 201, 20, 1986.

5. **Danpure, C.J.** Recent advances in the understanding, diagnosis and treatment of primary hyperoxaluria type 1. *J. Inherited Metab. Dis.* 12, 210, 1989.

6. **Danpure, C.J.** Molecular and clinical heterogeneity in primary hyperoxaluria type 1. *Am. J. Kidney Dis.* 17, 366, 1991.

7. **Danpure, C.J. and Jennings, P.R.** Further studies on the activity and subcellular distribution of alanine:glyoxylate aminotransferase in the livers of patients with primary hyperoxaluria type 1. *Clin. Sci.* 75, 315, 1988.

8. **Danpure, C.J., Jennings, P.R., Fryer, P., Purdue, P.E., and Allsop, J.** Primary hyperoxaluria type 1 : genotypic and phenotypic heterogeneity. *J. Inherit. Metab. Dis.* 17, 487, 1994.

9. **Cooper, P.J., Danpure, C.J., Wise, P.J., and Guttridge, K.M.** Immunocytochemical localization of human hepatic alanine: glyoxylate aminotransferase in control subjects and patients with primary hyperoxaluria type 1. *J. Histochem. Cytochem.* 36, 1285, 1988.

10. **Purdue, P.E., Takada, Y., and Danpure, C.J.** Identification of mutations associated with peroxisome-to-mitochondrion mistargeting of alanine/glyoxylate aminotransferase in primary hyperoxaluria type 1. *J. Cell Biol.* 111, 2341, 1990.

11. **Danpure, C.J., Cooper, P.J., Wise, P.J., and Jennings, P.R.** An enzyme trafficking defect in two patients with primary hyperoxaluria type 1: peroxisomal alanine/glyoxylate aminotransferase rerouted to mitochondria. *J. Cell Biol.* 108, 1345, 1989.

12. **Danpure, C.J., Purdue, P.E., Fryer, P. et al.** Enzymological and mutational analysis of a complex primary hyperoxaluria type 1 phenotype involving alanine:glyoxylate aminotransferase peroxisome-to-mitochondrion mistargeting and intraperoxisomal aggregation. *Am. J. Hum. Genet.* 53, 417, 1993.

13. **Purdue, P.E., Lumb, M.J., Fox, M., Griffo, G., Hamon Benais, C., Povey, S., and Danpure, C.J.** Characterization and chromosomal mapping of a genomic clone encoding human alanine:glyoxylate aminotransferase. *Genomics.* 10, 34, 1991.

14. **Takada, Y., Kaneko, N., Esumi, H., Purdue, P.E., and Danpure, C.J.** Human peroxisomal L-alanine: glyoxylate aminotransferase. Evolutionary loss of a mitochondrial targeting signal by point mutation of the initiation codon. *Biochem. J.* 268, 517, 1990.

15. **Purdue, P.E., Lumb, M.J., Allsop, J., and Danpure, C.J.** An intronic duplication in the alanine: glyoxylate aminotransferase gene facilitates identification of mutations in compound heterozygote patients with primary hyperoxaluria type 1. *Hum. Genet.* 87, 394, 1991.

16. **Danpure, C.J., Birdsey, G.M., Rumsby, G., Lumb, M.J., Purdue, P.E., and Allsop, J.** Molecular characterization and clinical use of a polymorphic tandem repeat in an intron of the human alanine:glyoxylate aminotransferase gene. *Hum. Genet.* 94, 55, 1994.

17. **Purdue, P.E., Allsop, J., Isaya, G., Rosenberg, L.E., and Danpure, C.J.** Mistargeting of peroxisomal L-alanine:glyoxylate aminotransferase to mitochondria in primary hyperoxaluria patients depends upon activation of a cryptic mitochondrial targeting sequence by a point mutation. *Proc. Natl. Acad. Sci. U. S. A.* 88, 10900, 1991.

18. **Rumsby, G., Jones, R., Danpure, C.J., and Samuell, C.T.** TaqI polymorphism of the alanine:glyoxylate aminotransferase (AGXT) locus. *Hum. Mol. Genet.* 1, 350, 1992.

19. **Purdue, P.E., Lumb, M.J., Allsop, J., Minatogawa, Y., and Danpure, C.J.** A glycine-to-glutamate substitution abolishes alanine:glyoxylate aminotransferase catalytic activity in a subset of patients with primary hyperoxaluria type 1. *Genomics.* 13, 215, 1992.

20. **Minatogawa, Y., Tone, S., Allsop, J., Purdue, P.E., Takada, Y., Danpure, C.J., and Kido, R.** A serine-to-phenylalanine substitution leads to loss of alanine:glyoxylate

aminotransferase catalytic activity and immunoreactivity in a patient with primary hyperoxaluria type 1. *Hum. Mol. Genet.* 1, 643, 1992.

21. **Nishiyama, K., Funai, T., Katafuchi, R., Hattori, F., Onoyama, K., and Ichiyama, A.** Primary hyperoxaluria type I due to a point mutation of T to C in the coding region of the serine:pyruvate aminotransferase gene. *Biochem. Biophys. Res. Commun.* 176, 1093, 1991.

22. **Nishiyama, K., Funai, T., Yokota, S., and Ichiyama, A.** ATP-dependent degradation of a mutant serine: pyruvate/alanine: glyoxylate aminotransferase in a primary hyperoxaluria type 1 case. *J. Cell Biol.* 123, 1237, 1993.

23. **Danpure, C.J., Jennings, P.R., and Watts, R.W.** Enzymological diagnosis of primary hyperoxaluria type 1 by measurement of hepatic alanine: glyoxylate aminotransferase activity. *Lancet.* 1, 289, 1987.

24. **Allsop, J., Jennings, P.R., and Danpure, C.J.** A new micro-assay for human liver alanine: glyoxylate aminotransferase. *Clin. Chim. Acta.* 170, 187, 1987.

25. **Wanders, R.J., Ruiter, J., van Roermund, C.W., Schutgens, R.B., Ofman, R., Jurriaans, S., and Tager, J.M.** Human liver L-alanine-glyoxylate aminotransferase: characteristics and activity in controls and hyperoxaluria type I patients using a simple spectrophotometric method. *Clin. Chim. Acta.* 189, 139, 1990.

26. **Toone, J.R. and Applegarth, D.A.** Micromethod for the assay of glutamate: glyoxylate aminotransferase and modifications of a micromethod for the assay of alanine: glyoxylate aminotransferase. Implications for the prenatal diagnosis of type I hyperoxaluria by fetal liver biopsy. *Clin. Chim. Acta.* 203, 105, 1991.

27. **Petrarulo, M., Pellegrino, S., Marangella, M., Cosseddu, D., and Linari, F.** High-performance liquid chromatographic microassay for L-alanine: glyoxylate aminotransferase activity in human liver. *Clin. Chim. Acta.* 208, 183, 1992.

28. **Thompson, J.S. and Richardson, K.E.** Isolation and characterization of glutamate: glycine transaminase from human liver. *Arch. Biochem. Biophys.* 117, 599, 1966.

29. **Wise, P.J., Danpure, C.J., and Jennings, P.R.** Immunological heterogeneity of hepatic alanine:glyoxylate aminotransferase in primary hyperoxaluria type 1. *FEBS Lett.* 222, 17, 1987.

30. **Rose, G.A., Arthur, L.J., Chambers, T.L., Kasidas, G.P., and Scott, I.V.** Successful treatment of primary hyperoxaluria in neonate. *Lancet.* 1, 1298, 1982.

31. **Leumann, E., Matasovic, A., and Niederwieser, A.** Primary hyperoxaluria type I: oxalate and glycolate unsuitable for prenatal diagnosis. *Lancet.* 2, 340, 1986.

32. **Leumann, E.P., Niederwieser, A., and Fanconi, A.** New aspects of infantile oxalosis. *Pediatr. Nephrol.* 1, 531, 1987.

33. **Danpure, C.J., Jennings, P.R., Penketh, R.J., Wise, P.J., and Rodeck, C.H.** Prenatal exclusion of primary hyperoxaluria type 1. *Lancet.* 1, 367, 1988.

34. **Danpure, C.J., Jennings, P.R., Penketh, R.J., Wise, P.J., Cooper, P.J., and Rodeck, C.H.** Fetal liver alanine: glyoxylate aminotransferase and the prenatal diagnosis of primary hyperoxaluria type 1. *Prenat. Diagn.* 9, 271, 1989.

35. **Danpure, C.J., Cooper, P.J., Jennings, P.R., Wise, P.J., Penketh, R.J., and Rodeck, C.H.** Enzymatic prenatal diagnosis of primary hyperoxaluria type 1: potential and limitations. *J. Inherited Metab. Dis.* 12 Suppl 2, 286, 1989.

36. **Illum, N., Lavard, L., Danpure, C.J., Horn, T., Aerenlund Jensen, H., and Skovby, F.** Primary hyperoxaluria type 1: clinical manifestations in infancy and prenatal diagnosis. *Child Nephrol. Urol.* 12, 225, 1992.

37. **Rumsby, G., Uttley, W.S., and Kirk, J.M.** First trimester diagnosis of primary hyperoxaluria type 1. *Lancet.* 344, 1018, 1994.

38. **Rumsby, G., Mandel, H., Avey, C., and Geraerts, A.** Polymorphisms in the human alanine:glyoxylate aminotransferase gene and their application to the prenatal diagnosis of primary hyperoxaluria type 1. *Nephrol. Dial. Transpl.* in press.

39. **Milliner, D.S., Eickholt, J.T., Bergstralh, E., Wilson, D.M., and Smith, L.H.** Primary hyperoxaluria: results of long-term treatment with orthophosphate and pyridoxine. *N. Engl. J. Med.* in press, 1994.

40. **Watts, R.W., Calne, R.Y., Rolles, K., Danpure, C.J., Morgan, S.H., Mansell, M.A., Williams, R., and Purkiss, P.** Successful treatment of primary hyperoxaluria type I by combined hepatic and renal transplantation. *Lancet.* 2, 474, 1987.

41. **Watts, R.W., Morgan, S.H., Danpure, C.J. et al.** Combined hepatic and renal transplantation in primary hyperoxaluria type I: clinical report of nine cases. *Am. J. Med.* 90, 179, 1991.

42. **Watts, R.W.E., Danpure, C.J., de Pauw, L., and Toussaint, C.** Combined liver-kidney and isolated liver transplantations for primary hyperoxaluria type 1. The European experience. *Nephrol. Dial. Transpl.* 6, 502, 1991.

43. **de Pauw, L., Gelin, M., Danpure, C.J., Vereerstraeten, P., Adler, M., Abramowicz, D., and Toussaint, C.** Combined liver-kidney transplantation in primary hyperoxaluria type 1. *Transplantation.* 50, 886, 1990.

44. **Toussaint, C., de Pauw, L., Vienne, A., Gevenois, P.A., Quintin, J., Gelin, M., and Pasteels, J.L.** Radiological and histological improvement of oxalate osteopathy after combined liver-kidney transplantation in primary hyperoxaluria type 1. *Am. J. Kidney Dis.* 21, 54, 1993.

45. **Danpure, C.J.** Scientific rationale for hepatorenal transplantation in primary hyperoxaluria type 1, in *Transplantation and Clinical Immunology* (vol 22), Touraine, J.L., Ed., Excerpta Medica., Amsterdam, 1991, 91.

46. **Danpure, C.J.** Advances in the enzymology and molecular genetics of primary hyperoxaluria type 1. Prospects for gene therapy. *Nephrol. Dial. Transpl.* in press.

Chapter 10

CELLULAR ABNORMALITIES OF OXALATE TRANSPORT IN NEPHROLITHIASIS

Bruno Baggio, M.D., D.Sc. and Giovanni Gambaro, M.D., Ph.D.

No less than 76% of all human kidney stones are made up of calcium oxalate (CaOx), and 80% of stone patients have the idiopathic form of nephrolithiasis[1]. The physico-chemical theory of lithogenesis explains stone formation by the precipitation, growth and crystalline aggregation of several lithogenic salts in the urine, and has contributed greatly to the understanding of the complex problem of the etiology and pathogenesis of calcium urolithiasis[2]. Calcium oxalate and calcium phosphate are potentially the most insoluble lithogenic salts under the ionic conditions normally present in urine; indeed, their supersaturation level even in the urine of non-stone forming subjects is very close to the point of spontaneous precipitation.

The risk of forming CaOx stones is determined by the degree of calcium salt supersaturation, and the activity level of factors stabilizing (inhibitors) and destabilizing (promotors) their nucleation, crystal-growth and crystal-aggregation[3]. Idiopathic calcium nephrolithiasis (ICN) is currently interpreted as the consequence of an imbalance between these factors[3,4].

The principal determinants of urine calcium salt saturation are urine pH, calcium and oxalate excretion, and diuresis. Most studies addressed the role of calcium ion in promoting ICN. In the last decade, however, the pathogenetic role of oxalate has been revaluated in the light of physico-chemical considerations and clinical findings[5]. Indeed, minimal changes in urinary oxalate concentration produce increase in CaOx supersaturation than equal variations in calcium concentration[6]. Moreover, severity of disease and entity of crystalluria are well correlated with oxaluria, rather than calciuria[7]. The causes that lead to an imbalance between saturation and urinary inhibition might reside in systemic, metabolic, or renal anomalies. The latter are certainly very important in the idiopathic forms of nephrolithiasis; in fact, primary tubular loss of calcium[8], reduced urinary citrate excretion due in part to partial defect in renal acidification[9], and altered excretion of renal tubular epithelium constituents, such as renal enzymes[10,11], glycosaminoglycans (GAGs)[4,12], and Thamm-Horsfall protein[13] have been often reported. Several studies addressed whether the tubular anomalies were primitive and therefore pathogenetically important, or only a consequence of stone formation, but no firm conclusion was reached[14];

0-8493-7673-4/95/$0.00+$.50

nonetheless, there is evidence that at least some anomalies are primitive in nature.

Considering the co-existence of familial occurence, tubular disease associated with ICN, and anomalies in the systemic handling of oxalate, a cellular hypothesis of ICN was proposed. The first observation of an abnormal transmembrane flux rate of oxalate in red blood cell (RBC) in a group of patients with idiopathic CaOx nephrolithiasis was presented in 1984[15]. Two years later, this finding was confirmed in a larger number of patients; it was also reported evidence that the defect is characteristic of primary CaOx stone disease, is not present in secondary forms of calcium nephrolithiasis, can be corrected by diuretics, and is determined genetically as an autosomal, monogenic, dominant trait with complete penetration and variable expressivity[16]. Narula et al[17], in a group of Indian nephrolithiasic patients, Jenkins et al[18] and Motola et al[19] in renal stone patients in the United States, and Takahiro et al[20] in Japanese patients confirmed the existence of an abnormal RBC oxalate fllux.

RBCs have long been a favourite model for the study of membrane structure and function, and many transport systems were first identified in these cells; it is therefore conceivable that a transport pathway present in RBC may also be found in other cells. Indeed, in renal cortical and papillary epithelial cells[21,22], in colon epithelium[23] the oxalate transport shares many features with the RBC flux. So it is not surprising that, when in looking for a pathogenetical link between the RBC defect and renal stone disease, a faster intestinal absorption[16], and a higher renal clearance (unpublished) of oxalate were shown to be associated with increased RBC oxalate flux.

Albeit the oxalate flux rate is not related with calciuria, uricuria and a family history of nephrolithiasis, it is associated with increased oxalate renal clearance (unpublished) and the recurrence of urolithiasis[24]. These findings suggest that the cellular oxalate anomaly might be considered a marker of recurrent idiopathic CaOx nephrolithiasis. This hypothesis was confirmed by a 7-year follow-up study in the 5 families previously described[16], disclosing 5 new cases of renal lithiasis and only in those subjects with the cellular anomaly of oxalate transport[24].

This cellular anomaly is not restricted to oxalate, but involves the transport of other anions and cations. In fact, the evaluation of the erythrocyte urate transport in 67 idiopathic CaOx stone formers revealed the presence of an abnormal urate self-exchange in 30% of the patients[25]. The urate flux was found to be abnormal in two young stone free sons of two stone former probands with a high RBC flux, which suggests that the urate flux abnormality, like the oxalate defect, might be genetically determined. In stone formers, a direct correlation between transmembrane urate flux and 24 hr urinary excretion of uric acid was observed, and patients with the RBC urate self-exchange anomaly showed a more intense activity, in terms of disease recurrence. A morphazinamide test in 6 nephrolithiasic patients with abnormal urate flux promoted a significant fall in urate excretion in control subjects, but had no effect in the renal stone patients. These findings suggest that hyperuricosuria, which is very common ICN and might play a pathogenetic role in CaOx stone formation, could be due to a cellular defect in transmembrane urate transport. Furthermore, this study provides evidence for a physiopathological link between the erythrocyte and tubular anomalies, and thus between an abnormal RBC anion flux and renal stone disease.

Recently, in about 60% of renal stone formers, an abnormal kinetics of erythrocyte Na/K/2Cl cotransport has been reported, suggesting a reduced affinity of the carrier for sodium[26]. Since the cotransport for sodium is inhibited by loop diuretics, we performed a furosemide test to verify the hypothesis that stone formers have an abnormal renal cotransport. The administration of furosemide induced a smaller increase in the fractional excretion of sodium in patients, compared to controls, suggesting a decreased Na/K/2Cl cotransport activity at the renal tubular level as well. The inhibited activity of the Na/K/2Cl cotransport in the thick ascending limb of the Henle's loop, by reducing the transepithelial potential difference, should decrease the paracellular tubular reabsorption of calcium, thus leading to hypercalciuria. This is supported by the negative correlation observed between 24 h urine calcium excretion and the RBC cotransport activity. These observations, demonstrating an abnormal Na/K/2Cl cotransport activity in RBC and kidney of nephrolithiasic patients, suggest that this anomaly might be relevant for stone formation, by hampering renal calcium reabsorption in the distal nephron, and thus determining critical physical-chemical conditions for CaOx crystallization.

This cellular scenario, which involves both anion and cation transport, further complicated the picture of the pathogenesis of nephrolithiasis, and prompted to modify the original hypothesis of a cellular defect restricted to oxalate transport. The updated opinion is that these findings, as a whole, support the idea that ion transport anomalies might be the expression of a still unknown cellular defect in idiopathic calcium stone formers. The crucial problem is to discern whether these cellular alterations had a single matrix, and, if this is true, which is the primary defect.

A possible way to reach this objective is to consider that ion transport is generally accepted as an energy dependent process, and is modulated by the phosphorylation state of many erythrocyte membrane carrier proteins. In this regard, it is worth recalling that in the initial studies on the transmembrane oxalate transport, evidence of an anomaly in some membrane proteins at the phosphorylation level, namely bands 2 (spectrin) and 3 (anion carrier) was presented[27]. Many observations support a functional correlation between membrane protein phosphorylation and the anion transport. Diuretics (hydrochlorothiazide and amiloride), which after oral administration restore the red cell oxalate exchange rate to nearly normal[16], *in vitro*, while inhibiting the oxalate exchange, reduce the band 3 phosphorylation level similarly to the anion transport inhibitor DIDS[28,29]; furthermore, the red cell oxalate transport seems to depend upon the intracellular energy level, as ATP depletion induces a reduction in oxalate flux[27]. Moreover, Pewitt has recently reported that the activity of Na/K/2Cl cotransport in avian erythrocytes is also regulated by the phosphorylation and dephosphorylation of a protein with a molecular weight of 150.000 Da[30].

Thus, to explain the origin of the erythrocyte anomalies associated with nephrolithiasis, 2 possible mechanisms may be forwarded: 1) a primary structural defect in some membrane proteins, leading to the exposure of additional sites for phosphorylation, or 2) an imbalance between protein kinase and phosphatase activity responsible of the phosphorylation level of the membrane proteins.

Regarding the first possibility, it is extremely unlikely that a primary defect could simultaneously involve several proteins, that are all abnormally phosphorylated and/or working in stone formers. Nonetheless, this aspect was addressed by evaluating the possible existence of a link between the increase in RBC oxalate flux and chromosome 17, where the band 3 gene is located; the absence of such a linkage would exclude band 3 primary anomalies as the culprit of the abnormal flux oxalate. Two three-generation families, in which the oxalate transport anomaly was present, were studied using different DNA polymorphic 17 q markers. No evidence of linkage with these markers was observed, suggesting that the abnormal erythrocyte self-exchange of oxalate is not linked with the band 3 gene, and thus supported the idea that a primary anomaly of this protein is not crucial for the expression of the oxalate anomaly[31].

The hypothesis of an imbalance between some proteinkinases and phosphatases seems more promising. Workers in this field are debating which kinases are involved in the phosphorylation of these proteins; in other words, whether the Serine/Threonine protein kinases cAMP dependent or the cAMP-independent, such as caseinkinase, or proteinkinase C or Ca^{2+}-calmodulin kinase are responsible for ^{32}P labeling in the band 2 and 3 proteins. Experiments to evaluate this aspect suggest that band 3 phosphorylation level and its carrier function are independent of cyclic AMP kinases and proteinkinase C activity[32].

This hypothesis was reinforced by a recent study demonstrating that, when human erythrocytes are treated with okadaic acid, a known inhibitor of casein and tyrosine phosphatases, they not only displayed an increased level of band 3 and 2 phosphorylation, but also faster transmembrane oxalate flux and Na/K/2Cl cotransport activity[33]. On the other hand, in a parallel study it has been shown that the intestinal oxalate uptake is not modulated by proteinkinase C, but by other proteinkinases, in particular by the Ca^{2+}-calmodulin kinase. Indeed, intestinal oxalate uptake is not modified by staurosporine, a specific inhibitor of proteinkinase C, while it is reduced by W-7, a known calmodulin inhibitor[34]. This last finding seems very interesting, because the Ca^{2+}-calmodulin kinase in human RBC is able to modulate both oxalate transport and Na/K/2Cl cotransport activity, suggesting that anion and cation carriers might depend on the same regulatory mechanism. In the light of these observations demonstrating the role of caseinkinases in the modulation of the phosphorylation of band 2 and 3 levels, and the anion transport, its activity was evaluated in RBC of stone forming patients and a significant higher caseinkinase activity in patients was observed (unpublished data).

The putative imbalance in kinase activities, demonstrated in RBC of renal stone formers, might either involve the enzymes themselves or modulating effectors of their activities. However, the recognition of differences in carrier protein isoforms between different tissues (in the kidney and the gut) and RBCs, precludes the assumption that a protein which is structurally abnormal in RBC would be correspondingly structurally abnormal in the kidney, for example. Therefore, it is intriguing to explain the systemic distribution of the ion trasport defects. The favourite idea is that the complex array of ion flux cell abnormalities observed in ICN is epiphenomenon of a still unknown primitive anomaly in the composition of cell membranes, the *milieu* in which carrier proteins work, possibly leading to anomalous functioning of one or more regulatory steps in the energizing mechanism of membrane carriers. In exploring this possibility, the attention was focused on cell membrane GAGs and lipids.

GAGs are considered important inhibitors of lithogenesis, and their urinary excretion is frequently reduced in nephrolithiasic patients[4,12]. On the other hand, GAGs are potent inhibitors of some proteinkinases, such as caseinkinase and tyrosinekinase[35]. Compared to control subjects, it is noteworthy that renal stone formers show a significantly lower erythrocyte GAG content, which is inversely correlated with both the transmembrane oxalate flux rate and the erythrocyte membrane protein phosphorylation rate[36]. Moreover the administration of some GAGs *in vitro* and *in vivo* promotes a significant reduction in erythrocyte oxalate self-exchange, a reduction in band 3 and 2 protein phosphorylation, and furthermore, a reduction in urinary oxalate excretion[35,37]

The rationale to study membrane lipids in stone formers is their recognized role as second cell messengers, and in the modulation of membrane protein functions[38-40]; furthermore an alteration in the urine excretion of PGE_2, a phospholipid metabolite, has been described in calcium nephrolithiasis[41-43]; finally, dietary fish oil supplements, which affect serum and cell membrane phospholipid fatty acid composition, reduce urine calcium and oxalate excretion by a still unknown mechanism[44,45]. Quite recently an anomalous arachidonic acid content in the plasma and erythrocyte membrane phospholipids of renal calcium stone patients was observed (unpublished). Moreover, a dietary interventional trial with n-3 polyunsaturated fatty acid ethyl esters, and *in vitro* experiments, in which arachidonic acid release in cell membrane was obtained with phospholipase A_2, disclosed that the membrane arachidonic acid has the potential to modulate the oxalate membrane transport.

These findings suggest that GAGs and lipids play an extensive role in nephrolithiasic patients and raise the possibility that a cellular derangement in their metabolism may be responsible for the cellular anomalies associated to renal stone disease.

In conclusion, trying to summarize the data concerning the cellular abnormalities observed in ICN, it is possible to propose that:

1) oxalate transport is not the only one anomaly discovered in RBCs of renal stone forming patients; alterations in the transport of other anions, like urate, and some cations are also present;

2) these ion transport anomalies might be the expression of a systemic cellular defect in idiopathic calcium stone formers;

3) the systemic cellular defect may have an important pathogenetical role in CaOx lithogenesis, by modifying the intestinal and renal handling of relevant ions, and thus leading to urine salt supersaturation;

4) there are reasons to believe that the pathogenesis of the erythrocyte anomalies involves most likely a primary defect in some membrane component.

REFERENCES

1. **Nordin, B.E.C., Hodgkinson, A., Peacock, M., Robertson, W.G.,** Urinary tract calculi, in *Nephrology*, Hamburger, J., Crosnier, J., Grunfeld, J.P., Eds., Wiley, New York and Paris, 1979, 1091.

2. **Fleisch, H.,** Inhibitors and promoters of stones, *Kidney. Int.*, 13, 361, 1978.

3. **Robertson, W.G., Peacock, M., Heyburn, P.J., Marshall, D.H., Clark, P.B.,** Risk factors in calcium stone disease of the urinary tract, *British J. Urol.*, 50, 449, 1978.

4. **Baggio, B., Gambaro, G., Oliva, O., Favaro, S., Borsatti, A.,** Calcium oxalate nephrolithiasis: an easy way to detect an imbalance between promoting and inhibiting factors, *Clin. Chim. Acta*, 124, 149, 1982.

5. **Smith, L.H.,** Hyperoxaluria, in Walker, V.R., Sutton, R.A.L., in *Urolithiasis*, Cameron, E.C.B., Pak, C.Y.C., Robertson, W.G., Eds., Plenum Press, New York, 1988, 405.

6. **Finlayson, B.,** Renal lithiasis in review, *Urol. Clin. N. Am.*, 1, 181, 1974.

7. **Robertson., W,G., Peacock, M.,** The cause of idiopathic calcium stone disease: hypercalciuria or hyperoxaluria?, *Nephron*, 26, 105, 1980.

8. **Pak, C.Y.C.,** Physiological basis for absorptive and renal hypercalciuria, *Am. J. Physiol.*, 237, F415, 1979.

9. **Minisola, S., Rossi, W., Pacitti, T.M., Scarnecchia, L., Bigi, F., Carnevale, V., Mazzuoli, G.,** Studies on citrate metabolism in normal subjects and kidney stone patients, *Miner. Electrol. Metab.*, 15, 303, 1989.

10. **Baggio, B., Gambaro, G., Ossi, E., Favaro, S., Borsatti, A.,** Increased urinary excretion of renal enzymes in idiopathic calcium oxalate nephrolithiasis, *J. Urol.*, 129, 1161, 1983.

11. **Khan, S.R., Shevock, P.N., Hackett, R.L.,** Urinary enzymes and calcium oxalate urolithiasis, *J. Urol.*, 142, 846, 1989.

12. **Baggio, B., Gambaro, G., Cicerello, E., Mastrosimone, S., Marzaro, G., Borsatti, A., Pagano,F.,** Urinary excretion of glycosaminoglycans in urological disease, *Clin. Biochem.*, 20, 449, 1987.

13. **Gambaro, G., Baggio, B., Favaro, S., Cicerello, E., Borsatti, A.,** Role de la mucoproteine de Tamm-Horsfall dans la lithogenese oxalique-calcique, *Nephrologie*, 5, 171, 1984.

14. **Jaeger, P., Portmann, L., Ginalsky, J.M., Jacquet, A.F., Temler, E., Burckardt, P.,** Tubulopathy in nephrolithiasis: Consequence rather than cause. *Kidney Int.*, 29, 563, 1986.

15. **Baggio, B., Gambaro, G., Marchini, F., Cicerello, E., Borsatti, A.,** Raised transmembrane oxalate flux in red blood cells in idiopathic calcium oxalate nephrolithiasis, *Lancet*, ii, 12, 1984.

16. **Baggio, B., Gambaro, G., Marchini, F., Cicerello, E., Tenconi, R., Clementi, M., Borsatti, A.,** An inheritable anomaly of red-cell oxalate transport in "primary" calcium nephrolithiasis correctable with diuretics, *N. Engl. J. Med.*, 314, 599, 1986.

17. **Narula, R., Sharma, S.H., Sidhu, H., Thind, .K., Nath, R,** Transport of oxalate in intact red blood cell can identify potential stone-formers, *Urol. Res.*, 16, 193, 1988.

18. **Jenkins, A.D., Langley, M.J., Bobbitt, M.W.,** Red blood cell oxalate flux in patients with calcium urolithiasis, *Urol. Res.*, 16, 209, 1988.

19. **Motola, J.A., Urivetsky, M., Molia, L., Smith, A.D.,** Transmembrane oxalate exchange: its relationship to idiopathic calcium oxalate nephrolithiasis, *J. Urol.*, 147, 549, 1992.

20. **Takahiro, K., Yamakawa, K., Kawamura, J.,** Erythrocyte oxalate self-exchange rate in Japan, in *Proc. 7th Int. Symp. on Urolithiasis,* Cairns, Australia, 1992, 21.

21. **Sigmon, D., Sanjaya, K., Carpenter, M.A., Miller, T., Menon, M., Scheid, C.,** Oxalate transport in renal tubular cells from normal and stone-forming animals, *Am. J. Kidney Dis.*, 17, 376, 1991.

22. **Wandzilak, T.R., Calò, L., D'Andre, S., Borsatti, A., Williams, H.E.**, Oxalate transport in cultured porcine renal epithelial cells, *Urol. Res.*, 20, 341, 1992.

23. **Hatch, M., Freel, R.W., Vaziri, N.D.**, Mechanisms of oxalate absorption and secretion across the rabbit distal colon, *Pflügers Arch.*, 426, 101, 1994.

24. **Gambaro, G., Marchini, F., Vincenti, M., Budakovic, A., Nardellotto, A., Baggio, B.**, Clinical features of idiopathic calcium-oxalate nephrolithiasis associated with the anomalous erythrocyte self-exchange of oxalate, *Proc. 5th Eur. Urolithiasis Symp.*, Manchester, U.K., 1994, 67.

25. **Gambaro, G., Vincenti, M., Marchini, F., D'Angelo, A., Baggio, B.**, Abnormal urate transport in erythrocytes of idiopathic calcium nephrolithiasis: a possible link with hyperuricosuria, *Clin. Sci.*, 85, 41, 1993.

26. **Baggio, B., Gambaro, G., Marchini,F., Vincenti, M., Ceolotto, G., Pessina, A.C., Semplicini, A.**, Abnormal erythrocyte and renal frusemide-sensitive sodium transport in idiopathic calcium nephrolithiasis, *Clin. Sci.*, 86, 239, 1994.

27. **Baggio, B., Clari,G., Marzaro, G., Gambaro, G., Borsatti, A., Moret, V.**, Altered red blood cell membrane protein phosphorylation in idiopathic calcium oxalate nephrolithiasis, *IRCS Med. Sci.*, 14, 368, 1986.

28. **Clari, G., Baggio, B., Marzaro, G., Gambaro, G., Borsatti, A., Moret, V.**, Phosphorylation of band 3 protein in nephrolithiasis, *Ann. N. Y. Acad. Sci.*, 488,533,1986.

29. **Baggio, B., Marzaro, G., Gambaro, G., Marchini, F., Borsatti, A., Clari, G.**, Effect of thiazides and amiloride on the phospho-rylation status of red cell membrane anion carrier, in *Diuretics: Basic, Pharmacological, and Clinical Aspects,* Andreucci, V.E. and Dal Canton, A., Eds., Martinus Nijhoff Publishing, Boston, 1987, 65.

30. **Pewitt, E.B., Hegde, R.S., Palfrey, H.C.**, (^{3}H)bumetanide binding to avian erythrocyte membranes. Correlation with activation and deactivation of Na/K/2Cl cotransport, *J. Biol. Chem.*, 265, 20747, 1990.

31. **Gambaro, G., Danieli, G.A., Borsatti, A., Marchini, F., Baggio, B.**, Band 3 gene is not linked with the abnormal RBC oxalate self-exchange in nepholithiasis, in *Proc. 12th Int. Cong. Nephrol.*, Jerusalem, Israel, 1993, 624.

32. **Baggio, B., Bordin, L., Gambaro, G., Piccoli, A., Marzaro, G., Clari, G.**, Evidence of a link between erythrocyte band 3 phosphorylation and anion transport in patients with "idiopathic" calcium oxalate nephrolithiasis, *Miner. Electrolyte Metab.*, 19, 17, 1993.

33. **Baggio., B., Bordin, L., Clari, G., Gambaro, G., Moret, V.,** Functional correlation   between   the Ser/Thr-phosphorylation of band-3 and band-3 mediated transmembrane anion transport in human erythrocytes, *Biochim. Biophys. Acta,* 1148, 157, 1993.
34. **Gambaro, G., Vincenti, M., Baggio, B., Calderaro, E.,** Oxalate transport in the intestinal epithelium and in the erythrocyte, *J. Am. Soc. Nephrol.,* 14, 707, 1993.
35. **Baggio. B., Gambaro, G., Marzaro, G., Marchini, F., Borsatti, A., Crepaldi, G.,** Effects of the oral administration of glycosaminoglycans on cellular abnormalities associated with idiopathic calcium oxalate nephrolithiasis, *Eur. J. Clin. Pharmacol.,* 40, 237, 1991.
36. **Baggio, B., Marzaro, G., Gambaro, G., Marchini, F., Williams, H.E., Borsatti, A.,** Glycosaminoglycan content, oxalate self-exchange and protein phosphorylation in erythrocytes of patients with "idiopathic" calcium oxalate nephrolithiasis, *Clin. Sci.* 79, 113, 1990.
37. **Baggio, B., Gambaro, G., Marchini, F., Marzaro, G., Williams, H.E., Borsatti, A.,** Correction of erythrocyte abnormalities in idiopathic calcium oxalate nephrolithiasis and reduction of urinary oxalate by oral glycosaminoglycans, *Lancet,* 338, 403, 1991.
38. **Pelech, S.L, Vance, D.E.,** Signal transduction via phosphatidylcholine cycles, *TIBS,* 14, 28-30, 1989.
39. **Liscovitch, M., Cantley, L.C.,** Lipid second messengers, *Cell, 77,* 329, 1994.
40. **Exton, J.H.** Phosphatidylcholine breakdown and signal transduction, *Biochim. Biophys. Acta* 1212, 26, 1994.
41. **Buck, A.C., Lote, C.J., Sampson, W.F.,** The influence of renal prostaglandins on urinary calcium excretion in idiopathic urolithiasis, *J. Urol.* 129, 421, 1983.
42. **Hirayama, H., Ikegami, K., Shimomura, T., Soejima, H., Yamamoto, T.,** The possible role of prostaglandins E_2 in urinary stone formation, *J.Urol.* 139, 549, 1988.
43. **Henriquez-La Roche, C., Rodriguez-Iturbe,   B., Parra, G.,** Increased urinary excretion of prostaglandin E2 in patients with idiopathic hypercalciuria is a primary phenomenon, *Clin. Sci.,* 83, 75, 1992.
44. **Buck, A.C., Davies, R.L., Harrison, T.,** The protective role of eicosapentaenoic acid (EPA) in the pathogenesis of nephrolithiasis, *J. Urol.* 146, 188, 1991.
45. **Rothwell, P.J.N., Green, R., Blacklock, N.J., Kavanagh, J.P.,** Does fish oil benefit stone formers?, *J. Urol.* 150,1391, 1993.

Chapter 11

OXALATE TRANSPORT ACROSS INTESTINAL
AND RENAL EPITHELIA

Marguerite Hatch and Robert W. Freel

Division of Nephrology, Department of Medicine,
University of California at Irvine, Irvine, CA 92717

The epithelial membranes of the gastrointestinal tract and kidneys are the principal interfaces for the exchange of oxalate between an organism and its environment.   As such, these organ systems play a major role in the maintenance of the mass balance of oxalate in the absence of metabolism of this organic acid.[1] In the present chapter we consider the transport systems of these interfacial epithelia in terms of their contribution to oxalate homeostasis in health and disease and possible transepithelial mechanisms for oxalate absorption and secretion.

I. INTESTINAL HANDLING OF OXALATE

The proportion of oxalate in urine that is derived from dietary sources is highly variable and studies show that it can range from 2% to 50% of the total dietary load.[2-7]   While the oxalate content of the diet can be estimated, and restricted if necessary, it is difficult to quantitate and control the concentration of intraluminal oxalate that is available for absorption along the length of the alimentary tract. Oxalate released upon digestion of food can form insoluble salts and it can bind non-specifically to other intraluminal compounds. Many investigators have addressed the relationship between dietary calcium and oxalate absorption and the renal excretion of both ions.[5,7-14] Invariably calcium restriction and hyperabsorption of calcium were associated with increased oxalate excretion. Dietary magnesium was similarly found to influence urinary oxalate excretion.[3,15] These observations, together with the fact that calcium oxalate stone-formers do not have significantly higher dietary intakes of calcium, magnesium, or oxalate[16-19]   underscore the relevance of the physicochemical environment of intraluminal contents in terms of oxalate solubility and availability.

Although oxalate can be absorbed by all segments of the intestine, most of the oxalate from food is absorbed from the upper part of the intestinal tract.[3,5,6] Timed absorption studies demonstrate maximum excretion of ^{14}C-oxalate two to four hours after an oral dose of the tracer.[3,5,12] In addition, patients with total colectomy appear to absorb and excrete the same amount of urinary oxalate as individuals with an intact colon.[6] Oxalate absorption does occur across the normal large intestine and this is also the location where greatly enhanced oxalate absorption occurs in patients with enteric hyperoxaluria .[2,6,20-31]

A. STOMACH

Little attention has been given to oxalate absorption by the mucosa of the stomach. A recent study examined urinary excretion of oxalate following gastric administration of various oxalate loads under conditions where gastric emptying was blocked.[32] The results showed a linear increase in urinary oxalate excretion with increasing gastric loading time. With a gastric loading time of 6 hours, over 60% of the 5 mM load was absorbed. Absorption of oxalate by gastric mucosa appears to be quantitatively significant and most likely occurs *via* non-ionic diffusion of the free acid along a concentration gradient.

B. SMALL INTESTINE

Initial studies of oxalate absorption by everted gut sacs of small intestine showed no evidence for carrier-mediated transport across rabbit[33] or rat[34] duodenum, jejunum and ileum. Since oxalate flux appeared to be strictly concentration-dependent and not affected by metabolic inhibitors,[33] it was concluded that intestinal absorption occurred by passive diffusion. In contrast, the involvement of active transport in small intestinal movement of oxalate was suggested by acute *in situ* experiments which demonstrated that the jejunum of nephrectomized rats secreted oxalate into the lumen against a concentration gradient.[35] More recent studies of oxalate transport across isolated, intact sheets of small intestine from rats[36] and rabbits[37, 38] clearly showed a net secretion of oxalate. In the absence of an electrochemical gradient the serosal to mucosal flux of oxalate was significantly greater in the jejunum compared to the ileum in rabbits,[38] whereas in rats, net secretion of oxalate was much greater in the ileum when compared to the jejunum.[36]

Because of the dietary contribution to urinary oxalate and the studies demonstrating oxalate absorption in the small intestine, the existence and relevance of intestinal oxalate movements in the opposite direction have not been seriously considered until recently.[39-42] It is noteworthy here that in "leaky" epithelia, such as the small intestine, the major route for the passive transepithelial movements of solutes is paracellular because the resistance of tight-junctions to solute flow is relatively low.[43] Hence, the contribution of the paracellular, passive flux of oxalate moving along its electrochemical gradient will undoubtedly predominate when intraluminal oxalate concentrations are high and consequently net absorption of oxalate will occur. While the foregoing studies provide few mechanistic details regarding the absorptive and secretory components of net oxalate flux across the small intestine, studies using membrane vesicle preparations have identified various anion exchanger systems that may mediate transepithelial oxalate movements across intestinal tissues.

The first indication that an oxalate uptake mechanism was present in intestinal brush border membrane vesicles, beyond passive, non-mediated uptake,[44] was the observation that oxalate *cis*-inhibited OH gradient-driven SO_4 uptake in rabbit ileal brush border membrane vesicles.[45] Additional studies using this membrane preparation revealed at least two distinct pathways (Table 1) whereby oxalate can participate in anion exchange: $Ox(SO_4)$-OH and Ox-Cl.[46,47] The former transport system was DIDS sensitive and was characterized as an $Ox(SO_4)$-OH exchanger based upon the fact that outwardly directed OH gradients stimulated Ox and SO_4 uptake and because both of the divalent anions *cis*-inhibited and *trans*-stimulated one another.[46] Uptake through this system was not affected by imposed electrical gradients (ie., an electroneutral exchange process).[47] Oxalate uptake into brush border membrane vesicles from pig jejunum has also been shown to occur by an Ox-SO_4 exchange process.[48] However, unlike the rabbit ileum, oxalate also *cis*-inhibited the Na-dependent uptake of SO_4, implying that oxalate is a substrate for this cotransport process.[48]

The second transporter characterized in the rabbit ileal brush border membrane vesicles was one mediating the exchange of oxalate for chloride (Table 1). In this system, both Ox and Cl *cis*-inhibited and *trans*-stimulated the counterflow of one another, the exchange process was also reported to be electroneutral, and DIDS sensitive.[47] The Cl-Ox exchanger was distinguished from the $Ox(SO_4)$-OH exchanger on the basis of unequal *cis*-inhibition by SO_4 or Cl and by the fact that the Ox-Cl system was less sensitive to probenecid.[46] It is noteworthy that, while electroneutral Cl-HCO_3 exchange has also been identified in rabbit ileal brush border membranes,[49] this exchanger does not appear to exhibit an affinity for the oxalate ion nor is the Cl-HCO_3 exchanger as strongly inhibited by DIDS as the Ox-Cl exchanger.[46]

At this time, only one mechanism (a $SO_4(Ox)$-HCO_3 exchange similar to renal tubule) has been provided to explain the movement of oxalate across the basolateral membrane of ileal enterocytes (Table 1). In rabbit ileal basolateral membranes, an outwardly directed HCO_3 gradient stimulated a DIDS sensitive, electroneutral, uptake of SO_4 which was *cis*-inhibited by the oxalate anion with a $K_i = 1.1$ mM.[50] An outward gradient of HCO_3 also promoted the DIDS sensitive uptake of Ox and Ox *trans*-stimulated SO_4 uptake, suggesting Ox and SO_4 share a common carrier.[50] A Cl-SO_4 exchanger which

Table 1

Oxalate exchange systems identified in apical and basolateral membrane vesicles from intestinal and renal epithelia

Animal-Tissue	Vesicle Fraction	Exchanger	Other Modes[a]	Charge Transfer	Inhibitors[b]	Reference
Rabbit- Renal cortex	Brush Border	$Ox^= - Cl^-$	$Ox^= - Formate^-$ / $Ox^= - OH^-$	Electrogenic	DIDS $K_i \cong .03$ mM / DIDS>>Furosemide> Prob / also SITS>DADS>DPC	99, 100, 104
Rabbit-Renal cortex	Brush Border	$Ox^= - SO_4^=$	$Ox^= - HCO_3^-$ / $SO_4^= - HCO_3^-$	Electroneutral	SITS>DPC>DADS	102, 104
Rabbit-Renal cortex	Basolateral	$Ox^= - SO_4^=$	$Ox^= - HCO_3^-$ / $SO_4^= - HCO_3^-$	Electroneutral	DIDS $K_i \cong .01$ mM	107
Rat-Renal cortex	Brush Border	$Ox^= - SO_4^=$ Na^+-dependent			low DIDS sensitivity in presence of Na $K_i = .35$ mM	103
Rat-Renal cortex	Brush Border	$Ox^= - OH^-$	$Ox^= - Cl^-$ / $Ox^= - Formate^-$	Electroneutral		101
Rat-Renal cortex	Basolateral	$Ox^= - SO_4^=$ Na^+-dependent ?			high DIDS sensitivity $K_i = .2$ μM	103, 108
Rabbit- Ileum	Brush Border	$Ox^= - OH^-$	$Ox^= - SO_4^=$ / $SO_4^= - OH^-$	Electroneutral	DIDS>Prob>Bum K^+-inhibits	45, 46
Rabbit- Ileum	Brush Border	$Ox^= - Cl^-$	$Ox^= - Formate^-$ / $Cl^- - Formate^-$ / $Ox^= - Ox^=$	Electroneutral	DIDS>Bum>Prob no K^+ effect	46, 47
Rabbit- Ileum	Basolateral	$Ox^= - HCO_3^-$	$Ox^= - SO_4^=$ / $SO_4^= - HCO_3^-$	Electroneutral	DIDS	50

a Only documented modes relevant to this discussion are presented.
b DIDS (4-4'-diisothiocyanostilbene-2,2'-disulfonic acid); SITS (4-aceto-4'-isothiocyano-2,2'-disulfonic stilbene); DADS (4-amino-4'-amino-2,2'- disulfonic stilbene); DPC (diphenylamine-2- carboxylate); Furo (furosemide); Prob (probenicid); Bum (bumetanide).

was strongly *cis*-inhibited by oxalate was also identified in these basolateral membrane vesicles, however it was suggested that oxalate was not actually transported by this carrier.[51] Thus, of the two anion exchangers identified to date (Table 1),only the $SO_4(Ox)$-HCO_3 exchanger appears to participate in oxalate movement across rabbit ileal basolateral membrane.

As noted previously, net oxalate secretion by rat and rabbit small intestine has been observed,[36,37,38] but the mechanistic details are largely unexplored but may be similar to those governing cAMP-induced chloride secretion.[52] In one preliminary account,[53] it was shown that an exogenous cAMP analog also stimulates oxalate secretion in isolated, short-circuited segments of the rabbit distal ileum. Ileal brush border membrane vesicles were shown to possess conductive pathways for oxalate and chloride that exhibited some of the characteristics of a pore or channel (ie., activation energies of a simple, non-mediated diffusion process and poorly saturable). Hence, the cAMP-induced ileal secretion of oxalate, like chloride,[52] may be limited by the apical membrane anion conductance of the secretory enterocyte. The rabbit colon also secretes oxalate by a mechanism that resembles that of chloride, as considered in the following section.

C. LARGE INTESTINE

Oxalate transport across the large intestine was also considered to be *via* passive diffusion since initial studies using everted gut sacs did not indicate involvement of a carrier-mediated process.[33,34] Later studies, which directly addressed the importance of calcium in maintaining the integrity of the transporting epithelium, clearly demonstrated that oxalate was actively transported by the large intestine of rats[54] and rabbits.[55] In *in vivo* studies with rats, the degree of oxalate absorption by rat caecum was shown to correlate with luminal pH.[56] With an intraluminal pH of 4.9 compared to 7.17, urinary oxalate excretion was significantly elevated consistent with an increase in caecal transport due to diffusion of the free acid ($pK_a = 3.83$[1]).

Segmental differences in oxalate transport were also demonstrated within the colon.[39,40] In the isolated, short-circuited proximal colon of the rabbit oxalate transport was found to be dependent upon the presence of other major ions.[39] Replacement of Na with n-methyl-D-glucamine or Cl with gluconate in the bathing solutions abolished net oxalate secretion and in the nominal absence of HCO_3 a net absorptive oxalate flux was unmasked. Further evidence for partial coupling of oxalate fluxes to Na-H and Cl-HCO_3 exchangers in the proximal colon was indicated by reductions in the unidirectional fluxes of oxalate following the mucosal addition of amiloride at doses that inhibit Na-H exchange and following bilateral application of DIDS. In addition, oxalate fluxes across the proximal colon were found to be sensitive to absorptive and secretory stimuli.[39] Epinephrine, which enhanced Na transport across this segment, also increased the absorptive flux of oxalate and net oxalate secretion was changed to net absorption. The addition of cAMP further enhanced the basal net secretion of oxalate across this segment. It is significant that intestinal secretory pathways for oxalate have been identified, since they can potentially provide an important extra-renal route for oxalate excretion. Furthermore, as demonstrated in both the proximal and distal (see below) segments of the colon, it is remarkable that the direction of net oxalate flux can be regulated by neuro-hormones which are known to affect enteric handling of major plasma electrolytes.

In the rabbit distal colon, where oxalate is absorbed under short-circuit conditions,[36,38,40,55] the movements and sensitivities of oxalate paralleled those of chloride.[38,40] Net movement of oxalate across this tissue was found to be Na-dependent but Cl-independent.[40] In the latter study it was proposed that oxalate enters the enterocyte *via* a SITS sensitive Cl-HCO_3 exchanger in the apical membrane and exits the cell *via* a DIDS sensitive base-anion exchanger which is coupled to an amiloride sensitive Na-H antiporter. The basolateral exit step was also shown to be sensitive to thiazide diuretics which reduced the absorptive flux of oxalate across rabbit colonic tissues.[38] Like the proximal part of the rabbit colon, net flux of oxalate across the distal colon was sensitive to secretory stimuli and cAMP reversed the net absorptive flux to a net secretory flux. Further experiments showed that the secretory pathways for oxalate and chloride had remarkably similar sensitivities to serosal furosemide and mucosal NPPB (a putative Cl conductance blocker). It was proposed that oxalate was accumulated intracellularly *via* a furosemide sensitive Na-K-2Cl cotransport system on the basolateral membrane and efflux across the apical membrane occurred *via* an NPPB sensitive pathway, possibly a Cl channel.[40] Again, this demonstration that the entire colon has the capability of regulating the secretion and absorption of oxalate implies that this segment may participate in the mass balance of oxalate and oxalate homeostasis. These oxalate secretory pathways that have been identified also underscore the importance of the substrate-specific, oxalate-degrading bacteria that have been isolated from the human large intestine.[57] It has been estimated from calculations based upon *in vitro* studies that these micro-organisms can make a significant contribution to intraluminal degradation of oxalate. In addition to scavenging oxalate from residual dietary sources, these colonic bacteria also contribute to the maintenance of a transepithelial gradient that is favorable for the passive, paracellular movement of oxalate from blood to lumen.

While there is a considerable body of information regarding oxalate transport in isolated colonic tissues, the transport systems proposed to account for net absorption or secretion of oxalate across the colon have not been directly tested using isolated membrane fractions. Exchange pathways similar to those described for the rabbit small intestine may be operative in colonocytes. In rat colonic apical membrane vesicles oxalate produced a modest *cis*-inhibition of HCO_3 or OH gradient driven ^{36}Cl uptake,[58] suggesting the possibility that

DIDS-sensitive Cl-OH or Cl-HCO_3 exchangers mediate oxalate absorption by this tissue. Oxalate was not a substrate for HCO_3 gradient driven butyrate uptake in these same vesicles.[59]

II. ALTERATIONS OF INTESTINAL OXALATE HANDLING IN DISEASE STATES

A. ENTERIC HYPEROXALURIA

Enteric hyperoxaluria specifically refers to hyperoxaluria resulting from enhanced absorption of oxalate secondary to a wide variety of gastrointestinal diseases,[20,22-26,60] ileal resection,[6,20,23] or jejuno-ileal bypass.[20,23,27-31] The common finding among these disorders causing hyperabsorption of oxalate is malabsorption of bile salts and fatty acids. The colon has been identified as the primary site of increased oxalate absorption[20,28,61,62] and two hypotheses have been extended to explain the mechanism. One model emphasizes the changes in the permeability of colonic mucosa[33,61,63-68] and the other suggests the importance of oxalate solubility and its availability in fecal contents.[6,23,26,30,62,67] Using a variety of techniques, including everted gut sac preparations[33], perfused loops of rat intestine,[61,67] and perfused colonic segments from humans,[68] many investigators concluded that dihydroxy bile salts and long chain hydroxy fatty acids increased oxalate absorption across this tissue. Low concentrations of bile salt applied to intact short-circuited sheets of rabbit colon resulted in large increases in the passive permeability of the mucosa to oxalate.[63] Using a similar experimental approach with rat colon, it was also demonstrated that bidirectional oxalate fluxes were increased following the addition of bile salts or fatty acids.[69] *In vitro* experiments examining oxalate solubility in the presence of fatty acids and calcium showed that the affinity of calcium for fatty acid was greater than that for oxalate and consequently, the concentration of free oxalate was increased.[33,70] The free concentration of oxalate was reduced by the resin cholestyramine,[33,71] which also binds bile salts .[24,71] Further confirmation that oxalate solubility in the lumen of the colon plays a key role in enteric hyperoxaluria is provided by numerous patient studies demonstrating that oral administration of calcium.[7,23,30,31,67,70,72] and cholestyramine[60,71] effectively reduces the hyperoxaluria. The mechanistic basis of enteric hyperoxaluria can be attributed to the complementary roles of enhanced mucosal permeability and increased oxalate solubility. Large increases in the paracellular passive flux of oxalate along its concentration results in enhanced absorption of oxalate in proportion to its intraluminal solubility. The relative contribution of the transcellular active movements of oxalate, which were described previously will be negligible under these circumstances.

B. ABSORPTIVE OR 'DIETARY' HYPEROXALURIA

Considerable attention has been given to the notion that some oxalate stone-formers have enhanced intestinal absorption of oxalate that is not secondary to steatorrhoea or bile salt malabsorption. Increased absorption of oxalate from the upper alimentary tract leading to hyperoxaluria is supported by a number of studies.[2,3,7,10,14,19,73,74] However, since most of these studies also involved the manipulation of dietary calcium, it is generally concluded that increased oxalate absorption is a direct consequence of oxalate bioavailability. Results of some investigations,[3,5,14,16] while acknowledging the importance of oxalate bioavailability, indicate that other factors may be involved. In studies of patients having an inheritable anomaly of red blood cell oxalate transport, it was suggested that the erythrocyte defect might also be present in enterocytes.[75,76] Thus, male stone-forming patients with the underlying red cell defect excreted significantly higher amounts of oxalate in urine following an oxalate load when compared to their brothers who were characterized as having normal red cell transport of oxalate.[75] It has also been reported that vitamin B_6 deficiency in rats is associated with an increase in oxalate uptake in intestinal rings[77] and isolated apical membrane vesicles.[78] Unfortunately, none of these investigative approaches permit a direct evaluation of *in vivo* intestinal transport mechanisms for oxalate in stone-forming patients because renal excretory mechanisms cannot be excluded and the dynamics of intraluminal oxalate availability cannot be definitively controlled in these types of studies. It is also possible that intestinal secretory pathways for oxalate may be functionally important in stone-formers who are presumed to hyperabsorb dietary oxalate. A reduction in the intestinal secretory component, rather than, or in addition to an enhancement of the absorptive component of oxalate transport could explain absorptive or 'dietary' hyperoxaluria.

C. INTESTINAL EXCRETION IN CHRONIC RENAL FAILURE

Until recently oxalate excretion was thought to occur exclusively *via* the kidneys but new information regarding alternative routes for the elimination of oxalate challenges this dogma. The extrarenal elimination of oxalate was confirmed in rats with chronic renal failure (CRF) in two separate studies which used different experimental approaches.[36,41] In one study, the fate of ^{14}C-oxalate, which was infused by a mini osmotic pump in normal rats and rats with experimentally induced chronic renal failure, was determined over a period of four days.[41] It was evident from the results that fecal excretion of the tracer was significantly higher in CRF rats when compared to normal which suggests that enteric elimination of oxalate plays an important role in conditions where renal function is compromised. A direct examination of the magnitude and direction of oxalate fluxes across isolated, short-circuited intestinal segments removed from a similar rat model demonstrated that intestinal transport of oxalate is altered in CRF.[36] The latter study showed that 6 weeks after 5/6 nephrectomy basal oxalate absorption across the colon, which was observed in control rats, was changed to secretion in rats with CRF. This study also showed that the basal secretory flux of oxalate across the jejunum and ileum was not affected by renal insufficiency. The results of these complementary investigations,[36,41] together with the observation that an increased population of oxalate-degrading microorganisms was found in the large intestine of patients with CRF,[79] provide evidence that colonic excretion of oxalate is potentially significant in CRF. While the regulation and mechanisms involved in the striking alterations in colonic oxalate transport in CRF have yet to be examined, extra-renal excretion of oxalate is an important consideration when interpreting studies which focus solely on urinary oxalate excretion measurements.

III. RENAL HANDLING OF OXALATE

A. CLASSICAL STUDIES

Renal handling of oxalate has been investigated in several species using a variety of experimental approaches. Based upon evidence provided in animal studies, there is a general agreement that oxalate is freely filtered by the glomerulus[80-84] and that oxalate undergoes bidirectional transport along the renal tubule.[83,85] However, there is conflicting information as to whether the net effect of bidirectional transport of oxalate results in tubular secretion or reabsorption. Contrary results were also evident from a number of clearance studies which examined the relationship between oxalate secretion and varying urinary flow rates. In one study, using rats, net fractional secretion of oxalate was found to be greater at low urinary flow rates.[81] In contrast, other studies which used dogs[80] and human subjects[8,14] found a significant positive correlation between urinary oxalate and urine flow. In two other studies carried out in rats[86] and sheep,[82] secreted oxalate was not related to urine flow.

Oxalate clearance studies employing ^{14}C-oxalate methods in chickens,[87] sheep,[82] dogs,[80] rats,[81,84,85,88,89] rabbits,[90] and man[83,91] have shown that oxalate clearance exceeds the clearance of an extracellular marker like inulin indicating net secretion of oxalate by the kidney. In marked contrast, oxalate clearance values determined by more non-invasive methods employing direct chemical measurements of plasma and urinary oxalate concentrations and urine production rates have shown net reabsorption of oxalate by the kidney. In the past, such differences were readily and correctly attributed to technical complications that led to an overestimation of plasma oxalate concentration.

In spite of the fact that analytical methods now yield similar results for plasma oxalate concentrations, recent studies on humans[16,92] and rats[36,41] (using this approach) show oxalate/creatinine clearance ratios still less than unity, implying renal reabsorption of oxalate in the unrestrained or unanesthetized organism. This result is also supported by a calculation of oxalate/creatinine clearance ratio which incorporates currently accepted normal values for 24 hr urine volume, urinary and serum oxalate levels, and creatinine clearance. If it is assumed that a healthy individual has an average 24 hr urine volume of 1.5 l (= 1.04 ml/min), an average oxalate excretion of 0.28 mmole/24 hr, and a plasma oxalate concentration of 2.5 mM, then renal clearance of oxalate is 77 ml/min. If a creatinine clearance of 120 ml/min is employed in this calculation, it follows that the estimated clearance ratio is 0.64, clearly indicating a net reabsorptive process in the kidney.

Studies of superficial renal cortical nephron segments using various micro techniques have most frequently indicated that the proximal tubule plays a major role in the renal handling of oxalate. Proximal tubular secretion of oxalate was first suggested by peritubular capillary microperfusion and tubular superperfusion experiments in rats.[81] These kinds of experiments showed that ^{14}C-oxalate appeared in the proximal tubule lumen in excess of the extracellular marker ^{3}H-inulin. Subsequent free-flow microperfusion studies have generally confirmed the existence of an oxalate secretory process in the proximal tubular epithelium.[84,86,89] However, in one study only reabsorption was reported along the rat proximal tubule and it was argued that the secretory component was located in nephrons or nephron segments that were inaccessible.[88] Net secretory fluxes of ^{14}C-oxalate have also been measured in isolated, perfused proximal tubular segments from the rabbit kidney.[90] In this study, the observed rates of oxalate secretion were markedly different between proximal tubules from superficial and juxtamedullary nephrons indicating that in this species (as opposed to rat) there is heterogeneity of oxalate handling within the kidney.

The proximal tubular secretory flux of oxalate also exhibits some of the characteristics of an active, carrier-mediated transport process. In rats, simultaneous perfusion of the proximal tubule and the peritubular capillary indicated that oxalate secretion was a saturable process that was reduced by sodium cyanide and, at capillary concentrations less than 4 mM oxalate, was inhibited by the organic anion transport inhibitor probenecid.[89] In rabbits the secretory flux of oxalate in isolated proximal tubular segments was temperature dependent and proceeded against a lumen negative transepithelial potential in the absence of a chemical gradient for oxalate.[90]

The reabsorptive component of the bidirectional renal transport of oxalate is also considered to occur in the proximal tubule. Several groups, using different microperfusion techniques in rats, have concluded that some 7% to 30% of the filtered load of oxalate is reabsorbed along the early to mid portions of the proximal tubule.[81,84,88] In one of these studies,[81] increasing perfusate oxalate concentrations up to 0.5 mM did not saturate the reabsorptive component, suggesting that this process is a passive, non-carrier mediated process. In contrast, results of a recent study suggest that the reabsorptive component of oxalate transport in the human kidney exhibits a sensitivity to the loop diuretic bumetanide and also to the carbonic anhydrase inhibitor acetazolamide.[93] In isolated, perfused rabbit proximal tubules the net flux of oxalate was only slightly larger than the bath to lumen flux indicating the absorptive flux was relatively small in this preparation.[90]

In situ studies using stop-flow peritubular capillary or tubular microperfusion methods have revealed several possible transport avenues for oxalate across luminal and contraluminal membranes of the proximal tubule.[94] Using these techniques it was demonstrated that oxalate interacts with a luminal, Na-dependent SO_4 cotransport system in the rat proximal tubule.[95] At the opposing membrane it was shown that inclusion of oxalate in peritubular capillary perfusates inhibited SO_4 uptake as well, but in a Na-independent manner, suggesting a SO_4-Ox exchange process on the contraluminal membrane.[96] This SO_4-anion exchanger only accepted aliphatic dicarboxylates with closely positioned COO⁻ groups, like oxalate and maleate.[97] Interestingly, oxalate was not a suitable substrate for the contraluminal Na-dependent dicarboxylate cotransporter, which is mediates the uptake of many Krebs cycle intermediates.[94,96]

Oxalate is but one of the many anions that is reabsorbed or secreted by the renal proximal tubule[94,98,99] and it had been frequently suggested that oxalate might share a common secretory pathway with other organic anions like urate or para-aminohippurate (PAH).[80,83,85,86,89] Even in the earlier oxalate clearance studies in dogs, the depression in oxalate excretion following PAH infusion was taken as evidence for a common transport mechanism.[80] However, as described below and elsewhere,[94,98,100] studies of oxalate uptake into isolated membrane vesicles would suggest that oxalate and urate/PAH transporters are distinct entities but may share common transport partners like OH or Na.

B. MEMBRANE VESICLE STUDIES

There have been relatively few vesicle studies that specifically focus on the mechanism of oxalate transport, *per se*, in the proximal tubule (Table 1). In rabbit renal microvillus (luminal) membranes there are at least two distinct anion exchangers that have been described that accept oxalate as one of the counterions (the Ox-Cl exchanger and the SO_4-HCO_3 exchanger) and each exhibits several exchange modes. The Cl-Ox system was identified in rabbit renal brush borders by the fact that outwardly directed gradients of oxalate (and formate) promoted the uptake of ^{36}Cl, indicating a Cl-Ox exchange process in these vesicles.[101] Inhibitor and substrate specificity studies further indicated that formate stimulation of Cl uptake actually occurred by two distinct carriers: the Ox(formate)-Cl and a Cl-formate exchanger.[101] A preliminary account has further suggested that the Ox-Cl exchanger also accepts OH.[102] Thus, three modes of oxalate transport across the luminal membrane of the rabbit proximal tubule are possible *via* the Cl-Ox exchanger: Ox-Cl, Ox-formate, and Ox-OH.

An Ox-OH(Cl) exchanger has also been identified in rat renal cortical brush borders, yet this exchanger was reported to have a different stoichiometry since the exchange process was electroneutral.[103]

Luminal oxalate exchange *via* a SO_4-HCO_3 exchanger has also been proposed for the rabbit, although the experimental details for this have only been presented in review format[104] and differ with some published studies using rat luminal membrane vesicles.[105] In the rabbit this SO_4-HCO_3 exchanger is reported to be electroneutral, has little affinity for Cl, and is capable of transporting oxalate in two modes (Ox-SO_4 or Ox-HCO_3).[104] A SO_4-HCO_3 exchanger has also been identified in rat renal microvillus membranes, but the only study evaluating oxalate as a substrate for this exchanger also reported a significant Na-dependence,[105] hence this system would not appear to be a simple anion exchanger.

These oxalate-anion exchange systems and the Cl-formate exchanger of rabbit renal brush border membranes have been compared with respect to their sensitivities to compounds that are known inhibitors of erythrocyte band 3-mediated exchange processes.[106] It is notable that the inhibitor profiles for each of these renal anion exchangers were distinct from each other and different from band 3 profiles. Such pharmacological distinctions presumably reflect structural differences in the binding sites of the transport proteins and suggest that the oxalate exchangers in the kidney are structurally distinct from band 3 proteins. How these observations impact the hypothesis[75,107,108] that defects in band 3-mediated oxalate uptake in erythrocytes of stone-formers may be also present and defective in the renal epithelia of these individuals remains to be determined.

Oxalate transport across rabbit renal basolateral membrane vesicles has also been shown to occur *via* a SO_4-HCO_3 exchanger. In this species, outwardly directed gradients of HCO_3 (but not OH^-) were shown to stimulate $^{35}SO_4$ or ^{14}C-Ox uptake by a transport system that was highly DIDS sensitive and independent of the presence of sodium ion.[109] Furthermore, the HCO_3-driven accumulation of Ox and SO_4 was potential-independent, indicating electroneutral exchange of $2HCO_3$ (or CO_3) for one divalent anion. In addition to Ox-HCO_3 exchange, Ox-SO_4, and Ox-S_2O_3 exchange modes were also reported, but other anions of physiological interest (Cl, formate, urate, PO_4) failed to *cis*-inhibit HCO_3-driven Ox or SO_4 uptake which indicates a rather narrow spectrum substrates.[109] While oxalate has been shown to *trans*-stimulate SO_4 uptake into rat renal basolateral membrane vesicles, this transport process was markedly different in exhibiting a Na dependence and a wider range of accepted anions (including Cl, PO_4).[105,110] Whether these differences are species specific or due to technical differences has not been resolved.

C. OTHER TRANSPORT MODELS

While most *in vitro* investigations of oxalate transport have employed isolated tissues or membrane vesicles, there have been several recent reports that suggest that the transport systems may also be examined in isolated cells and in established cell lines. In one such report, possible alterations in renal oxalate transport systems were addressed using suspensions of cells dispersed from renal cortex and papilla from normal and stone-forming rats.[74] Both cortical and papillary cells from normal rats exhibited time-dependent, pH- and DIDS-sensitive uptake of oxalate. In stone-forming rats, oxalate uptake by cortical cells was depressed, whereas oxalate uptake by papillary cells was enhanced, relative to control animals. It was concluded that these alterations in renal cell oxalate transport may contribute to the process of renal stone formation in this animal model.

Several groups have addressed some of the characteristics of oxalate transport in cultured LLC-PK1 cells,[111-114] an established renal cell line derived from the pig proximal tubule. In one study, where the sidedness of ^{14}C-Ox uptake was evaluated, an asymmetry in the distribution of oxalate exchange systems was reported.[112] Oxalate uptake across the apical membrane was DIDS inhibitable, Cl sensitive, but independent of luminal SO_4 and HCO_3 and membrane potential.[112] Oxalate uptake from the basolateral side of the monolayer was also inhibited by DIDS, but sensitive to bathing solution SO_4 and HCO_3 (not Cl). While these results suggest the participation of some of the oxalate exchange systems established using different models, others have reported that LLC-PK1 cells do not exhibit the ability to take up oxalate across either face of the monolayer.[113] As the methodologies were similar in these two reports, it would appear that expression of oxalate transport systems may be variable and dependent upon lineage or culture conditions.

D. FINAL COMMENTS

The existence of, what would appear to be, relatively specific exchange mechanisms for oxalate transport in both renal and intestinal epithelia suggests the possibility that these transport systems may subserve functions that are not directly related to oxalate homeostasis. One such function follows from the hypothesis that oxalate[101] (and formate[115]) exchange in the proximal tubule may participate in electroneutral, NaCl reabsorption by the mammalian proximal tubule. Since Na-H and Ox-Cl exchangers exist in parallel in the luminal membrane of proximal tubule cells, it was suggested that oxalate secreted from the cell into the lumen *via* the Cl-Ox exchange system could provide a luminal membrane uptake mechanism for Cl. Recycling of Ox across the luminal membrane *via* Ox-OH exchange would produce the net thermodynamic effect of electroneutral NaCl uptake, since the OH secreted in Ox-OH exchange would neutralize the proton secreted by the luminal Na-H exchanger. The implication that oxalate and its transport systems may be important in the reabsorption of NaCl and obliged water is supported by microperfusion studies wherein physiological concentrations (mM) of oxalate (or formate) were shown to stimulate a DIDS-sensitive reabsorption of Cl and volume in the proximal[116,117] and early distal tubules of rats.[118]

How the apical and basolateral oxalate transport pathways coordinate the vectorial translocation of oxalate across intestinal and renal interfacial epithelia is still a matter interest and of conjecture because the electrochemical potential differences for the solutes involved have not been experimentally established. However, consideration of the likely activities of the transport partners for oxalate in the rabbit proximal tubule has led to proposal that the electrochemical potential of intracellular oxalate is greater than the surrounding luminal and peritubular environments.[109] This condition could then result in efflux of oxalate across both luminal and contraluminal membranes of the proximal tubule cell (ie., bidirectional transport). The direction of net transcellular transport would then depend on the relative magnitudes of the apical and basolateral transport processes.[109]

REFERENCES

1. Hodgkinson, A., *Oxalic Acid in Biology and Medicine*, Academic Press, London, 1977, 159.
2. Finlayson, B., Calcium stones: Some physical and clinical aspects, in *Calcium Metabolism in Renal Failure and Nephrolithiasis*, David, D. S., John Wiley & Sons, New York, 1977, 337.
3. Barilla, D. E., Notz, C., Kennedy, D. and Pak, C. Y., Renal oxalate excretion following oral oxalate loads in patients with ileal disease and with renal and absorptive hypercalciurias. Effect of calcium and magnesium, *Am. J. Med.*, 64, 579, 1978.
4. Rampton, D. S., Kasidas, G. P., Rose, G. A. and Sarner, M., Oxalate loading test: a screening test for steatorrhoea, *Gut*, 20, 1089, 1979.
5. Marangella, M., Fruttero, B., Bruno, M. and Linari, F., Hyperoxaluria in idiopathic calcium stone disease: further evidence of intestinal hyperabsorption of oxalate, *Clin. Sci.*, 63, 381, 1982.
6. Earnest, D. L., Johnson, G., Williams, H. E. and Admirand, W. H., Hyperoxaluria in patients with ileal resection: an abnormality in dietary oxalate absorption, *Gastroenterology*, 66, 1114, 1974.
7. Lindsjö, M., Danielson, B. G., Fellström, B., Lithell, H. and Ljunghall, S., Intestinal oxalate and calcium absorption in recurrent renal stone formers and healthy subjects, *Scand. J. Urol. Nephrol.*, 23, 55, 1989.

8.   Zarembski, P. M. and Hodgkinson, A., Some factors influencing the urinary excretion of oxalic acid in man, *Clin. Chim. Acta*, 25, 1, 1969.

9.   Marshall, R. W., Cochran, M., Robertson, W. G., Hodgkinson, A. and Nordin, B. E., The relation between the concentration of calcium salts in the urine and renal stone composition in patients with calcium-containing renal stones, *Clin. Sci.*, 43, 433, 1972.

10.   Smith, L. H., Diet and hyperoxaluria in the syndrome of idiopathic calcium oxalate urolithiasis, *Am. J. Kidney Dis.*, 17, 370, 1991.

11.   Bataille, P., Pruna, A., Gregoire, I., Charransol, G., de, F. J., Ledeme, N., Finet, M., Coevoet, B., Fievet, P. and Fournier, A., Critical role of oxalate restriction in association with calcium restriction to decrease the probability of being a stone former: insufficient effect in idiopathic hypercalciuria, *Nephron*, 39, 321, 1985.

12.   Erickson, S. B., Cooper, K., Broadus, A. E., Smith, L. H., Werness, P. G., Binder, H. J. and Dobbins, J. W., Oxalate absorption and postprandial urine supersaturation in an experimental human model of absorptive hypercalciuria, *Clin. Sci.*, 67, 131, 1984.

13.   Massey, L. K. and Sutton, R. A., Modification of dietary oxalate and calcium reduces urinary oxalate in hyperoxaluric patients with kidney stones, *J. Am. Diet. Assoc.*, 93, 1305, 1993.

14.   Galosy, R., Clarke, L., Ward, D. L. and Pak, C. Y. C., Renal oxalate excretion in calcium urolithiasis, *J. Urol.*, 123, 320, 1980.

15.   Berg, W., Bothor, C., Pirlich, W. and Janitzky, V., Influence of magnesium on the absorption and excretion of calcium and oxalate ions, *Eur. Urol.*, 12, 274, 1986.

16.   Hatch, M., Oxalate status in stone-formers. Two distinct hyperoxaluric entities, *Urol. Res.*, 21, 55, 1993.

17.   Griffith, H. M., O'Shea, B., Kevany, J. P. and McCormick, J. S., A control study of dietary factors in renal stone formation, *Br. J. Urol.*, 53, 416, 1981.

18.   Robertson, W. G., Dietary factors important in calcium stone formation, in *Urolithasis and Related Clinical Research*, Schwille, P. O., Smith, L. H., Robertson, W. G. and Vahlensieck, W., Plenum Press, New York, 1985, 61.

19.   Hodgkinson, A., Evidence of increased oxalate absorption in patients with calcium-containing renal stones, *Clin. Sci. Mol. Med.*, 54, 291, 1978.

20.   Dobbins, J. W. and Binder, H. J., Importance of the colon in enteric hyperoxaluria, *N. Engl. J. Med.*, 296, 298, 1977.

21.   Stauffer, J. O., Stewart, R. J. and Bertrand, G., Acquired hyperoxaluria: Relationship to dietary calcium content and severity of steatorrhea, *Gastroenterology*, 66, 783, 1974.

22.   McDonald, G. B., Earnest, D. L. and Admirand, W. H., Hyperoxaluria correlates with fat malabsorption in patients with sprue, *Gut*, 18, 561, 1977.

23.   Stauffer, J., Hyperoxaluria and intestinal disease, *Dig. Dis.*, 22, 921, 1977.

24.   Chadwick, V. S., Modha, K. and Dowling, R. H., Mechanism for hyperoxaluria in patients with ileal dysfunction, *N. Engl. J. Med.*, 289, 172, 1973.

25.   Dowling, R. H., Rose, G. A. and Sutor, D. J., Hyperoxaluria and renal calculi in ileal disease, *Lancet*, 1, 1103, 1971.

26.   Andersson, H. and Jagenburg, R., Fat-reduced diet in the treatment of hyperoxaluria in patients with ileopathy, *Gut*, 15, 360, 1974.

27.   Hylander, E., Jarnum, S., Jensen, H. J. and Thale, M., Enteric hyperoxaluria: dependence on small intestinal resection, colectomy, and steatorrhoea in chronic inflammatory bowel disease, *Scand. J. Gastroenterol.*, 13, 577, 1978.

28.   Hofmann, A. F., Laker, M. F., Dharmsathaphorn, K., Sherr, H. P. and Lorenzo, D., Complex pathogenesis of hyperoxaluria after jejunoileal bypass surgery. Oxalogenic substances in diet contribute to urinary oxalate, *Gastroenterology*, 84, 293, 1983.

29.   Lindsjö, M., Danielson, Fellström, B., Lithell, H. and Ljunghall, S., Intestinal absorption of oxalate and calcium in patients with jejunoileal bypass, *Scand. J. Urol. Nephrol.*, 23, 283, 1989.

30.   Hylander, E., Jarnum, S., Kempel, K. and Thale, M., The absorption of oxalate, calcium, and fat after jejunoileal bypass, *Scand. J. Gastroenterol.*, 15, 343, 1980.

31.   Hylander, E., Jarnum, S. and Nielsen, K., Calcium treatment of enteric hyperoxaluria after jejunoileal bypass for morbid obesity, *Scand. J. Gastroenterol.*, 15, 349, 1980.

32.   Hautman, R. E., The stomach: a new and powerful oxalate absorption site in man, *J. Urol.*, 149, 1401, 1993.

33.   Binder, H. J., Intestinal oxalate absorption, *Gastroenterology*, 67, 441, 1974.

34.   Caspary, W. F., Intestinal oxalate absorption. I. Absorption in vitro, *Res. Exp. Med. (Berl.)*, 171, 13, 1977.

35.   Dobson, D. M. and Finlayson, B., Oxalate transport from plasma to intestinal lumen in the rat, *Urol. Surg.*, 540, 1973.

36.   Hatch, M., Freel, R. W. and Vaziri, N. D., Intestinal excretion of oxalate in chronic renal failure, *J. Am. Soc. Nephrol.*, 5, 1339, 1994.

37.   Hatch, M. and Vaziri, N. D., Segmental differences in intestinal oxalate transport, *FASEB J.*, 5, A1138, 1991.

38. Hatch, M. and Vaziri, N. D., Do thiazides reduce intestinal oxalate absorption?: a study in vitro using rabbit colon, *Clin. Sci.*, 86, 353, 1994.
39. Hatch, M., Freel, R. W. and Vaziri, N. D., Characteristics of the transport of oxalate and other ions across rabbit proximal colon, *Pflügers Arch.*, 423, 206, 1993.
40. Hatch, M., Freel, R. W. and Vaziri, N. D., Mechanisms of oxalate absorption and secretion across the rabbit distal colon, *Pflügers Arch.*, 426, 101, 1994.
41. Costello, J. F., Smith, M., Stolarski, C. and Sadovnic, M. J., Extrarenal clearance of oxalate increases with progression of renal failure in the rat., *J. Am. Soc. Nephrol.*, 3, 1098, 1992.
42. Hatch, M. and Vaziri, N. D., Enhanced enteric excretion of urate in rats with chronic renal failure, *Clin. Sci.*, 86, 511, 1994.
43. Frizzell, R. A. and Schultz, S. G., Ionic conductances of extracellular shunt pathway in rabbit ileum, *J. Gen. Physiol.*, 59, 318, 1972.
44. Menon, M. and Mahle, C. J., Oxalate metabolism and renal calculi, *J. Urol.*, 127, 148, 1982.
45. Schron, C. M., Knickelbein, R. G., Aronson, P. S., Della, P. J. and Dobbins, J. W., Effects of cations on pH gradient-stimulated sulfate transport in rabbit ileal brush-border membrane vesicles, *Am. J. Physiol.*, G614, 1985.
46. Knickelbein, R. G., Aronson, P. S. and Dobbins, J. W., Substrate and inhibitor specificity of anion exchangers on the brush border membrane of rabbit ileum, *J. Membr. Biol.*, 88, 199, 1985.
47. Knickelbein, R. G., Aronson, P. S. and Dobbins, J. W., Oxalate transport by anion exchange across rabbit ileal brush border, *J. Clin. Invest.*, 77, 170, 1986.
48. Wolffram, S., Grenacher, B. and Scharrer, E., Transport of selenate and sulphate across the intestinal brush-border membrane of pig jejunum by two common mechanism, *Q. J. Exp. Physiol.*, 73, 103, 1988.
49. Knickelbein, R., Aronson, P. S., Schron, C. M., Seifter, J. and Dobbins, J. W., Sodium and chloride transport across rabbit ileal brush border. II. Evidence for Cl-HCO3 exchange and mechanism of coupling, *Am. J. Physiol.*, 249, G236, 1985.
50. Knickelbein, R. G. and Dobbins, J. W., Sulfate and oxalate exchange for bicarbonate across the basolateral membrane of rabbit ileum, *Am. J. Physiol.*, G807, 1990.
51. Schron, C. M., Knickelbein, R. G., Aronson, P. S. and Dobbins, J. W., Evidence for carrier-mediated Cl-SO4 exchange in rabbit ileal basolateral membrane vesicles, *Am. J. Physiol.*, 253, G404, 1987.

52. Halm, D. R. and Frizzell, R. A., Intestinal chloride secretion, in *Textbook of Secretory Diarrhea*, Lebenthal, E. and Duffey, M., Raven Press, Ltd., New York, 1990, 47.

53. Freel, R. W., Hatch, M. and Vaziri, N. D., Conductive pathways for oxalate (Ox) and chloride in rabbit ileal brush border membrane vesicles (BBMV)., *FASEB J*, 9, A87, 1995.

54. Freel, R. W., Hatch, M., Earnest, D. L. and Goldner, A. M., Oxalate transport across the isolated rat colon. A re-examination, *Biochim. Biophys. Acta*, 600, 838, 1980.

55. Hatch, M., Freel, R. W., Goldner, A. M. and Earnest, D. L., Oxalate and chloride absorption by the rabbit colon: sensitivity to metabolic and anion transport inhibitors, *Gut*, 25, 232, 1984.

56. Diamond, K. L., Fox, C. C. and Barch, D. H., Role of cecal pH in intestinal oxalate absorption in the rat, *J. Lab. Clin. Med.*, 112, 352, 1988.

57. Allison, M. J., Cook, H. M., Milne, D. B., Gallagher, H. and Clayman, R. V., Oxalate degradation by gastrointestinal bacteria from humans, *J. Nutr.*, 116, 455, 1986.

58. Rajendran, V. M. and Binder, H. J., Cl-HCO$_3$ and Cl-OH exchanges mediate Cl uptake in apical membrane vesicles of rat distal colon, *Am. J. Physiol.*, G874, 1993.

59. Mascolo, N., Rajendran, V. M. and Binder, H. J., Mechanism of short-chain fatty acid uptake by apical membrane vesicles of rat distal colon, *Gastroenterology*, 101, 331, 1991.

60. Smith, L. H., Fromm, H. and Hofmann, A. F., Acquired hyperoxaluria, nephrolithiasis, and intestinal disease. Description of a syndrome, *N. Engl. J. Med.*, 286, 1371, 1972.

61. Dobbins, J. W. and Binder, H. J., Effect of bile salts and fatty acids on the colonic absorption of oxalate, *Gastroenterology*, 70, 1096, 1976.

62. Modigliani, R., Labayle, D., Aymes, C. and Denvil, R., Evidence for excessive absorption of oxalate by the colon in enteric hyperoxaluria, *Scand. J. Gastroenterol.*, 13, 187, 1978.

63. Hatch, M., Freel, R. W., Goldner, A. M. and Earnest, D. L., Effect of bile salt on active oxalate transport in the colon, in *Colon and Nutrition*, Kasper, H. and Goebell, H., MTP Press Ltd., Lancaster, 1982, 299.

64. Earnest, D. L., Enteric hyperoxaluria, *Adv. Intern. Med.*, 24, 407, 1979.

65. Bright-Asare, P. and Binder, H. J., Stimulation of colonic secretion of water and electrolytes by hydroxy fatty acids, *Gastroenterology*, 64, 81, 1973.

66.   Gaginella, T. S., Lewis, J. C. and Phillips, S. F., Ricinoleic acid effects on rabbit intestine: an ultrastructural study, *Mayo Clin. Proc.*, 51, 569, 1976.

67.   Saunders, D. R., Sillery, J. and McDonald, G. B., Regional differences in oxalate absorption by rat intestine: evidence for excessive absorption by the colon in steatorrhoea, *Gut*, 16, 543, 1975.

68.   Fairclough, P. D., Feest, T. G., Chadwick, V. S. and Clark, M. L., Effect of sodium chenodeoxycholate on oxalate absorption from the excluded human colon--a mechanism for 'enteric' hyperoxaluria, *Gut*, 18, 240, 1977.

69.   Kathpalia, S. C., Favus, M. J. and Coe, F. L., Evidence for size and charge permselectivity of rat ascending colon. Effects of ricinoleate and bile salts on oxalic acid and neutral sugar transport, *J. Clin. Invest.*, 74, 805, 1984.

70.   Earnest, D. L., Williams, H. E. and Admirand, W. H., A physicochemical basis for treatment of enteric hyperoxaluria, *Trans. Assoc. Am. Physicians*, 88, 224, 1975.

71.   Stauffer, J. Q., Humphreys, M. H. and Weir, G. J., Acquired hyperoxaluria with regional enteritis after ileal resection. Role of dietary oxalate, *Ann. Intern. Med.*, 79, 383, 1973.

72.   Earnest, D. L., Perspectives on incidence, etiology, and treatment of enteric hyperoxaluria, *Am. J. Clin. Nutr.*, 30, 72, 1977.

73.   Marshall, R. W., Cochran, M. and Hodgkinson, A., Relationships between calcium and oxalic acid intake in the diet and their excretion in the urine of normal and renal-stone-forming subjects, *Clin. Sci.*, 43, 91, 1972.

74.   Sigmon, D., Kumar, S., Carpenter, B., Miller, T., Menon, M. and Scheid, C., Oxalate transport in renal tubular cells from normal and stone-forming animals, *Am. J. Kidney Dis.*, 17, 376, 1991.

75.   Baggio, B., Gambaro, G., Marchini, F., Cicerello, E., Tenconi, R., Clementi, M. and Borsatti, A., An inheritable anomaly of red-cell oxalate transport in "primary" calcium nephrolithiasis correctable with diuretics, *N. Engl. J. Med.*, 314, 599, 1986.

76.   Borsatti, A., Calcium oxalate nephrolithiasis: defective oxalate transport, *Kidney Int.*, 39, 1283, 1991.

77.   Farooqui, S., Mahmood, A., Nath, R. and Thind, S. K., Nutrition & urolithiasis: Part I-intestinal absorption of oxalate in vitamin B6 deficient rats, *Indian J. Exp. Biol.*, 19, 551, 1981.

78.   Gupta, R., Sidhu, H., Rattan, V., Thind, S. K. and Nath, R., Oxalate uptake in intestinal and renal brush-border membrane vesicles (BBMV) in vitamin B6-deficient rats, *Biochem. Med. Metab. Biol.*, 39, 190, 1988.

79.   Comici, M., Balestri, P. L., Lupetti, S., Colizzi, V. and Falcone, G., Urinary excretion of oxalate in renal failure, *Nephron*, 30, 269, 1982.

80.   Cattell, W. R., Spencer, W. G., Taylor, A. G. and Watts, R. W. E., The mechanism of the renal excretion of oxalate in the dog, *Clin. Sci.*, 22, 43, 1962.

81.   Greger, R., Lang, F., Oberleithner, H. and Deetjen, P., Handling of oxalate by the rat kidney, *Pflügers Arch.*, 374, 243, 1978.

82.   McIntosh, G. H. and Belling, G. B., An isotopic study of oxalate excretion in sheep, *Aust. J. Exp. Biol. Med. Sci.*, 53, 479, 1975.

83.   Greger, R., Renal transport of oxalate, in *Renal Transport of Organic Substances*, Greger, R., Lang, F. and Silbernagel, S., Springer-Verlag, Berlin, 1981, 224.

84.   Weinman, E. J., Frankfurt, S. J., Ince, A. and Sansom, S., Renal tubular transport of organic acids. Studies with oxalate and para-aminohippurate in the rat, *J. Clin. Invest.*, 61, 801, 1978.

85.   Greger, R., Lang, F., Oberliethner, H. and Sporer, H., Renal handling of oxalate, *Renal Physiol., Basel*, 2, 57, 1980.

86.   Knight, T. F., Senekjian, H. O. and Weinman, E. J., Effect of para-aminohippurate on renal transport of oxalate, *Kidney Int.*, 15, 38, 1979.

87.   Tremaine, L. M., Bird, J. E. and Quebbemann, A. J., Renal tubular excretory transport of oxalate in the chicken, *J. Pharmacol. Exp. Ther.*, 233, 7, 1985.

88.   Hautmann, R. and Osswald, H., Renal handling of oxalate. A micropuncture study in the rat, *Arch. Pharmacol.*, 304, 277, 1978.

89.   Knight, T. F., Sansom, S. C., Senekjian, H. O. and Weinman, E. J., Oxalate secretion in the rat proximal tubule, *Am. J. Physiol.*, 240, F295, 1981.

90.   Senekjian, H. O. and Weinman, E. J., Oxalate transport by proximal tubule of the rabbit kidney, *Am. J. Physiol.*, 243, F271, 1982.

91.   Osswald, H. and Hautmann, R., Renal elimination kinetics and plasma half-life of oxalate in man, *Urol. Int.*, 34, 440, 1979.

92.   Kasidas, G. P., Assaying of oxalate in plasma, in *Oxalate metabolism in relation to urinary stone*, Rose, G. A., Springer, Berlin, 1988, 26.

93.   Boer, P., Beutler, J. J., van Rijn, H. J. M., Berckmas, R. J., Koomans, H. A. and Dorhout Mees, E. J., Urinary oxalate excretion during intravenous infusion of diuretics in man, *Nephron*, 54, 187, 1990.

94.   Ullrich, K. J., Specificity of transporters for 'organic anions' and 'organic cations' in the kidney., *Biochim. Biophys. Acta*, 1197, 45, 1994.

95.   David, C. and Ullrich, K. J., Substrate specificity of the luminal Na(+)-dependent sulphate transport system in the proximal renal tubule as compared to the contraluminal sulphate exchange system, *Pflügers Arch.*, 421, 455, 1992.

96.   Ullrich, K. J., Rumrich, G., Fritzsch, G. and Klöss, S., Contraluminal para-aminohippurate (PAH) transport in the proximal tubule of the rat kidney. II. Specificity: aliphatic dicarboxylic acids, *Pflügers Arch.*, 408, 38, 1987.

97.   Ullrich, K. J., Rumrich, G. and Klöss, S., Contraluminal sulfate transport in the proximal tubule of the rat kidney. III. Specificity: disulfonates, di- and tri-carboxylates and sulfocarboxylates, *Pflügers Arch.*, 404, 300, 1985.

98.   Kinne, R. K. H., Selectivity and direction: plasma membranes in renal transport, *Am. J. Physiol.*, 260, F153, 1991.

99.   Wright, E. M., Transport of carboxylic acids by renal membrane vesicles, *Annu. Rev. Physiol.*, 47, 127, 1985.

100.   Aronson, P. S. and Kuo, S.-M., Heterogeneity of anion exchangers mediating chloride transport in the proximal tubule, *Ann. N.Y. Acad. Sci.*, 574, 96, 1989.

101.   Karniski, L. P. and Aronson, P. S., Anion exchange pathways for Cl-transport in rabbit renal microvillus membranes, *Am. J. Physiol.*, 253, F513, 1987.

102.   Kuo, S.-M. and Aronson, P. S., Oxalate-OH exchange in rabbit renal microvillus membrane vesicles, *FASEB J.*, 2, A753, 1988.

103.   Yamakawa, K. and Kawamura, J., Oxalate:OH exchange across rat renal cortical brush border membrane, *Kidney Int.*, 37, 1105, 1990.

104.   Aronson, P. S., The renal proximal tubule: a model for diversity of anion exchangers and stilbene-sensitive anion transporters, *Annu. Rev. Physiol.*, 51, 419, 1989.

105.   Bästlein, C. and Burchhardt, G., Sensitivity of rat renal luminal and contraluminal sulfate transport systems to DIDS, *Am. J. Physiol.*, 250, F226, 1986.

106.   McConnell, K. R. and Aronson, P. S., Effects of inhibitors on anion exchangers in rabbit renal brush border membrane vesicles, *J. Biol. Chem.*, 269, 21489, 1994.

107.   Baggio, B., Gambaro, G., Marchini, F., Cicerello, E. and Borsatti, A., Raised transmembrane oxalate flux in red blood cells in idiopathic calcium oxalate nephrolithiasis, *Lancet*, 2, 12, 1984.

108.   Baggio, B., Bordin, L., Clari, G., Gambaro, G. and Moret, V., Functional correlation between the Ser/Thr-phosphorylation of band-3 and band-3-mediated transmembrane anion transport in human erythrocytes, *Biochim. Biophys. Acta*, 1148, 157, 1993.

109.   Kuo, S.-M. and Aronson, P. S., Oxalate transport via the sulfate/HCO$_3$ exchanger in rabbit renal basolateral membrane vesicles, *J. Biol. Chem.*, 263, 9710, 1988.

110.   Löw, I., Friedrich, T. and Burckhardt, G., Properties of an anion exchanger in rat renal basolateral membrane vesicles, *Am. J. Physiol.*, 246, F334, 1984.

111.   Ebisuno, S., Koul, H., Menon, M. and Scheid, C., Oxalate transport in a line of porcine renal epithelial cells--LLC-PK1 cells, *J. Urol.*, 152, 237, 1994.

112.   Koul, H., Ebisuno, S., Renzulli, L., Yanagawa, M., Menon, M. and Scheid, C., Polarized distribution of oxalate transport systems in LLC-PK1 cells, a line of renal epithelial cells, *Am. J. Physiol.*, 266, F266, 1994.

113.   Verkoelen, C. F., Romijn, J. C., de, B. W., Boeve, E. R., Cao, L. C. and Schroder, F. H., Absence of a transcellular oxalate transport mechanism in LLC-PK1 and MDCK cells cultured on porous supports, *Scanning Microsc.*, 7, 1031, 1993.

114.   Wandzilak, T. R., Calo, L., D'Andre, S., Borsatti, A. and Williams, H. E., Oxalate transport in cultured porcine renal epithelial cells, *Urol. Res.*, 20, 341, 1992.

115.   Karniski, L. P. and Aronson, P. S., Chloride/formate exchange with formic acid recycling: a mechanism of active chloride transport across epithelial membranes, *Proc. Natl. Acad. Sci. U.S.A.*, 82, 6362, 1985.

116.   Wareing, M. and Green, R., Effect of formate and oxalate on fluid reabsorption from the proximal convoluted tubule of the rat, *J. Physiol.*, 477, 347, 1994.

117.   Wang, T., Giebisch, G. and Aronson, P. S., Effects of formate and oxalate on volume absorption in rat proximal tubule, *Am. J. Physiol.*, 263, F37, 1992.

118.   Wang, T., Agulian, S. K., Giebisch, G. and Aronson, P. S., Effects of formate and oxalate on chloride absorption in rat distal tubule, *Am. J. Physiol.*, 264, F730, 1993.

Chapter 12

OXALATE MEASUREMENT IN BIOLOGICAL FLUIDS

Martino Marangella and Michele Petrarulo

I. HISTORICAL OVERVIEW

Calcium oxalate was first described as a constituent of certain renal stones in the early nineteenth century, and implausible, primitive procedures for its determination in urines were based on gravimetry [1]. During the thirties of the present century, a titrimetric procedure using potassium permanganate was introduced for both blood and urine measurements [2], and its use was still popular in the early fifties. Apart from being tedious and time consuming, this method gave poorly reliable plasma values, in the order of hundreds of $\mu mol/L$. Starting from the sixties, the purification of specific enzymes [3,4], and successively, the availability of highly sensitive chromogen systems [5,6] brought about a decided improvement in urine measurements.

In recent years, the advent of liquid chromatographic techniques has further improved the reliability and feasibility of oxalate assays [7,8]. The fact that all pre-treatment steps to separate oxalate from the original matrix can now be omitted has dramatically improved the accuracy and precision of urine measurements. For plasma assays, the progressive narrowing of differences between direct methods and indirect *in vivo* isotopic procedures has constituted definite progress, which was only possible after it had been recognized that proper handling of blood samples is crucial to prevent *in vitro* oxalogenesis as a source of unreliable results.

The issue of oxalate measurements in plasma and urine and other body fluids will be addressed in more detail in this chapter.

II. CHEMICAL AND PHYSICO-CHEMICAL PROPERTIES OF OXALATE

Oxalate is a strong dicarboxylic acid, 90 g/mol molecular weight, and dissociation constants ($pKa_1 = 1.25$ and $pKa_2 = 4.27$) which indicate that it is almost entirely dissociated at pHs higher than 5, typical of the majority of the biological fluids. In these conditions oxalate behaves as a strong ligand of calcium. Oxalate also binds to other relevant cations, such as magnesium, sodium, potassium and ammonium, as indicated by the stability constants of some of the resulting complex species listed in Table 1 [9]. The thermodynamic solubility product of calcium oxalate monohydrate is very low ($K_{sp,th} = 2.5 \times 10^{-9}$ $mol^2 \cdot L^{-2}$) It increases with increasing ionic strength: at usual ionic strength of plasma (0.160 mol/L) K_{sp} is 2.3×10^{-8} $mol^2 \cdot L^{-2}$ [10].

Table 1.
Stability   Constants   of   Relevant   Cation-Oxalate
Complexes at   0.15 mol/L Ionic Strength

Complex Species	Stability Constants (log ß)
H(Ox)⁻	3.98
Ca(Ox)	2.36
Mg(Ox)	2.67
Na(Ox)⁻	0.51
K(Ox)⁻	0.32
NH_4(Ox)⁻	0.65

III.   OXALATE DISTRIBUTION IN BODY FLUIDS

As outlined below, the kidney is highly efficient in removing oxalate, therefore both plasma levels and miscible pool size are very low, in the order of 1-2 µmol/L and 10-50 µmoles, respectively. In plasma there is virtually no protein binding, oxalate is freely filtered at the glomerulus and is thought to distribute uniformly in the extracellular fluid.

Most of the information on the distribution spaces of oxalate comes from isotopic dilution studies, which are generally performed by using the [14C] labelled oxalate. From these studies it has been unequivocally shown that the distribution volume of oxalate exceeds that of extracellular fluid by a factor of about 1.5 [11], and is only 65-70% of total body water [12]. This means on the one hand, that oxalate is not freely diffusible in all the body water and, on the other that there is an as yet unknown space outside the extracellular fluid where oxalate is located.

An increase in the miscible pool of oxalate is induced by renal insufficiency, alone or associated with increased oxalate biosynthesis. Even when oxalate pool size is expanded many times, the distribution space does not increase [11,13]. Whereas plasma oxalate has been studied by both indirect and direct procedures, intracellular oxalate levels have so far been determined only rarely. In a recent study Hatch has measured oxalate concentration spectrophotometrically in both serum and whole blood from normal individuals [14]. The fact that she found no difference between the two sets of determinations would suggest that oxalate concentration in red blood cells is similar to that of serum. However, we still do not know whether any change of oxalate in extracellular fluids is paralleled by concurrent changes in intracellular fluids. In the setting of relentless oxalate retention, plasma clearance has been reported to exceed renal clearance, resulting in a tissue accumulation of oxalate, starting at 20 or 40 mL/min GFR in patients with renal failure unrelated to or associated with primary hyperoxaluria, respectively [15]. These findings have been interpreted as indicating the constitution of a new poorly exchangeable pool, represented by tissue deposition of calcium oxalate crystals. This hidden pool is not accessible with non invasive procedures and can only be detected by measurements of oxalate in body tissues, such as bone (see below).

IV.   OXALATE DETERMINATION IN URINE AND BLOOD

A.   OXALATE DETERMINATION IN URINE
1.   Sampling,   Storage and Processing
Samples from daily urinary excretion are generally recommended for investigating oxalate dismetabolisms. Spot urines may suitably be used in pediatric settings or for circadian studies. Irrespective of the type of urine collection, the correct pre-analytical management of samples, i.e., harvesting, storage and pre-treatment, is so crucial for reliability of the results, that improper sample processing in the pre-analytical phase may frustrate the adoption of even highly accurate assaying techniques. Urine should be collected in the presence of preservatives, and calcium oxalate formation should be prevented by acidifying and warming before assaying [16,17]. The most important interference in oxalate determination involves ascorbate, which may affect oxalate assay in different ways during both pre-analytical and analytical operations. In neutral and alkaline environments, ascorbate degrades spontaneously and produces oxalate and threonate [18], thereby increasing the measurable oxalate [19]. The implications of this finding on urine assays have been fully realized since the early eighties [20-25]. Subsequently it has been suggested that urine should be collected and stored frozen in the presence of mineral acid (i.e. 0.1 mol/L HCl, final conc.) in order to prevent the ascorbate to oxalate conversion [21].

In considering a biological sample for analysis of oxalate, it must also be taken into account that:
- the alkalinization may give rise to abnormally high results due to the ascorbate promoted oxalogenesis [19];
- the preconcentration and purification of oxalate by precipitation should be avoided because it is generally incomplete and poorly reproducible [21].

2.   Chemical Techniques
a. Atomic Absorption Spectrometry
Calcium oxalates can be separated by precipitation from urines at pH between 4 and 6 in the presence of an excess of calcium ions. This allows urinary oxalate to be determined indirectly, by measuring calcium by means of atomic absorption spectrometry. The difference between total (added and endogenous) and residual (present in the supernatant after precipitation) calcium is related to the oxalate content [26,27].

b. Colorimetry-Spectrophotometry
Several non-enzymatic spectrophotometric procedures for the determination of urine oxalate were proposed, up to the late seventies. Preliminary purification of oxalate by solvent extraction or precipitation as the calcium salt was invariably needed, because of poor sensitivity and specificity of the chromogenic systems. Oxidation with permanganate or cerium was used for the titrimetric determination of urine oxalate [28,29]. Glyoxylate produced by the reduction of oxalate was reacted with resorcinol [30] or phenylhydrazine/ferricyanide [31] to obtain

fluorescent or uv-absorbing derivatives. Oxalate was reduced to glycolate before reaction with chromotropic acid and subsequent formation of a colored adduct [32,33]. Baadenhuijsen and Jansen measured the decrease in absorbance induced by urine oxalate on an uranium(IV)-4-(2-pyridilazo)resorcinol complex [34].

More recently, other spectrophotometric procedures based on the phase-transfer of oxalate ligand complexes have been described [35,36]. Concurrently, more simple and accurate enzymatic or ion chromatographic techniques have been developed, and these are now generally considered to better meet the requirements of routinely usable methods.

3.   Enzymatic Procedures

a. Oxalate Oxidase

The analytical use of specific oxalate-degrading enzymes (now commercially available) has simplified sample processing and enhanced reliability. Oxalate oxidase from moss, barley or other vegetable sources, at pH around 4, catalyzes the oxidation of oxalate into carbon dioxide and hydrogen peroxide, and the monitoring of the products of degradation can be related to the original oxalate content. In earlier procedures the enzymatic reaction was monitored by measuring the carbon dioxide produced[37].

The availability of the Trinder chromogenic system, which is highly sensitive for detecting hydrogen peroxide, has made this a marker of choice. This system is based on the peroxidase catalyzed activation of hydrogen peroxide with production of hydroxide radicals, capable of coupling proper chromogen substrates to form highly absorbing quinoneimine dyes [5]. This chromogenic system, however, was unsuitable for assaying unprocessed urine, essentially because of the presence of reducing substances, which react with the free radicals, decreasing the color yield [38,39]. This drawback has now been overcome by the use of simple pre-treatments, such as charcoal adsorption or selective oxidation, which eliminate ascorbate and other reducing interferences [23, 40-42]. Various procedures that use Trinder-like reaction have been reported using either immobilized [20,42,43] or soluble enzyme [40,41,44,45].

It will be recognized that, in addition to accomplishing reliable and simple low determinations of oxalate in urine [46], oxalate oxidase is now successfully used for blood plasma determinations, which, in normals, means at oxalate concentrations two orders of magnitude lower [43,47].

b. Oxalate Decarboxylase

Oxalate decarboxylase from fungi cultures (*Collybia velutipes*) was found to catalyze the decarboxylation of oxalate with formation of carbon dioxide and formic acid with optimum pH round 3 [4]. Since the early sixties this enzyme has been applied to urine oxalate assay after oxalate precipitation [48], urine pre-concentration [49], or no pre-treatment [50]. However, the measurement of the carbon dioxide evolved, performed either manometrically or by measurement of pH modification, was poorly sensitive and precise, as well as tedious and time-consuming.

The use of formic acid as an analytical marker has represented an important step forward in the analytical use of this enzyme [51,52]. In the presence of formate dehydrogenase and NAD, the formate produced by the oxalate oxidase reaction is oxidized with concurrent formation of NADH, which is detected spectrophotometrically. The assay on unprocessed urines gives poor analytical recoveries [51]. It has been suggested that the addition of EDTA could overcome this drawback [53]. Nevertheless, the simplicity and the reliability of the procedure make it currently suitable for routine purposes [54].

4. Chromatographic Procedures

a. Gas Chromatography

Applications of gas-liquid chromatography (GLC) in the determination of oxalate in urine have been reported since the early seventies [55-60]. Compared with the complexity of other procedures at that time, GLC made a valuable contribution to the reliability of measurements of urine oxalate. Different versions were proposed; fundamentally, oxalate had to be extracted and esterified in an anhydrous environment before the chromatographic assay. Advances in separation and detection technologies have enhanced sensitivity and accuracy. Capillary GLC can be considered as a valuable tool to assess oxalate concentration in body fluids [61,62]. Unfortunately, tedious sample management and the requirement of both highly technical specialization and special equipment hinder a widespread adoption of this technique.

b. Liquid Chromatography

Early liquid chromatographic approaches to the study of acidurias by using uv-detection were unsuitable for oxalate assay because of the low specific absorption of oxalate [63,64]. The use of electrochemical detection after precipitation of oxalate and cation-exchange/ion-pair chromatographic separation provided the first, though poorly reproducible, liquid chromatographic procedure for the determination of oxalate in urines [65]. Subsequent works using similar techniques seemed to better fulfill the need for simplicity and reproducibility [66]. A reversed-phase/ion-pair separation with uv-detection, based on solid-phase extraction clean-up before chromatography, was proposed. However, this quick and simple procedure was affected by negative ascorbate interference and non-linear response [67].

1,2-Diaminobenzene or 9-anthryldiazomethane have been used for the derivatization of oxalate before the chromatographic separation of the resulting uv-absorbing [68,69] or fluorescent [70] derivative. The requirement of sample processing coupled with the need for special instrumentation have limited the diffusion of these sensitive techniques.

Anion exchange-chromatography with suppressed conductivity detection has signified a new and important opportunity for oxalate assay [71]. Suppression of eluent conductivity greatly enhances the sensitivity of conductimetric detection, which, coupled with chromatographic separation using selective and low capacity resins, is particularly suitable for the quantification of small anions. This technique seemed to be applicable to urinary oxalate assay without any sample pre-treatment, and consequently quick as well as precise [7,8]. However, the use of alkaline eluents promotes the in-column ascorbate-to-oxalate conversion; the presence of exceedingly high concentrations of ascorbate, as in the case of huge intakes, may affect the oxalate assay. It has been suggested that overestimation can be prevented by either buffering the sample with boric acid [22], or eliminating ascorbate [72], or lowering the eluent pH [73].

The current opinion is that ion chromatography is one of the most simple and reliable techniques for the determination of oxalate in urines. Unfortunately, the need for dedicated and automatized instrumentation still limits its diffusion.

B. OXALATE DETERMINATION IN BLOOD
1. Indirect Radionuclide-Dilution Procedures

A major issue in the field of plasma oxalate measurements has been the discrepancy between indirect and chemical direct procedures. For a long time oxalate estimates obtained by means of indirect isotopic-dilution methods were taken as reference values [11,74-77]. With these models plasma oxalate is calculated by the formula:

$$Plasma\ Oxalate = Urine\ Oxalate/[^{14}C]Oxalate\ Clearance$$

which is derived from the general formula for renal clearance calculation. Urine oxalate is measured directly, whereas oxalate clearance is assessed by *in vivo* radioisotopic dilution procedures. A number of authors have used this approach, reporting reference ranges between 0.6 and 2.9 µmol/L, in close agreement with each other. Table 2 lists some of these results.

Table 2
Reference Ranges of Plasma Oxalate Concentration
Assessed by Indirect *in vivo* Isotopic Dilution Studies

Author, Ref.	Oxalate Concentration (µmol/L)	
	Mean	Range
Pinto, 74	1.39	0.39 - 2.90
Constable, 75	1.04	0.75 - 1.40
Prenen, 76	1.39	1.04 - 1.78
Watts, 11	0.95	0.65 - 1.45
Linari, 77	2.1	1.30 - 2.40

The use of indirect procedures is however hampered by a number of disadvantages. First, the use of radionuclide-labelled oxalate is required. Second, considerable amounts of blood are necessary for a complete study to be carried out. Third, the direct measurement of oxalate in urine may be a further source of inaccuracy. Fourth, the infusion technique and the use of one or two-compartment model may affect reliability of the results. Concerning this latter point, differences have been found when single injection or continuous infusion of [^{14}C]oxalate are used, single injection being shown to overestimate renal clearances by as much as 30 to 52%, with consequent underestimation of plasma levels of oxalate [76].

These *in vivo* procedures cannot be used routinely and are now substituted by modern direct assays of plasma oxalate. Apart from investigative applications, they are still of historical importance, in that in consequence of these results laboratories have been compelled to optimize direct procedures for plasma oxalate.

2. Direct Determination: Sampling, Storage and Processing

A number of chemical, enzymatic and chromatographic methods, applied for the blood oxalate assay have been experimented during the last fifteen years. Whereas no particular difficulties have been encountered with pathologically high concentrations, these have produced conflicting orders-of-magnitude ranges for normal subjects. The reasons for the aforementioned discrepancy between *in vivo* isotopic methods and direct assays remained unexplained until the early eighties [78]. Today it is widely accepted that this was caused by spontaneous oxalogenesis that takes place in blood soon after sample collection. Oxalogenesis is essentially favoured by the low concentration ratio between plasma oxalate and potential interfering substances and precursors [79,80]. This phenomenon was overlooked for long time, but once it was focused, it provoked many efforts to achieve a pre-analytical blood sample stabilization [61,79]. Highly accurate sample processing has progressively narrowed discrepancies with indirect *in vivo* assays, to the extent that many currently available procedures can be considered sufficiently accurate [43,47,81-84].

In general, in order to stabilize the original oxalate content even after prolonged frozen storage, eparinized blood should be managed in the cold, and the separated plasma should be promptly acidified and deproteinized. In addition, as mentioned above for urines, all alkalinization during the analytical process should be avoided because this would promote the endogenous ascorbate to oxalate conversion, which would result in falsely manifold increases in oxalate concentration.

The earlier approach to plasma oxalate determination aimed at preventing misleading increases in oxalate was proposed in 1980 [79]. Blood was collected in the presence of specific inhibitors of enzymes supposed to produce oxalate and plasma was analyzed using an enzymatic method after deproteinization with hydrochloric acid and treatment at 100°C. Although the ascorbate interference was neglected the acidification prevented it from degrading and helped to produce acceptable results (0-5.4 μmol/L). However, even omitting these enzyme inhibitors, the ultrafiltration of plasma and precipitation of oxalate as calcium oxalate at neutral pH, associated with enzyme assays, gave similar [45,80] or slightly higher [85] results. Lower levels were found when ultrafiltrates were promptly acidified before precipitation [81].

Ultrafiltration of HCl-acidified plasma (at pH around 4) supplemented with sodium nitrite yielded concentrations close to 2 μmol/L. This treatment appeared to be suitable for short term stabilization of oxalate [86]. However, this procedure was affected by some decrease of ultrafilterability of oxalate due to binding to proteins at that pH. The soundness of these results has therefore been questioned [43,81,82,87].

Ultrafiltration of more strongly HCl-acidified plasma (pH<2) yielded complete recoveries, was suitable for long frozen storage of samples and provided mean oxalate concentrations of 2.5 µmol/L, when coupled with an immobilized oxalate oxidase assay [43].

Plasma samples deproteinized/acidified with sulfosalicylic acid were found to be stabilized during frozen storage, and assays performed with soluble oxalate oxidase after charcoal treatment gave similarly low concentrations in normal individuals (0.6-2.9 µmol/L) [47].

Reliable oxalate concentrations in normal subjects (1-2 µmol/L) have been observed by means of enzyme/bioluminescence assays after acidification of plasma, using an acid exchange resin and omitting sample deproteinization. These authors have claimed that this procedure is useful in stabilizing oxalate during the frozen storing of samples from normal individuals [6,88].

Other enzymatic procedures in which the sample was treated under alkaline conditions, which means neglecting the problem of spontaneous ascorbate-to-oxalate degradation, gave exceedingly high values [89-94].

Prompt acidification of plasma with HCl has been found to induce some stabilization of oxalate before capillary GLC assay, and reference values ranging between 1.3-5.3 µmol/L [61] or between 2.7-6.0 µmol/L [62] have been reported. However, no convincing data supporting the suitability of this pre-treatment for frozen storage have been produced.

Anion exchange chromatography has also been used for determination of oxalate in blood. Several reports have been concurrently published with the use of this technique after different sample preparation. Samples were immediately ultrafiltered and stored at -80°C [95], or else stored frozen and ultrafiltered [96], or strongly acidified and ultrafiltered [82], or deproteinized by acidificating and heating [97] before being assayed. The results obtained have been conflicting, in that, whereas some have reported acceptably low reference ranges (0.7-2.9) [95], others have been unable to do so. Ultrafiltration followed by treatment of the ultrafiltrate with a silver cation exchanger has been shown to give reliable results [83]. We have recently reexamined the ion chromatographic determination of oxalate and have shown that acidification and acetonitrile extraction of plasma before assay provided similarly acceptable results, and stabilization during frozen storage [84].

The analytical details of some of the above procedures will be discussed in the following sections.

3.   Enzymatic Techniques
a. Oxalate Oxidase
The use of oxalate oxidase coupled with hydrogen peroxide measurement has so far been extensively challenged. These techniques are substantially derived by urinary applications. Apart from an unsuccessful attempt with the use of the catalase/aldehyde dehydrogenase/NADP reaction [93], hydrogen peroxide has been most often detected by using the peroxidase/Trinder color reaction, which is much

more sensitive. The enzyme has been used either in the immobilized [43,86] or in the soluble form [45,47,87,92,94] after different sample processings.

As mentioned above, reliability of the methods crucially depends both on avoiding alkaline conditions and eliminating reducing interferences (i.e., ascorbate, etc.). Immobilized oxalate oxidase has been described in a continuous flow system in which ascorbate interference was prevented by nitrites [86]. Recently, this method has been modified by using a more sensitive detector and a more suitable sample preparation; since, in view of the pH dependence of oxalate ultrafilterability, plasma is strongly acidified before ultrafiltration. These modifications have produced reliable results, even in normal subjects [43].

Reaction with the enzyme in the soluble form has been applied after plasma ultrafiltration and precipitation of oxalate; despite the need for internal standardization and the complexity of sample handling, acceptably low concentrations were obtained in the normal range (0.4-3.7 μmol/L) [45]. In a further application, in which oxalate precipitation was avoided and ascorbate was eliminated enzymatically, the determination was performed on samples from hemodialyzed patients: the poor recoveries obtained were improved by lowering the sample-to-reagent ratio [87]. As mentioned above, with the use of soluble oxalate oxidase after deproteinization with sulfosalicylic/acid and purification with charcoal, we were able to furtherly increase the sample-to-reagent ratio, obtaining a proportional increase in sensitivity; low recoveries in samples from uremic patients were corrected by the charcoal treatment [47].

a. Oxalate Decarboxylase

The proper handling of samples is an essential prerequisite for achieving acceptable results. The following methods have been set up, taking this concern into account.

Earlier oxalate decarboxylase methods, which were affected by the lack of a sensitive analytical signal, required either preconcentration of oxalate by precipitation or large sample volumes for analysis [79,81]. The use of this enzyme, for urinary assay coupled with measurement of pH changes in an alkaline buffer which fixed the carbon dioxide produced by oxalate degradation, has already been mentioned [50]. Its adaptation to plasma required 50 mL of sample, tedious reagent preparation and prolonged incubation time; the authors themselves recognized the insuitability of their procedure for routine purposes [79]. A radioenzymatic isotope dilution assay has been described in which [14C]oxalate was added, as a tracer, to a 15 mL plasma sample. After ultrafiltration, precipitation as calcium oxalate, lyophilization and extraction of the precipitate, the sample was incubated with oxalate decarboxylase and the radioactive carbon dioxide measured by scintillation counting. A mean oxalate concentration of 4 μmol/L was reported for normals [85].

The oxalate decarboxylase/formate dehydrogenase reaction has been experimented in plasma assays, and in this case, too, the reported applications were derived from those originally set up for urines [51]. Ultrafiltration of 15 mL of plasma sample, precipitation as calcium oxalate, lyophilization and extraction of the precipitate before the double-enzyme assay provided mean values for normals averaging 3 μmol/L. This method required isotopic dilution to account for possible oxalate losses during the sample processing [80]. In a subsequent modification, the ultrafiltrate was collected directly into HCl in order to acidify promptly: mean oxalate concentration in normal plasma averaged 1.25 μmol/L [81].

A more sensitive procedure, requiring 2 mL of plasma, uses the above double enzyme reaction coupled with a commercially available enzyme/bioluminescent assay for the NADH generated. The plasma is simply acidified with an acid exchange resin before being stored at -20°C or assayed [6]. In a further modification the bioluminescence enzymes, i.e. NADH:FMN oxidoreductase and luciferase, are co-immobilized to provide a cheaper reusable reagent suitable for continuous flow assay [88].

4.   Chromatographic Techniques
Apart from the specific sample handling described above, the anion exchange chromatographic and GLC techniques for plasma oxalate do not substantially differ from those for urines. It must be added that, whereas reliable anion chromatographic applications for plasma oxalate in normals yield acceptable reference values [83,84,95], concentrations obtained by the more sophisticated GLC techniques [61,98] are invariably higher than those given by indirect procedures.

V.   OXALATE MEASUREMENT IN CLINICAL PRACTICE

A.   CLINICAL SIGNIFICANCE OF URINE AND BLOOD ASSAYS
The relevance of oxalate measurements in human pathology stems from the peculiar physicochemical properties of calcium oxalates, which have an extremely low solubility in aqueous solution and, hence, in biological fluids. In human beings oxalate is an end-product of metabolism and the kidney is virtually the only route by which it can be removed from body fluids. In normal subjects the renal clearance is exceptionally effective, so that oxalate levels in plasma, and in other extracellular fluids, are very low. Oxalate levels of intracellular spaces could be even lower. Therefore, in normal conditions, body fluids are virtually always undersaturated with calcium oxalates [10]. This advantage is however offset by the fact that urine concentrations often reach supersaturation [99,100].

The fact that oxalate is found in more than 60% of all kidney stones indicates that, in urine, it represents a major risk factor in calcium oxalate stone disease [101]. Therefore, measurements of oxalate in urine have been universally included in the protocols for metabolic evaluation of patients with calcium nephrolithiasis. By examining a wide range of cases, it has been shown that disorders in oxalate excretion occur in a substantial proportion of such patients (Table 3).

Table 3.
Prevalence of Hyperoxaluria in Calcium
Oxalate Nephrolithiasis

Author, Ref.	Prevalence (%)
Robertson & Peacock, 99	13 - 23
Baggio et al., 102	50
Jaeger et al., 103	11 - 22
Bek-Jensen & Tiselius, 104	17
Yendt & Cohanim, 105	21
Marangella, 106	32

In addition to clinical conditions in which hyperoxaluria is a primary defect, untoward derangements of urine oxalate may be caused by improper dietary or pharmacological maneuvers. Table 4 lists the main causes of abnormal oxalate excretion.

Table 4.
Clinical Applications of Measurements of
Oxalate in Urine

Defined defects of oxalate excretion
Mild idiopathic hyperoxaluria
Primary hyperoxaluria type I, II (III)
Enteric hyperoxaluria
Dietary hyperoxaluria
Excessive intakes of bioavailable oxalate
Low calcium diets
Animal Protein-rich diets (?)
Pyridoxine deficiency
Drug-induced Hyperoxaluria
Ascorbate supplementation (?)
Pyridoxilate
Xylitol infusion
Ethilen glycol poisoning

The question of the dietary dependence of oxalate excretion is very complex and is worthy of attention. Dietary hyperoxaluria may arise directly from excessive intakes of bioavailable oxalate [107]. Low calcium diets are known to increase intestinal absorption of oxalate [108], namely in patients with absorptive hypercalciuria [103] or elevated levels of vitamin D [109]. Diet-related hyperoxaluria may also be endogenously caused, in patients on total parenteral nutrition [110], upon xylitol administration [111], in pyridoxine deficiency [112], or in the still controversial cases of animal protein-rich diets [113,114] and ascorbate supplementation [115].

Plasma oxalate determination is not routinely included in the first-visit evaluation of patients with calcium stone disease, because these patients do not exhibit notable abnormalities in blood oxalate. However, when abnormally high levels of urine oxalate are found in baseline exams it is advisable to consider taking plasma measurements. Updated assays of plasma oxalate now ensure considerable sensitivity and reliability, and hence even subtle changes can be detected. Plasma oxalate tends to increase in conditions associated with increased oxalate metabolic generation. Conversely, recent reports have suggested that patients with mild idiopathic hyperoxaluria have an increase in the renal clearance of oxalate [116]; this might result in decreases in plasma oxalate detectable with modern procedures.

Renal insufficiency is a source of misleading results, because, even in the early phases, reduction in oxalate excretion induces oxalate retention [75,117]. In this subset results and reference ranges for both plasma and urine measurements should be re-established. Oxalate retention represents a threat for the development of systemic oxalosis, featured by extra-renal deposition of calcium oxalate crystals. This is a potentially useful application of oxalate measurement in plasma, and applies to patients with end-stage renal failure due to either primary hyperoxaluria or oxalosis-unrelated renal diseases. Measurements of plasma concentrations are helpful tools for detecting and differentiating hyperoxaluria syndromes in the setting of end-stage renal failure, and for planning dialysis and transplantation strategies. Main clinical applications of plasma oxalate measurement are listed in Table 5.

Table   5.
Clinical   Applications of   Measurements   of
Oxalate in   Plasma

Idiopathic hyperoxaluria (Renal Hyperoxaluria)
Primary hyperoxalurias
Enteric hyperoxaluria
Oxalosis-unrelated renal failure
Primary or Secondary oxalosis

B.   ASSESSMENT OF REFERENCE RANGES

Correct reference ranges are crucial for the interpretation of both urine and plasma measurements. Reference ranges must be fixed using well-defined procedures for sample collection and sample handling, and should be applied to series matched for sex, age and renal function.

1.   Age and Sex Profiles.
a. Urine Oxalate in Adults

Urine oxalate excretion in normal adults shows a normal or log-normal distribution of values. Means ±(SD) and reference ranges for males and females obtained in our laboratory using an ion chromatographic procedure are shown in Table 6.

Table 6.
Sex-matched   Reference Ranges of Urine Oxalate
in Normal Individuals

	All (80)	Males (41)	Females (39)	p
mmol/24h				
Mean ± SD	0.32 ± 0.14	0.33 ± 0.15	0.30 ± 0.14	n.s.
Range	0.10 - 0.68	0.10 - 0.68	0.10 - 0.66	n.s.
µmol/mmol uCr				
Mean ± SD	26.1 ± 12.5	25.3 ± 11.5	26.9 ± 11.6	n.s.
Range	7.1 - 72.3	7.1 - 59.1	8.5 - 72.3	n.s.

Males do not differ from females in oxalate excretion expressed as both mmol/24h and µmol/mmol urine creatinine. Similarly, no gender differences for urine oxalate in normal individuals have been reported by others,
using different procedures [83,118,119]. Indeed, oxalate excretion was independent of age in our normal adults aged twenties through seventies. Upper limits for normal oxalate excretion are fixed worldwide at 0.5 mmol/24h or 50 µmol/mmol urine creatinine [101,104,105,119].

b. Urine Oxalate in Infants and Children
The question of reference ranges is intriguing in the case of infants and children, and controversy still exists concerning the way results should be normalized, that is, body surface area, body weight, urine creatinine. In these patients daily urine collection are often difficult to perform, oxalate measurements on spot urines have therefore been proposed to obviate this.
A number of interesting papers have recently been produced on this item. A large group of normal infants and children was studied for oxalate excretion by Leumann et al. using oxalate oxidase [120]. They found that oxalate to creatinine ratios, assessed on non-fasting morning urine, tended to increase during the first few days of life and then decreased linearly between 6 months and 4.9 years of age. Children between 5 and 16 years of age behaved in a similar way, though with slighter differences. Fasting and non-fasting samples did not differ significantly. Barratt et al. performed a similar study on 137 normal children aged between less than 1 year and more than 12 years [121]. They confirmed that oxalate to creatinine molar ratios decreased with age. Morgenstern et al. made a longitudinal study of oxalate (and glycolate) excretion during the first year of life. The oxalate to creatinine ratio was distributed log-normally, but there was no significant decrease in individual patients during this span of time [122]. However, the range was about ten times higher than that of normal individuals

from their laboratory. Von Schnakenbourg et al. measured oxalate excretion on spot urine by ion-chromatography in 169 healthy children aged 1 day to 13 years, and found than means decreased from 131 ± 57 in infants to 42 ± 31 μmol/mmol creatinine in children over 10 [123]. They attributed this decrease to the progressive gain in muscle mass and hence in creatinine production, with age.

Some conclusions can be drawn from the aforementioned studies: first, spot urines can be used instead of 24-h urines provided age-matched reference values obtained with a given procedure be set. Second, oxalate to creatinine ratio decreases with age and appears to stabilize after adolescence: this can possibly be accounted for by an increase in creatinine production and excretion with age, and not by an actual decrease in oxalate excretion. Third, there are many clinical settings in which oxalate measurements in infants and children are of interest, including primary hyperoxalurias, formula-fed infants [122], premature infants receiving parenteral nutrition [110].

c. Plasma Oxalate

Age and sex profiles of plasma oxalate concentrations have received relatively less attention. The definition of reference ranges has been crucially affected by difficulties in obtaining reliable assays. The more recent ion-chromatographic or enzymatic procedures report normal values close to those found by indirect radioisotopic dilution methods. In three independent studies using anion exchange chromatography Schwille et al. [95], Hagen et al. [83] and we ourselves [84] found no difference in the reference intervals between males and females. Similar results were obtained using oxalate oxidase assays [43, 47]. Data from different series are listed in Table 7.

Table 7.
Means and Reference Intervals of Plasma Oxalate (μmol/L)
in Males and Females from Five Independent Studies

Author, Ref.	Method	Males		Females	
Schwille, 95	Anion Chrom	1.98	(1.4-2.5)	1.78	(0.7-2.9)
Wilson, 43	Ox Oxidase	2.8	-	2.3	-
Hagen, 83	Anion Chrom	2.0	(0.8-3.2)	1.8	(1.0-2.6)
Petrarulo, 84	Anion Chrom	1.89	(0.8-3.0)	1.91	(0.9-3.4)
Petrarulo, 47	Ox Oxidase	1.32	(0.6-2.7)	1.24	(0.6-2.4)

There are no available data concerning age-related differences of plasma oxalate in adult individuals. Barratt et al. reported a geometric mean of plasma oxalate of 1.53 μmol/L (range 0.78-3.02), in normal children, which was independent of age [121].

2. Circadian and Seasonal Variations of Oxalate

The dietary influences on metabolic generation and intestinal absorption of oxalate account for its great variability in the day-to-day determinations. This would also explain diurnal and seasonal variations of oxalate excretion. Because this variability may restrict the reliability of oxalate measurements, it seems worth trying to understand these trends.

In London, Hallson et al. reported a significant increase in oxalate excretion in calcium stone-formers during summer, and attributed this to both an increase in a dietary supply of oxalate-rich foods and to a decrease in dietary calcium [124]. Assuming a link between vitamin D and oxalate excretion [109] their observations could be tentatively explained with enhanced vitamin D activation due to an increase in hours of sunshine during summer.

Hargreave el al. described diurnal variations of oxalate, which peaked following breakfast and lunch, and decreased during the waking hours [125]. They attributed these changes to intestinal absorption of dietary oxalate, but failed to confirm this in a subsequent study on evening urinary oxalate from stone formers [126]. It has been suggested that the oxalate excretion of normal subjects follows a circadian rhythm, not found in stone-formers [127]. Adult individuals were found to have a higher oxalate to creatinine ratio on 24-h urine than on morning urine [83]. Finch et al. found that in normal subjects day-to-day variations of urine oxalate were low on oxalate-controlled diets, whereas great variability was observed on unrestricted diets [128]. This led them to conclude that the oxalate endogenously produced is very constant from day to day and that fluctuations are almost entirely due to diet. More recently, Holmes et al. have observed considerable variability in day to day oxalate excretion despite a control in calory and protein intake in normal subjects [129].

All these data substantially point to possible pitfalls related to sampling of urine for oxalate measurements. In fact, oxalate excretion exhibits day-to-day and intra-day variations and the fine mechanisms of these fluctuations are as yet poorly understood. This may result in underscoring of the oxalate-related risk of stone formation and justifies the advice of measuring oxalate repeatedly on different days and on fractionated collections.

3.   The Case of Impaired Renal Function

The occurrence of renal insufficiency involves a number of items. First, new reference ranges to detect cases of altered oxalate excretion. Second, changes in renal handling of oxalate induced by or associated with renal failure. Third, evaluation of the residual risk of stone formation in the course of renal failure. Fourth, evaluation of the risk of systemic oxalosis resulting from chronic oxalate retention.

Because the kidney is virtually the only route for removing oxalate from body fluids, renal insufficiency is expected to induce a decrease in urine excretion and an increase in plasma levels of oxalate. Theoretically, renal insufficiency could affect the handling of oxalate, as it does for other anions such as phosphate and citrate [130]. Urine and plasma oxalate have so far been measured only rarely in patients with impaired renal function. More evidence is available from indirect *in vivo* studies of renal clearance of oxalate [15,75,76]. Renal failure impairs the ability to excrete oxalate and this leads to an early increase in plasma concentrations, as emerges from the close correlation between plasma oxalate and creatinine [10,15]. Oxalate to creatinine ratios for both urine and plasma may be used to establish new reference values. Tables 8 and 9 list the results obtained in our Institution in different groups of renal disease and renal function.

Table 8.
Oxalate Excretion in Different Groups of Renal Function

	GFR (μmL/min)	uOx (mmol/24h)	uOx/uCr (μmol/mmol)
Normal GFR	114 ± 22	0.28 ± 0.08	30 ± 16
Primary Hyperoxaluria	46 ± 39	1.53 ± 0.74	359 ± 188
Enteric Hyperoxaluria	64 ± 27	0.96 ± 0.49	110 ± 47
CRF-Oxalate unrelated	30 ± 16	0.27 ± 0.10	25 ± 13

Table 9.
Plasma Levels of Oxalate in Different Groups of Renal Function

	GFR mL/min	POx μmol/L	uOx/Cr μmol/mmol
Normal GFR	114 ± 18	1.3 ± 0.7	26.4 ± 13.8
Primary Hyperoxaluria	46 ± 39	53.7 ± 52	247 ± 137
CRF-Oxalate Unrelated	30 ± 16	20.1 ± 11.6	79 ± 42
Non-PH Dialysis patients	0	44.0 ± 6.9	62 ± 21
PH patients on RDT	0	168 ± 14	227 ± 112

From these tables it appears that renal failure tends to blunt differences in oxalate excretion especially in cases of hyperoxaluric syndrome, whereas the oxalate to creatinine molar ratios maintain or amplify these differences and cancel overlapping between subjects with and without hyperoxaluria. The same applies to plasma levels of oxalate: renal failure tends to level-off differences between patients with and without hyperoxaluria, while the creatinine ratio has a higher discriminating capacity.

4.   Oxalate-related Hyperacidurias
The discrimination between different types of hyperoxaluria may rely on assessment of associated abnormalities in generation and urine excretion of metabolically related substances. Mostly relevant in this regard, is the case of primary hyperoxalurias, which can be differentiated from the accompanying hyperglycolic or hyper-L-glyceric acidurias, in type 1 and 2 respectively. Conversely, the absence of related hyperacidurias suggests an exogenous origin of hyperoxaluria. Methods for measuring these acids were complex and poorly reliable until a few years ago. Recently, some procedures have been proposed, which have proved accurate, highly sensitive, and powerful to make differential diagnosis. These include HPLC [131,132], ion chromatographic [83], and GLC [98,133,134] methods, detailed description of which is beyond the bounds of this chapter.

Both 24-h or random samples can be used, to carry out urine measurements and results expressed as excretions or creatinine molar ratios. Glycolate reference ranges for urine and plasma from normal individuals obtained with some recent procedures are listed in Table 10.

Table 10.
Reference Ranges for Plasma and Urine Glycolate in Adult Normal Subjects Obtained with IC or HPLC Assays

Author, Ref.	Plasma	Urine
	µmol/L	µmol/24h
HPL, 131	4.5 - 13.6	185 - 761
IC, 83	1.4 - 7.4	0 - 1400
GLC, 98	0.8 - 6.6	-

VI. OXALATE MEASUREMENT IN OTHER BIOLOGICAL FLUIDS AND TISSUES

Oxalate has seldom been measured in biological fluids other than plasma and urine. A major application of these measurements is amniotic fluid. Barratt et al. [121] assayed oxalate concentration in amniotic fluid from 63 uncomplicated pregnancies, reporting concentrations of 19 ± 4.3 µmol/L. In a similar study Wandzilak et al. analyzed 70 amniotic samples by ion chromatography and reported reference values of 18.9 ± 8.6 µmol/L, with oxalate to creatinine molar ratio of 0.29 ± 0.14, similar to that found in urine of infants less than 1 year [136].

A major application of these measurements would be pre-natal diagnosis of primary hyperoxaluria, but some authors have claimed that amniotic fluid is unsuitable for this purpose [137]. The recent introduction of specific probes to identify defects of alanine:glyoxylate aminotransferase gene in fetal cells appears more promising in this regard, though it can not be excluded that the application of updated procedures for oxalate and glycolate on amniotic samples could achieve valuable results.

Other applications of oxalate measurements may be required to detect extra-renal deposition of calcium oxalate crystals, as in the case of systemic oxalosis. Oxalosis as a complication of hyperoxaluria has so far been detected histologically. Recently, we have developed a procedure for measuring oxalate content in bone specimens from biopsies of the iliac crest [138]. Basically, the procedure uses 10 to 80 mg specimens, which are cleaned of non-osseous material and treated for 24 hours at 37°C with a working mixture containing collagenase. The slice is then rinsed with water and dehydrated by mixing with 100% ethanol. After being weighed carefully, the fragment undergoes acidic digestion with concentrated hydrochloric acid for 48-72 hours at room temperature. The resulting solution is diluted with water and filtered through 0.22 µm filters. Oxalate and phosphate are measured on these final solutions using ion chromatography with suppressed conductivity. The results are expressed as µmol/g bony tissue and µmol/mmol phosphate, in order to obviate weighing errors. This procedure appears highly sensitive, accurate and reproducible, in that oxalate concentrations can be detected even in just a few µmol/g bony tissue. The reference range from 10 individuals without metabolic bone disease is reported in Table 11 and compared with values from dialysis patients with or without oxalosis.

Table 11.
Measured Bony Content of Oxalate

	Oxalate	Oxalate/Phosphate
	µmol/g bone	µmol/mmol
Control Subjects	0.54 - 1.15	0.17 - 0.42
Non-Oxalotic RDT	0.75 - 14.6	0.21 - 5.7
PH patients on RDT	4.2 - 907	1.4 - 436

The table shows that patients with systemic oxalosis are clearly differentiated from both control subjects and dialysis patients without hyperoxaluric syndromes.

REFERENCES

1. **Schultzen, O.**, Quantitative Bestimmung des Oxalsäuen Kalkes in Harn, *Arch. Anat. Physiol.*, 6, 719, 1868.
2. **Merz, K. W., Maugeri, S.**, Über das Vorkommen und die Bestimmung de Oxalsäure im Blut, *Physiol. Chem.*, 201, 31, 1931.
3. **Chiriboga, J.**, Some properties of an oxalic oxidase purified from barley seedlings, *Biochem. Biophys. Res. Comm.*, 11(4), 277, 1963.
4. **Shimazono, H., Hayaishi, O.**, Enzymatic decarboxylation of oxalic acid, *J. Biol. Chem.*, 227, 151, 1957.
5. **Gochman, N., Schmitz, J. M.**, Automated determination of uric acid, with use of a urease-peroxidase system, *Clin. Chem.*, 17, 1154, 1971.
6. **Parkinson, I. S., Kealey, T., Laker, M.F.**, The determination of plasma oxalate concentrations using an enzyme/bioluminescent assay, *Clin. Chim. Acta*, 152, 335, 1985.
7. **Mhale, C. J., Menon, M.**, Determination of urinary oxalate by ion chromatography: preliminary observation, *J. Urol.*, 127, 159, 1982.
8. **Robertson, W. G., Scurr, D. S., Smith, A., Orwell, R. L.**, The determination of oxalate in urine and urinary calculi by a new ion-chromatographic technique, *Clin. Chim. Acta*, 126, 91, 1982.
9. **Daniele, P. G., Marangella, M.**, Ionic equilibria in urine: a computer model system improved by accurate stability constants values, *Annali di Chimica*, 72, 25, 1982.
10. **Marangella, M., Petrarulo, M., Vitale, C., Daniele, P. G., Sammartano, S., Cosseddu, D., Linari, F.**, Serum calcium oxalate saturation in patients on maintenance haemodialysis for primary hyperoxaluria or oxalosis-unrelated renal diseases, *Clin. Sci.*, 81, 483, 1991.
11. **Watts, R. W. E., Veall, N., Purkiss, P.**, Sequential studies of oxalate dynamics in primary hyperoxaluria, *Clin. Sci.*, 65, 627, 1983.

12. **Hautmannn, R., Osswald, H.**, Pharmacokinetic studies of oxalate in man, *Invest. Urol.*, 16, 395, 1979.
13. **Marangella, M., Petrarulo, M., Mandolfo S., Vitale C., Cosseddu, D., Linari, F.**, Plasma profiles and dialysis kinetics of oxalate in patients receiving hemodialysis, *Nephron*, 60, 74, 1992.
14. **Hatch, M.**, Spectrophotometric determination of oxalate in whole blood, *Clin. Chim. Acta*, 193, 199, 1990.
15. **Morgan, S. H., Purkiss, P., Watts, R. W. E., Mansell, M. A.**, Oxalate dynamics in chronic renal failure. Comparison with normal subjects and patients with primary hyperoxaluria, *Nephron*, 46, 253, 1987.
16. **Hodgkinson, A.**, Sampling errors in the determination of urine calcium and oxalate: solubility of calcium oxalate in HCl-urine mixtures, *Clin. Chim. Acta*, 109, 239, 1981.
17. **Braiotta, E. A., Buttery, J. E., Ludvigsen, N.**, The effects of pH, temperature and storage on urine oxalate, *Clin. Chim. Acta*, 147, 31, 1985.
18. **Herbert, R. W., Hirst, E. L., Percival, E. G. V., Reynolds, R. J. W., Smith, F.**, The constitution of ascorbic acid, *J. Chem. Soc.*, 1270, 1933.
19. **Harris, A. B.**, Vitamin-C-iduced hyperoxaluria, *Lancet*, i, 399, 1976.
20. **Potezny, N., Bais, R., O'Loughlin, P. D., Edwards, J. B., Rofe, A. M., Conyers, A. J.**, Urinary oxalate determination by use of immobilized oxalate oxidase in a continuous-flow system, *Clin. Chem.*, 29, 16, 1983.
21. **Mazzachi, B. C., Teubner, J. K., Ryall, R. L.**, Factors affecting measurement of urinary oxalate, *Clin. Chem.*, 30, 1339, 1984.
22. **Robertson, W. G., Scurr, D. S.**, Prevention of ascorbic acid interference in the measurement of oxalic acid in urine by ion-chromatography, *Clin. Chim. Acta*, 140, 97, 1984.
23. **Kasidas, G. P., Rose, G. A.**, Spontaneous in vitro generation of oxalate from L-ascorbate in some assays for urinary oxalate and its prevention, in *Urolithiasis and Related Clinical Research*, Swille, P. O., Smith, L. H., Robertson, W. G., Vahlensieck, W, Eds., Plenum Press, New York, 1985, 653.
24. **Mazzachi, B. C., Teubner, J. K. Ryall R. L.**, The effect of ascorbic acid on urine oxalate measurement, in *Urolithiasis and Related Clinical Research*, Swille, P. O., Smith, L. H., Robertson, W. G., Vahlensieck, W, Eds., Plenum Press, New York, 1985, 649.
25. **Chalmers, A. H., Cowley, D. M., McWhimmey B. C.**, Stability of oxalate in urine: relevance to analyses for ascorbate and oxalate, *Clin. Chem.*, 31, 1703, 1985.
26. **Menaché R.**, Routine micromethod for determination of oxalic acid in urine by atomic absorption spectrophotometry, *Clin. Chem.*, 20, 1444, 1974.
27. **Koehl, C., Abecassis, J.**, Determination of oxalic acid in urine by atomic absorption spectrophotometry, *Clin. Chim. Acta*, 70, 71, 1976.
28. **Archer, H. E., Dormer, A. E., Scowen, E. F., Watts, R. W. E.**, Studies on the urinary excretion of oxalate by normal subjects, *Clin. Sci.*, 16, 405, 1957.
29. **Koch, G. H., Strong, F. M.**, Determination of oxalate in urine, *Anal. Biochem.*, 27, 162, 1969.
30. **Zarembski, P. M., Hodgkinson, A.**, The fluorimetric determination of oxalic acid in blood and other biological fluids, *Biochem. J.*, 96, 712, 1965.
31. **Hamelle, G., Bressole, F.**, Microdosage de l'acide oxalique dans les urines, *Travaux de la Societè de Pharmacie de Montpellier*, 35, 195, 1975.

32. **Dempsey, E. F., Forbes, A. P., Melick, R. A., Henneman, P. H.**, Urinary oxalate excretion, *Metabolism*, 9, 52, 1960.
33. **Hodgkinson, A., Williams, A.**, An improved colorimetric procedure for urine oxalate, *Clin. Chim. Acta*, 36, 127, 1972.
34. **Baadenhuijsen, H., Jansen, A. P.**, Colorimetric determination of urinary oxalate recovered as calcium salt, *Clin. Chim. Acta*, 62, 315, 1975.
35. **Salinas, F., Martínez-Vidal, J. L., Gonzáes-Murcia, V.**, Extraction-spectrophotometric determination of oxalate in urine and blood serum, *Analyst*, 114, 1685, 1989.
36. **Muñoz Leyva, J. A., Hernández Artiga, M. P., Aragón Mendez, M. M., Quintana Perez, J. J.**, Atomic absorption and uv-vis absorption spectrophotometric determination of oxalate in urine by ligand exchange extraction, *Clin. Chim. Acta*, 195, 47, 1990.
37. **Kohlbecker, G., Richter, L., Butz, M.**, Determination of oxalate in urine using oxalate oxidase: Comaparison with oxalate decarboxylase, *J. Clin. Chem. Clin. Biochem.*, 17, 309, 1979.
38. **Bais, R., Potezny, N., Edwards, J. B., Rofe, A. M., Conyers, A. J.**, Oxalate determination by immobilized oxalate oxidase in a continuous flow system, *Anal. Chem.*, 52, 508, 1980.
39. **White-Stevens, R. H.**, Interference by ascorbic acid in test systems involving peroxidase. I. reversible indicators and the effects of copper, iron, and mercury, *Clin. Chem.*, 28, 578, 1982.
40. **Laker, M. F., Hofmann, A. F., Meeuse B. J. D.**, Spectrophotometric determination of urinary oxalate with oxalate oxidase, *Clin. Chem.*, 26, 827, 1980.
41. **Obzansky, D. M., Richardson K. E.**, Quantification of urinary oxalate with oxalate oxidase from beet stems, *Clin. Chem.*, 29, 1815, 1983.
42. **Kasidas, G. P., Rose, G. A.**, Continuous flow assay for urinary oxalate using immobilised oxalate oxidase, *Ann. Clin. Biochem.*, 22, 412, 1985.
43. **Wilson, D. M., Liedtke, R. R.**, Modified enzyme-based colorimetric assay of urinary and plasma oxalate with improved sensitivity and no ascorbate interference: reference values and sample handling procedures, *Clin. Chem.*, 37, 1229, 1991.
44. **Buttery, J. E., Ludvigsen, N., Braiotta, E. A., Pannall, P. R.**, Determination of urinary oxalate with commercially available oxalate oxidase, *Clin. Chem.*, 29, 700, 1983.
45. **Berckmans, R. J., Boer, P.**, An inexpensive method for sensitive enzymatic determination of oxalate in urine and plasma, *Clin. Chem.*, 34, 1451, 1988.
46. **Sigma Diagnostics**, Oxalate: quantitative, enzymatic determination of oxalate in urinr at 590 nm, in *Technical bulletin, procedure no.591*, Sigma Chemical Co. U.S.A., St. Louis MO, 1989, 1.
47. **Petrarulo, M., Cerelli, E., Marangella, M, Cosseddu, D., Vitale., C., Linari, F.**, Assay of plasma oxalate with soluble oxalate oxidase, *Clin. Chem.*, 40, 2030, 1994.
48. **Mayer, G. G., Markow, D., Karp, F.**, Enzymatic oxalate determination in urine, *Clin. Chem*, 9, 334, 1963.
49. **Ribeiro, M. E., Elliot, J. S.**, Direct enzymatic determination of urinary oxalate, *Invest. Urol.*, 2, 78, 1964.
50. **Hallson, P. C., Rose G. A.**, A simplified and rapid enzymatic method for determination of urinary oxalate, *Clin. Chim. Acta*, 55, 29, 1974.

51. **Costello, J., Hatch, M. Bourke, E.,** An enzymic method for the spectrophotometric determination of oxalic acid, *J. Lab. Clin. Med.,* 87, 903, 1976.

52. **Yriberri, J., Posen, S.,** A semi-automatic enzymic method for estimating urinary oxalate, *Clin. Chem.,* 26, 881, 1980.

53. **Cannon, G. D., Eaton, R. H., Glen, A. C. A.,** Enzymic determination of urinary oxalate, with EDTA added to improve recovery, *Clin. Chem. (letter),* 29, 1855, 1983.

54. **Boeringher Mannheim Biochemica,** Oxalic acid. uv-Method, in *Methods in biochemical analysis and food analysis,* Boehringer Mannheim GmBH Biochemica D, Mannheim, 1986, 86.

55. **Duburque, M. T., Melon, J. M., Thomas, J., Thomas, E., Pierre, R., Charransol, G., Desgrez, P.,** Dosage et identification de l'acide oxalique dans les milieux biologiques, *Ann. Biol. Clin.,* 28, 95, 1970.

56. **Chalmers, R. A., Watts, R. W. E.,** The quantitative extraction and gas liquid chromatographic determination of organic acids in urine, *Analyst,* 97, 958, 1972.

57. **Tocco, D. J., Duncan, A. E. W., Noll, R. M., Duggan, D. E.,** An electron-capture gas chromatographic procedure for the estimation of oxalic acid in urine, *Anal. Biochem.,* 94, 470, 1979.

58. **Farrington, C. J., Chalmers, A. H.,** Gas-chromatographic estimation of urinary oxalate and its comparison with a colorimetric method, *Clin. Chem,* 25, 1993, 1979.

59. **Gelot, M. A., Lavoue, G., Belleville, F., Nabet, P.,** Determination of oxalates in plasma and urine using gas chromatography, *Clin. Chim. Acta,* 106, 279, 1980.

60. **Dicorcia, A., Samperi, R., Vinci, G., D'ascenzo, G.,** Simple, reliable chromatographic measurement of oxalate in urine, *Clin. Chem.,* 28, 1457, 1982.

61. **Wolthers, B. G., Hayer, M.,** The determination of oxalic acid in plasma and urine by means of capillary gas chromatography, *Clin. Chim. Acta,* 120, 87, 1982.

62. **Lopez, M., Tuchman, M., Scheinman, J. I.,** Capillary gas chromatography measurement if oxalate in plasma and urine, *Kidney. Int.,* 28, 82, 1985.

63. **Molnár, I., Horváth, C.,** Rapid separation of urinary acids by high-performance liquid chromatography, *J. Chromatogr.,* 143, 391, 1977.

64. **Buchanan, D. N., Thoene, J. G.,** Dual-column high performance liquid chromatographic urinary organic acid profiling, *Anal. Biochem.,* 124, 108, 1982.

65. **Mayer, W. J., McCarthy, J. P., Greenberg, M. S.,** The determination of oxalic acid in urine by high performance liquid chromatography with electrochemical detection, *J. Chromatogr. Sci.,* 17, 656, 1979.

66. **Kok, W. T., Groenendijk, G., Brinkman, U. A. T., Frei, R. W.,** Determination of oxalic acid in biological matrices by liquid chromatography with amperometric detection, *J. Chromatogr.,* 315, 271, 1984.

67. **Larsson, L., Libert, B., Asperud, M.,** Determination of urinary oxalate by reversed-phase ion-pair "high-performance" liquid chromatography, *Clin. Chem.,* 28, 2272, 1982.

68. **Hughes, H., Hagen, L., Sutton, R. A. L.,** Determination of urinary oxalate by high-performance liquid chromatography, *Anal. Biochem.,* 119, 1, 1982.

69. **McWhinney, B. C., Cowley, D. M., Chalmers, A. H.**, Simplified column liquid chromatographic method for measuring urinary oxalate, *J. Chromatogr.*, 383, 137, 1986.

70. **Imaoka, S., Funae, Y., Sugimoto, T., Hayahara, N., Maekawa, M.**, Specific and rapid assay of urinary oxalic acid using high-performance liquid chromatography, *Anal. Biochem.*, 128, 459, 1983.

71. **Stevens, T. S., Davis, J. C.**, Hollow fiber ion-exchange suppressor for ion chromatography, *Anal. Chem.*, 53, 1488, 1981.

72. **Petrarulo, M., Marangella, M., Bianco, O., Marchesini, A., Linari, F.**, Preventing ascorbate interference in ion-chromatographic determinations of urinary oxalate: four methods compared, *Clin. Chem.*, 36, 1642, 1990.

73. **Dionex Corporation**, Determination of oxalate in urine by ion chromatography, in *Application note no.36*, Dionex Corporation U.S.A., Sunnyvale CA, 1992, 1.

74. **Pinto, B., Crespi, G., Sole-Barcells F., Barcelo, P.**, Patterns of oxalate metabolism in recurrent oxalate stone formers, *Kidney Int.*, 5, 285, 1974.

75. **Constable, A. R., Joekes, A. M., Kasidas G. P., O'Regan, P., Rose, G. A.**, Plasma levels and renal clearance of oxalate in normal subjects and in patients with primary hyperoxaluria or chronic renal failure, *Clin. Sci.*, 56, 299, 1979.

76. **Prenen, J. A. C., Boer, P., Dorhout Mees, E. J., Endeman, H. J., Spoor, S. M., Oei, H. Y.**, Renal clearance of [^{14}C]oxalate: comparison of constant-infusion with single-injection techiniques, *Clin. Sci.*, 63, 47, 1982.

77. **Linari, F., Marangella, M.**, Oxalate nephropathy. Pathophysiology and biochemical features, in *Tubulo-Interstitial Nephropaties*, Amerio, A., Coratelli, P., Massry, S. G., Eds., Kluwer Acad. Pub., Boston, 1991, chap. 9.

78. **Hodgkinson, A.**, Determination of oxalic acid in biological material, *Clin. Chem.*, 16, 547, 1970.

79. **Ackay, T., Rose, G. A.**, The real and apparent plasma oxalate, *Clin. Chim. Acta*, 101, 305, 1980.

80. **Maguire, M., Fituri, N., Keog, B., Costello, J.**, The effect of storage on serum oxalate values, in *Urolithiasis, Clinical and Basic Research*, Smith, L. H., Robertson, W. G., Finlayson, B., Eds., Plenum Press, New York, 1981, 963.

81. **Costello, J., Landwehr, D. M.**, Determination of oxalate concentration in blood, *Clin. Chem.*, 34, 1540, 1988.

82. **Petrarulo, M., Bianco, O., Marangellla, M., Pellegrino, S., Linari, F.**, Ion chromatographic determination of plasma oxalate in healthy subjects, in patients with chronic renal failure and in cases of hyperoxaluric syndromes, *J. Chromatogr.*, 511, 223, 1990.

83. **Hagen, L., Walker, V. R., Sutton, R. A. L.**, Plasma urinary oxalate and glycolate in healthy sybjects, *Clin. Chem.*, 39, 134, 1993.

84. **Petrarulo, M., Cerelli, E., Marangella, M., Maglienti, F., Linari, F.**, Ion-chromatographic determinations of plasma oxalate reexamined, *Clin. Chem.*, 39, 537, 1993.

85. **Cole, F. E., Gladden, K. M., Bennett, D. J., Erwin, D. T.**, Human plasma oxalate concentration re-examined, *Clin. Chim. Acta*, 139, 137, 1984.

86. **Kasidas, G. P., Rose, G. A.**, Measurement of plasma oxalate in healthy subjects and in patients with chronic renal failure using immobilised oxalate oxidase, *Clin. Chim. Acta*, 154, 49, 1986.

87. **Rolton, H. A., McConnell, K. N., Modi, K. S., Macdougall, A. I.,** A simple, rapid assay for plasma oxalate in uraemic patients using oxalate oxidase, which is free from vitamin C interference, *Clin. Chim. Acta*, 182, 247, 1989.

88. **Parkinson, I. S,, Channon, S. M., Tomson, C. R. V., Adonai, L.R., Ward, M.K., Laker, M.F.,** The determination of plasma oxalate concentrations using an enzyme/bioluminescent assay. 2. Co-immobilisation of bioluminescent enzymes and studies of in vitro oxalogenesis, *Clin. Chim. Acta*, 179, 97, 1989.

89. **Hatch, M., Bourke, E., Costello, J.,** New enzymic method for serum oxalate determination, *Clin. Chem.*, 23, 76, 1977.

90. **Samsoondar, J., More, R. W., Kellen, J. A.,** Enzymatic determination of oxalates, *Enzymes*, 30, 273, 1983.

91. **Bennett, D. J., Cole, F. E., Frohlich, E. D., Erwin, D. T.,** A radioenzymatic isotope-dilution assay for oxalate in serum or plasma, *Clin. Chem.*, 25, 1810, 1979.

92. **Sugiura, M., Yamamura, H., Hirano, K., Ito, Y., Sasaki, M., Morikawa, M., Inoue, M., Tsuboi, M.,** Enzymic determination of serum oxalate, *Clin. Chim. Acta*, 105, 393, 1980.

93. **Kohlbecker, G., Butz, M.,** Direct spectrophotometric determination of serum and urinary oxalate with oxalate oxidase, *J. Clin. Chem. Clin. Biochem.*, 19, 1103, 1981.

94. **Borland, W. W., Payton, C. D., Simpson, K., MacDougall, A. I.,** Serum oxalate in chronic renal failure, *Nephron*, 45, 119, 1987.

95. **Schwille, P. O., Manoharan, M., R_menapf, G., W_lfel, G., Berens, H.,** Oxalate measurement in the picomol range by ion chromatography: values in fasting plasma and urine of controls and patients with idiopathic calcium urolithiasis, *J. Clin. Chem. Clin. Biochem.*, 27, 87, 1989.

96. **Politi, L., Chiaraluce, R., Consalvi, V., Cerulli, N., Scandurra, R.,** Oxalate, phosphate and sulphate determination in serum and urine by ion chromatography, *Clin. Chim. Acta*, 184, 155, 1989.

97. **Skogerboe, K. J., Felix-Slinn, T., Synovec, R. E.,** Ion chromatographic determination of oxalate in plasma: correlation study with an enzymatic method, *Anal. Chim. Acta*, 237, 299, 1990.

98. **France, N. C., Holland, P. T., Wallace, M. R.,** Contribution of dialysis to endogenous oxalate production in patients with chronic renal failure, *Clin. Chem.*, 40, 1544, 1994.

99. **Robertson, W. G., Peacock, M., Nordin, B. E. C.,** Activity products in stone-forming and non-stone-forming urine, *Clin. Sci.*, 34, 579, 1968.

100. **Pak, C. Y. C., Holt, K.,** Nucleation and growth of brushite and calcium oxalate in urine of stone formers, *Metabolism*, 25, 665, 1976.

101. **Robertson, W. G., Peacock, M.,** The cause of idiopathic calcium stone disease: Hypercalciuria or hyperoxaluria?, *Neprohn*, 26, 105, 1980.

102. **Baggio, B., Gambaro, G., Favaro, S., Borsatti, A.,** Prevalence of hyperoxaluria in idiopathic calcium oxalate kidney stone disease, *Nephron*, 35, 11, 1983.

103. **Jaeger, P., Portmann, L., Jacquet, A. F., Burckhardt, P.,** Influence of the calcium content of the diet on the incidence of mild hyperoxaluria in idiopathic renal stone formers, *Am. J. Nephrol.*, 5, 40, 1985.

104. **Bek-Jensen, H., Tiselius, H. G.,** Stone formation and urine composition in calcium stone formers without medical treatment, *Eur. Urol.*, 16, 144, 1989.

105. **Yendt, E. R., Cohanim, M.,** Clinical and laboratory approaches for evaluation of nephrolithiasis, *J. Urol.,* 141, 764, 1989.

106. **Marangella, M.,** Diagnostic profiles of the metabolic abnormalities in idiopathic calcium nephrolithiasis, *It. J. Min. Electr. Met.,* 8, 103, 1994.

107. **Brinckley, L. J., Gregory, J., Pak, C. Y. C.,** A further study of oxalate bioavailability in foods, *J. Urol.,* 144, 94, 1990.

108. **Marangella, M., Fruttero, B., Bruno, M., Linari, F.,** Hyperoxaluria in idiopathic calcium stone disease: further evidence of intestinal hyperabsorption of oxalate, *Clin. Sci.,* 63, 381, 1982

109. **Giannini, S., Nobile, M., Castrignano, R., Pati, T., Tasca, A., Villi, G., Pellegrini, F., D'Angelo, A.,** Possible link between vitamin D and hyperoxaluria in patients with renal stone disease, *Clin. Sci.,* 84, 51, 1993.

110. **Campfield, T., Braden, G.,** Urinary oxalate excretion by very low birth weight infants receiving parenteral nutrition, *Pediatrics,* 84, 860, 1989.

111. **Nhu Uyen, N., Dumoulin, G., Henriet, M. T., Berthelay, S., Regnard, J.,** Carbohydrate metabolism and urinary excretion of calcium and oxalate after ingestion of polyol sweeteners, *J. Clin. Endocrinol. Metab.,* 77, 388, 1993.

112. **Ribaya, J. D., Gershoff, S. N.,** Effects of sugars and vitamin B6 on oxalate synthesis in rats, *J. Nutr.,* 112, 2161, 1984.

113. **Robertson, W. G., Heyburn, P. J., Peacock, M., Hanes, F. A., Swaminathan, R.,** The effect of high animal protein intake on the risk of calcium stone-formation in the urinary tract, *Clin. Sci.,* 57, 285, 1979.

114. **Marangella, M., Bianco, O., Martini, C., Petrarulo, M., Vitale, C., Linari, F.,** Effect of animal and vegetable protein intake on oxalate excretion in idiopathic calcium stone disease, *Br. J. Urol.,* 63, 348, 1989.

115. **Wandzilak, T. R., D'Andre, S. D., Davis, P. A., Williams, H. E.,** Effect of high dose vitaminc C on urinary oxalate levels, *J. Urol.,* 151, 834, 1994.

116. **Kasidas, G. P., Nemat, S., Rose, G. A.,** Plasma oxalate and creatinine and oxalate/creatinine clearance ratios in normal subjects and in primary hyperoxaluria. Evidence for renal hyperoxaluria, *Clin. Chim. Acta,* 191, 67, 1990.

117. **Tomson, C. R., Channon, S. M., Ward, M. K., Laker, M. F.,** Oxalate retention in chronic renal failure: tubular *vs* glomerular disease, *Clin. Nephrol.,* 32, 87, 1989.

118. **Parks, J. H., Coe, F. L.,** A urinary calcium-citrate index for the evaluation of nephrolithiasis, *Kidney Int.,* 30, 85, 1986.

119. **Ryall, R. L., Harnett, R. M., Hibberd, C. M., Mazzachi, B. C., Mazzachi, R. D., Marshall, V. R.,** Urinary risk factors in calcium oxalate stone disease. Comparison of men and women, *Br. J. Urol.,* 60, 480, 1987.

120. **Leumann, E. P., Dietl, A., Matasovic, A.,** Urinary oxalate and glycolate excretion in healthy infants and children, *Pediatr. Nephrol.,* 4, 493, 1990.

121. **Barratt, T. M., Kasidas, G. P., Murdoch, I., Rose, G. A.,** Urinary oxalate and glycolate excretion and plasma oxalate concentration, *Arch. Dis.Child.,* 66, 501, 1991.

122. **Morgenstern, B. Z., Milliner, D. S., Murphy, M. E., Simmons, P. S., Moyer, T. P., Wilson, D. M., Smith, L. H.,** Urinary oxalate and glycolate excretion patterns in the first year of life: A longitudinal study, *J. Pediatr.,* 123, 248, 1993.

123. **Von Schnakenburg, C., Byrd, D.J., Latta, K., Reusz, G. S., Graf, D., Brodhel, J.,** Determination of oxalate excretion in spot urines of healthy children by ion chromatography, *Eur. J. Clin. Chem. Biochem.,* 32, 27, 1994.

124. **Hallson, P. C., Kasidas, G. P., Rose, G. A.**, Seasonal variation in urinary excretion of calcium and oxalate in normal subjects in patients with idiopathic hypercalciuria, *Br. J. Urol.*, 49, 1, 1977.
125. **Hargreave, T. B., Sali, A., Mackay, C., Sullivan, M.**, Diurnal variation in urinary oxalate, *Br. J. Urol.*, 49, 597, 1977.
126. **Banks, J. G., Sullivan, M., Hargreave, B.**, Evening urinary oxalate excretion in stone formers, *Br. J. Urol.*, 51, 349, 1979.
127. **Touitou, Y., Touitou, C., Charransol, G.**, Alterations in circadian rhythmicity in calcium oxalate stone formers, *Int. J. Chronobiol.*, 8, 175, 1983.
128. **Finch, A.M., Kasidas, G. P., Rose, G. A.**, Urine composition in normal subjects after oral ingestion of oxalate-rich foods, *Clin. Sci.*, 60, 411, 1981.
129. **Holmes, R. P., Goodman, H. O., Hart, L. J., Assimos, D. G.**, Relationship of protein intake to urinary oxalate and glycolate excretion, *Kidney Int.*, 44, 366, 1993.
130. **Marangella, M.**, Renal handling of citrate in chronic renal insufficiency, *Nephron*, 57, 439, 1991.
131. **Petrarulo, M., Marangella, M., Linari, F.**, High-performance liquid chromatographic determination of plasma glycolic acid in healthy subjects and in cases of hyperoxaluria syndromes, *Clin. Chim. Acta*, 196, 17, 1991.
132. **Petrarulo, M., Marangella, M., Linari, F.**, High-performance liquid chromatographic assay for L-glyceric acid in body fluids. Application in primary hyperoxaluria type 2, *Clin. Chim. Acta*, 211, 143, 1992.
133. **Chalmers, R. A., Tracey, B. M., Mistry, J., Griffiths, K. D., Green, A., Winterborn, M. H.**, L-Glyceric aciduria (primary hyperoxaluria type 2) in siblings in two unrelated families, *J. Inher. Metab. Dis.*, 7(2), 133, 1984.
134. **Kamerling, J. P., Gerwig, G. J., Vliegenthart, J. F. G., Duran, M., Ketting, D., Wadman, S.K.**, Determination of the configurations of lactic and glyceric acids from human serum and urine by capillary gas-liquid chromatography, *J. Chromatogr.*, 143, 117, 1977.
135. **Marangella, M., Petrarulo, M., Vitale, C., Linari, F.**, Plasma and urine glycolate assays for differentiating the hyperoxaluria syndromes, *J. Urol.*, 148, 608, 1992.
136. **Wandzilak, T. R., Hanson, F. W., Williams, H. E.**, The quantitation of oxalate in amniotic fluid by ion-chromatography, *Clin. Chim. Acta*, 185, 131, 1989.
137. **Leumann, E., Niederwieser, A., Fanconi, A.**, New aspects of infantile oxalosis, *Pediatr. Nephrol.*, 3, 531, 1987.
138. **Marangella, M., Vitale, C., Petrarulo, M., Tricerri, A., Cerelli, E., Cadario, A., Portigliatti Barbos, M., Linari, F.**, Bony content of oxalate in patients with primary hyperoxaluria or oxalosis-unrelated renal failure, *Kidney Int.*, 1995, in press.

Chapter 13

URINARY MACROMOLECULES IN CALCIUM OXALATE STONE AND CRYSTAL MATRIX: GOOD, BAD, OR INDIFFERENT?

Rosemary L. Ryall and Alan M.F. Stapleton
Flinders Medical Centre, Bedford Park SA
Australia

I. INTRODUCTION

Increase in the total macromolecules of human urine inevitably accompanies injury to the kidney and urinary passages.
Boyce [1]

The controlling influence of macromolecules in the construction of healthy biomineralised tissues is undisputed.[2] It is now well recognised that the organic component of such tissues, which in animals include bone, cartilage, shell, dentin and enamel, is crucial to the biomineralisation process. Some macromolecules in these systems are responsible for initiating mineralisation, defining its physical limits and dictating its cessation, but others also provide the architectural framework upon which the inorganic salts are laid down. Less clearly defined, however, are the roles played by macromolecules in the formation of human uroliths, a process possessing all the hallmarks of uncontrolled biomineralisation.[2] That they perform *some* function might appear obvious, for they account for part of every stone, whether composed of struvite or apatite, calcium oxalate or uric acid.[3,4] This organic component, which by analogy with that of normal mineralised tissues, is known as matrix: matrix is as inevitably and integrally a part of the stone as the mineral itself. Although recognition of the presence of matrix in human calculi dates back several hundred years, its more recent history began with a series of studies by Boyce and his colleagues. It is not our intention here to recount this history in detail: it has been the subject of several reviews.[5,6] Instead, we will skim only briefly over the early discoveries and discuss in greater detail more recent findings, which, primarily as a result of significant advances in the technology of protein chemistry have, to some extent, rendered many of those of our predecessors obsolete.

Boyce and Garvey [3] pioneered the modern study of kidney stone matrix when they dissolved kidney stones using EDTA and analysed the resulting extract. Matrix was noted to represent approximately 2.5% of the dry weight of stones composed primarily of calcium or urate, 9-11% of cystine calculi and about 62% of the rare lucent "matrix stone" invariably associated with urinary infection. However, despite its relatively small contribution to the weight of most calculi, in most instances the organic material remaining after dissolution formed a cast of the original stone, indicating that it occupied much more space, especially in calcium oxalate stones, than was suggested by its proportional weight. In fact, matrix is distributed as a network throughout the entire structure of a stone [4,7] and plays a role in determining the architecture of calculi.[1,8] It is commonly present as a series of concentric layers associated with radial striations that appear ordered, rather than random.[3] For a detailed discussion of the inter-relations between the crystals and matrix of kidney stones, readers are referred to a series of excellent papers by Khan, Hackett and colleagues.[7,9-12] But while the physical features of stone ultrastructure have been reasonably amenable to direct microscopic examination, its chemical composition has proved more difficult to explore.

The organic component of stones has proved notoriously resistant to dissolution: only a portion of it is solubilized by EDTA treatment.[4,13] We therefore know much more about the physical structure of matrix than about its chemical constituents, and what knowledge we have managed to glean about the latter pertains only to that portion able to be dissolved - approximately 25% of the total weight.[14] Although Boyce [1] described the principal components of stone matrix released by EDTA dissolution, to our knowledge, few attempts to update Boyce's analyses have been made by later authors - mute testimony perhaps to the real difficulties inherent in studying matrix. And these are not confined to its obstinate insolubility. To this, as well as the complex chemical nature of stone matrix, must be added alterations in the molecular structure of its component macromolecules caused by chemical isolation procedures themselves.[4] The lag time between stone formation and analysis of the matrix may also influence the findings: storage of stones prior to analysis, sometimes for decades[15] may allow chemical cross-linking, dehydration and degradation of component macromolecules, rendering the gathered information inaccurate and, thereby, misleading. As if these difficulties were not enough reason for regarding early data with a healthy degree of circumspection, there is yet another factor that has further confused the issue.

As shown in figure 1, matrix constituents can originate from two sources. Firstly, normal urinary macromolecules can become trapped inside crystals of stone minerals as a result of their binding to the crystal surface and becoming engulfed by the advancing growth front.[16] Secondly, others not normally present in urine, but which are released by endothelial damage wrought by the buffeting action of large crystalline particles or developing stones within the renal collecting system can also become incorporated into the final stone structure.[16,17] Macromolecules may also be released into the urine by chemical damage resulting from increased oxalate concentrations prior to the initial formation of crystal nuclei.[18] The derivation of at least part of stone matrix from extra-urinary sources is beyond doubt, confirmed by demonstration of the occurrence in stone matrix of substances not significantly present in healthy urine, such as red blood cells,[19] erythrocyte

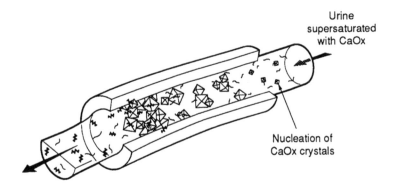

Figure 1. Stylized schema showing a renal tubule through which is flowing urine supersaturated with CaOx. Nucleation of small crystals of CaOx is depicted, and normal urinary macromolecules (∿) bind to the newly formed particles and become embedded within the architecture of the crystals as they grow in the supersaturated medium. Once sufficient growth and aggregation have occurred to cause retention of the stone nucleus within the tubule, release into the urine of abnormal macromolecules (∿) can occur as a result of damage to the urothelium by the inchoate stone. These can also become incorporated into the stone between existing crystals, and also within individual crystals as a result of further solute deposition or crystal nucleation. (Reprinted with permission from Current Opinion in Urology).

and mitochondrial ghosts, bacteria,[20] lipids,[21,22] and large serum proteins normally excluded by the glomerulus.[14,23] Furthermore, the contribution of this so-called "secondary" matrix[16] is consistent with the observation that the concentration of matrix increases significantly from the stone's core to its outer surface:[4] the larger the calculus, the greater its potential to cause injury. And any new medical graduate is well aware that stones are known to be associated with haematuria. Because the study of *stone* matrix will not permit us to distinguish between normal urinary macromolecules involved in the formation, growth and aggregation of crystals, macromolecular products of metabolic tissue damage that may have induced precipitation, or those released by physical injury to the urinary epithelium by the crystals themselves, some researchers have more recently turned to the organic matrix of *crystals* freshly precipitated from urine, since they are the precursors of stones. In this chapter we will review what is known about the macromolecular component of both stone and crystal matrix.

Nonetheless, though a number of factors have conspired to frustrate our attempts to define the role played by matrix in stone formation, few would argue that some role it undoubtedly must play.[7] A functional role, *prima facie*, seems self-evident, for matrix macromolecules are **there**. But why? Are they, like their healthy counterparts, agents that direct the mechanics of mineral deposition? Do they begin the process? Dictate when it will stop? Are they merely the structural scaffolding upon which the mineral is laid down? Are they promiscuous players, unavoidably caught up in a pathological event in which they have no real stake at all? Or perhaps they are just by-products of the whole, unhappy process; unwilling participants stripped off the walls of the urinary tract as the developing stone weaves its destructive course through the kidney and urinary collecting system, and ultimately becoming part of the final structure whose embryo caused their release from the healthy tissue it damaged along the way? Perhaps they are all of these. Answers to these questions, as has been reasoned so frequently in the past, must lie in identifying the individual components of matrix, for until we know what they **are**, we cannot possibly deduce what they **do**. But first it is necessary to review briefly what is known about the effects of urinary macromolecules on CaOx crystallization, because that will at least provide us with some clues about the reasons for their presence in stone matrix.

II. MATRIX COMPONENTS - PROMOTERS, INHIBITORS OR JUST ALONG FOR THE RIDE?

The most simple explanation for the occurrence of kidney stones would be that the urine of stone formers is supersaturated with stone minerals, which consequently precipitate in their urine, while their more salubrious fellows avoid this by excreting urine that is undersaturated - perhaps by drinking greater quantities of fluid. Although highly attractive, such an explanation is not, unfortunately, borne out by empirical observations. Under everyday conditions, urine is commonly supersaturated with respect to CaOx, even in people who have never formed a stone;[24] all of us occasionally pass crystals in our urine. Recurrent stone formers, however, tend to excrete such crystals in greater quantities and clustered into larger particles than those observed in healthy controls.[25,26] It may be inferred from these findings that any urinary molecule that affects the mass of CaOx deposited from urine, or the size of the crystal particles produced, has the potential to influence the likelihood that crystalline particles are retained in the renal collecting system, and thereby, the development of stone disease. This rationale has formed the basis of the so-called inhibitor theory of stone formation, which reasons that the urine of normal individuals contains efficient inhibitor molecules that reduce the likelihood of crystal nucleation; but should nucleation occur, these inhibitors proscribe their subsequent growth or aggregation into large particles by binding to the surfaces of the newly formed crystal embryos. However, although these inhibitors may sometimes completely prevent these processes, their more common effect is simply to retard them, so that once bound to the crystal surfaces, they can become buried as inclusions within the structure as additional solute ions are deposited and engulf them. When this occurs, they can actually increase the size of the crystals, even when the amount of mineral deposited is the same.[27] However, although this provides a simple explanation for the presence of macromolecules in stones, we must also consider the possibility that the macromolecules themselves may have caused the crystals to nucleate.

Levels of supersaturation required for homogeneous (spontaneous) CaOx crystal nucleation rarely occur *in vivo*, so it is generally accepted that precipitation in urine results from heterogeneous nucleation. The matrix theory of stone formation argues that macromolecules in urine are heterogeneous nucleators, actively inducing precipitation of stone salts by sequestering high, localized concentrations of mineral solute ions, causing them to precipitate at lower levels of urinary saturation than are normally required for homogeneous precipitation. What happens to these macromolecules? Paradoxically, the answer to this question is precisely the same as it would be if they had acted as inhibitors - they become trapped within the stone architecture. There are, of course, other possibilities. Urinary macromolecules might just be innocent joyriders, binding adventitiously to the crystal surfaces, not affecting the crystallization process at all, but still becoming part of the final stone structure. And the possibility cannot be discounted that some macromolecules do not bind, but might still interfere with crystallization.

It may come as no surprise that there is evidence to suggest that urinary macromolecules can do all of these. Some promote CaOx crystal nucleation in inorganic solutions [eg[28]] and in concentrated whole urine [eg[29]]. others like albumin appear to exert no significant effect on crystallization in urine [30], but are nonetheless found in stones. Many studies, far too numerous to document, have shown that a variety of urinary proteins inhibit CaOx crystal growth [eg[31,32]] and aggregation [eg[30,33]]. More surprisingly, some can actually do more than one of these: Tamm-Horsfall mucoprotein can inhibit CaOx crystal aggregation, but can also act as a promoter of crystal deposition, depending upon experimental conditions.[34] It has also been demonstrated that polyelectrolytes and proteins that inhibit crystallization in solution, can act as promoters when they are immobilized on to surfaces.[35] And the issue is further complicated by the fact that the potency of urinary macromolecules increases inversely in relation to the prevailing ionic strength.[36] Is it any wonder then, that inhibitors have been shown to act quite differently depending upon whether their activity is measured in inorganic metastable solutions or in urine itself? For instance, human serum albumin[37] and chondroitin sulphate[38] inhibit both CaOx crystal growth and aggregation when tested in an inorganic crystallization system, yet have no material effect at all on CaOx crystallization occurring in undiluted human urine.[30] In the light of this confusion, how can we possibly deduce what effects they might have *in vivo*? For the most part, this question is unanswerable, since it is apparent that their effects are manifold, unpredictable, and paradoxical. Above all, they are certainly not amenable to generalization: knowledge of their roles in stone formation will therefore be obtained only by laboriously teasing out the information for each individual macromolecule. And in the following sections, that is what we will attempt to do.

III. THE CHEMICAL COMPOSITION OF MATRIX

The contributions of Boyce and his coworkers to the study of matrix were monumental, yet they were able to identify only a few individual matrix components, their efforts effectively limited by the technology available to them at the time. One can only wonder what we might now know about matrix had Boyce and his team had access to the advanced tools of protein physicochemistry and molecular biology that we enjoy today. The main components of the EDTA-soluble portion of matrix were described by Boyce[1] as 64% protein, 9.6% non-amino sugars, 5% hexosamine as glucosamine, 10% bound water, and the remainder, "inorganic ash" - principally calcium and phosphate. Although not detected by Boyce,[1] lipids have also been shown to be significant components of stone matrix.[21] Nonetheless, proteins comprise the major part of matrix, an observation confirmed more recently by Sugimoto et al[39] using high-pressure liquid chromatography (HPLC). It was inevitable therefore that proteins would tend to dominate the study of matrix, and consequently, considerably more is known about them than the other two principal groups, namely lipids and glycosaminoglycans (GAGs). We will therefore delay discussion of matrix proteins and begin with what is known about lipids.

A. LIPIDS

Lipids have been identified in stones composed principally of CaOx, struvite or uric acid,[22,40] with phospholipids accounting for 8.6% of the total lipid, which in turn represents approximately 10.25% of stone matrix obtained by sonication and demineralization with EDTA.[22,40] In other mineralization systems lipids are thought to play an important role, contributing by weight 7 - 14% of bone, 2 - 6% of dentin and 12 - 22% of newly mineralized enamel matrices.[41] It is the acidic phospholipids, particularly phosphatidyl serine and phosphatidyl inositol, which are known to bind cations, that appear to be involved in the calcification process. The lipids identified in stone matrix are similar to those commonly found in cell walls;[21,22,40] their presence in stone matrix may therefore result from passive incorporation of sloughed cells of the urinary tract. Others, however, have suggested that lipids play a more active role, since cell membranes[42,43] and the lipids of CaOx stone matrix[21] can catalyse the nucleation of CaOx from a metastable solution. These findings are consistent with the deposition, during pathological calcification, of calcium on to cell membranes, and the integral involvement of phospholipids in the complexation of calcium in a diverse range of biological systems, most notably, blood coagulation. Compared with GAGs and proteins, study of the role of lipids in matrix formation is in its infancy and a great deal more information must yet be obtained, including studies involving their effect on CaOx crystallization in whole human urine, before it will be possible to assign to them a definitive role in stone pathogenesis.

B. GLYCOSAMINOGLYCANS (GAGS)

The presence of GAGs in stone matrix was inferred by Boyce and Garvey as long ago as 1956[3]. More recently, Nishio et al[44] reported that between 0.19 and 0.58% of a stone's weight consists of GAGs, implying that they must comprise a significant portion of the matrix. If we relate this to the fact that matrix represents approximately 2.5% of the dry weight of a CaOx stone,[3] it appears that GAGs may account for up to 20% of the weight of matrix. This alone is sufficient to suggest that they may fulfil some function in stone formation.

GAGs occurring in the urine of normal individuals typically consist of 55% chondroitin sulphate (ChS), 20% heparan sulphate (HS), 11% low-sulphated ChS and 4 - 10% hyaluronic acid (HA).[45] It is therefore somewhat surprising that only HS, and to a lesser extent HA, have been reported to be present in CaOx stones,[44,46] while a more recent study could detect only HS.[47] ChS, the most abundant GAG in urine, though present in small amounts in magnesium ammonium phosphate[44] and apatite stones,[44] has not been detected in CaOx stones.[44,46,47] These results strongly suggest that the incorporation of GAGs into stones is a selective process, a notion supported by the finding that inclusion of GAGs into CaOx crystals freshly precipitated from human urine is also highly selective. Like stones, such crystals contain only HS.[47,48] ChS is incorporated into CaOx crystals only in the absence of HA, indicating that they probably compete for the same binding sites on the crystal surface.[48] Similar results have been reported for uric acid stones and crystals, with HS being the only GAG detected.[15]

However, the simple fact of their presence in stones or crystals tells us nothing about the mechanism by which GAGs came to be there. For many years it was assumed that urinary GAGs were inhibitors of stone formation, because a number of studies [see[48]] had reported that either ChS or heparin (not present in human urine) were potent inhibitors of CaOx crystallization in inorganic metastable solutions. However, such studies tell us virtually nothing about the possible effects of GAGs under physiological conditions, and the more recent studies of stone and crystal matrix must cast doubt on the inhibitory role of urinary GAGs. The absence of ChS from stone matrix, coupled with earlier observations that it inhibits CaOx crystallization in inorganic solutions, prompted Nishio et al[44] to conclude that HS and HA, which *are* present in the matrix, must be promoters of stone formation. However, Roberts and Resnick[46] on the basis of identical findings, suggested the opposite interpretation. They attributed the presence of HS and HA in stone matrix to their strong binding affinity for the surface of the crystals comprising the stone,[49] and since the magnitude of an inhibitor's effect is presumed to be dependent upon the efficiency of its binding to the crystal surface, they implied that HS and HA normally fulfil an inhibitory role in urine. This view was shared by Yamaguchi et al[47] who inferred from their results that HS must be an inhibitor of CaOx crystal growth and aggregation, since HS was the only GAG present in soluble stone and crystal matrices, which exhibited strong inhibitory effects in their study. Unpublished data from our laboratory have shown that HS also inhibits CaOx crystal aggregation in undiluted, ultrafiltered human urine. On the other hand, ChS has no effect in the same experimental system,[30] an observation consistent with its absence from stones and CaOx crystals. However, recent evidence suggests that urinary GAGs can fulfil both roles, promoting CaOx crystal nucleation and reducing the final size of the crystals produced,[50] thereby lessening their chance of retention within the urinary tract.

On the basis of present evidence, therefore, it would appear that ChS, though present in urine in larger quantities than HS, plays no significant role in CaOx crystallization or in stone formation. HS, on the other hand may act as an inhibitor, although confirmation of such a role must await the results of further studies. What is quite clear, however, is that the routine measurement of *total* urinary GAGs excretion is unlikely to be of any practical benefit in the diagnosis and management of urolithiasis. With the benefit of hindsight, it is not surprising that comparisons of urinary GAG excretion in stone formers and normal subjects have produced conflicting findings, with reports that stone formers had abnormally low excretion rates [eg[51]], even as recently as 1992,[52] being countered by papers demonstrating that the two groups do not differ [eg[53,54]]. Specific determination of individual GAGs, particularly HS, may throw some light on their true role in kidney stone formation; however, until this is achieved, the nature of the involvement of GAGs in stone formation, if any, must remain a matter for speculation.

C. PROTEINS

Mucoprotein or "uromucoid" was one of the first matrix components identified by Boyce and Garvey.[3] Both terms have been used synonymously to describe Tamm-Horsfall Glycoprotein (THG), the most abundant single protein found in human urine.[55] However, although it was reportedly present in matrix,[56] the finding was undermined somewhat by the absence from matrix of sialic acid, which forms the side chains of THG.[3,56-58] This apparent discrepancy was circumvented by hypothesising that the incorporation of THG into matrix involved desialylation, a notion that was soon dispelled by the unequivocal demonstration[59] of the presence of sialic acid in stones. These authors suggested that THG might be passively deposited, since it occurred in all stones, irrespective of mineral composition. No other urinary protein has featured as prominently in the stone literature as THG, and because of this, it will be discussed in more detail in the next section.

Using immunological techniques, Boyce et al[14] showed that some components of serum were readily identifiable in the matrix of stones passed by patients known to have had urinary tract infections, but found only infrequently in uninfected patients. Albumin and one or two α-globulins were the most consistent components from serum, while γ-globulins appeared in association with pyelonephritis.[14] Over a period of several years Boyce's group published a number of papers describing a highly insoluble glycoprotein that defied all attempts at purification, which they dubbed Matrix Substance A.[14] Although MSA featured in a number of papers, its study was ephemeral; interest in it waned when work suggested that it consisted not of one, but of several separate substances.[60] This, combined with the existence of similarities between MSA and several known calcium-binding proteins,[61] and the development of modern techniques of protein physicochemistry, effectively tolled the death knell for MSA and it is unlikely that its study will ever enjoy a reprise: reference to it in the literature has long since passed. However, while modern technology contributed to the effective despatch of MSA from the stone scene, it promised far more, since it enabled easier purification and identification of specific proteins in stone and crystal matrix.

In addition to nephrocalcin (NC),[62] whose prominence in the stone literature justifies its more detailed discussion in the following section, other proteins detected in soluble stone matrix include haemoglobin, neutrophil elastase, another serine protease of unknown origin[17] and transferrin.[63] However, the possible significance of the presence of all but NC has not been explored further. Another protein, isolated from CaOx and mixed struvite/calcium phosphate stones, was of special interest because it contained γ-carboxyglutamic acid (Gla), an unusual amino acid found in proteins involved in calcification processes and blood clotting, because of its extraordinary ability to bind calcium. Described by Lian et al[64] as having a molecular weight of 17 kDa and containing approximately 40 Gla residues per 1000 amino acids, the protein was presumed to be distinct from the F1 activation peptide of prothrombin which contains 61.3 Gla residues per 1,000 amino acids,[65] and unlikely to be related to prothrombin or osteocalcin. Lian et al[64] therefore surmised that the protein was representative of a new class of Gla proteins. Despite the potential importance of this protein to stone pathogenesis its study has never been pursued, and there are no available data refuting conclusively the possibility that it may, after all, be related to urinary prothrombin fragment 1 (UPTF1), which is excreted in normal human urine.[66] UPTF1, formerly known as Crystal Matrix Protein (CMP), will be discussed more extensively below because it is this protein that predominates in calcium oxalate crystals precipitated from human urine.[16] Another protein whose presence in crystals has been noted is α-1 microglobulin. Morse and Resnick[67] reported the presence of what they called α-2 microglobulin in 2D SDS-PAGE gels of proteins isolated from calcium oxalate crystals generated in urine, although in a personal communication to us Resnick later confirmed that it should have been identified as α-1 microglobulin. Study of α-1 microglobulin has not been extended, although its close genetic relationship to inter-α-trypsin inhibitor (α-1 microglobulin and the HI 30 light chain of inter-α-trypsin inhibitor represent different portions of a single translation product of the same gene on chromosome 9 [68]), which will be discussed more fully below, probably justifies its more extensive investigation in the future.

Using improved methods for dissolving stone matrix, Binette and Binette[69,70] have isolated a protein from the matrix of both renal calculi and gallstones; amino acid sequence analysis indicates that the protein is related to CD59 protein, also known as protectin. To date, the significance to stone formation of neither CD59, nor superoxide dismutase, which the same authors also isolated from stones,[71] has been established. Umekawa et al[72] detected α-1-anti-trypsin, the most abundant protease inhibitor in human serum, in extracts of calcium calculi, providing direct support for recent findings, detailed below, that serum protease inhibitors such as inter-α-trypsin inhibitor (ITI) may play some inhibitory role in stone formation.[73] Another relative newcomer on the stone matrix scene, also deserving of more extensive discussion (see below), is uropontin, a urinary form of the bone protein osteopontin.[74]

Despite the fact that stone matrix has been shown to contain an ever-increasing list of individual proteins, in most cases it is impossible to say with any certainty just why they might be there. One can speculate, of course. For instance, the presence of superoxide dismutase[71] may perhaps be explained by the fact that the enzyme acts as a protector of tissue damage by scavenging the toxic superoxide anion.[75] Could the enzyme's presence be explained by release of this radical in response to cellular injury caused by the kidney stone itself - an example of a macromolecule incorporated as a secondary matrix component? However, speculation alone will not unravel the mystery of why proteins are in stone matrix; but detailed study of individual proteins may. We will now discuss what is known about several urinary proteins that have been subjected to rigorous study because they have been found to occur in stones, or because they have been isolated from urine and shown to influence the crystallization of CaOx. And we will begin with Tamm-Horsfall glycoprotein because it has the longest historical association with stones, and as a consequence, has been subjected to the most intense experimental scrutiny.

IV. PROTEINS IMPLICATED IN STONE FORMATION

A. TAMM-HORSFALL GLYCOPROTEIN (THG)

Tamm-Horsfall glycoprotein (THG) enjoys the distinction of being the most extensively investigated urinary macromolecule in urolithiasis research, probably because it is the most abundant protein in human urine, and the fact that uromucoid, one of the several names by which it has been known since its discovery, was one of the first components of stone matrix to be identified by Boyce and Garvey.[3] Although first described by Morner in 1895 [cited in[76]], Tamm and Horsfall were the first to isolate it from urine when, in 1950, they reported that it was responsible for the inhibition of hemagglutination induced by myxomavirus.[77] The precise physiological function of THG, which is the predominant component of renal casts, remains to be defined, although there is some evidence for its preventing bacterial infection within the urinary tract.[78] THG is a renal protein of all placental invertebrates,[79] localized to the luminal aspect of epithelial cells of the distal convoluted tubules[80] and distributed throughout the epithelial cells of the thick ascending limb of the loops of Henlé.[81]

Known to have a monomeric molecular weight of 80 kDa, THG is present in urine in polymeric forms measuring up to several million Da, 20 - 200 mg of which is excreted daily.[82] Carbohydrate, which accounts for

30% of its weight,[83] is important for self-association of the molecule: polymerization is increased in the presence of calcium ions[84] and albumin,[85] and decreased with increasing urea concentrations[86] and alkalinity.[84] It is its extraordinary ability to self-associate in urine into large structures visible to the naked eye that probably account for its known effects on CaOx crystallization, and it is these effects to which we will now primarily confine our discussion.

Despite its abundance in urine, THG is found only sparingly in stone matrix, and is absent from CaOx crystals precipitated from whole urine,[16] which would seem to indicate that THG binds only weakly, if at all, to CaOx crystals. Since it has been accepted for some time that inhibitors act by binding to crystal surfaces, we might expect THG to be a poor inhibitor of CaOx crystallization - at least in urine, where stones form. Unfortunately, the situation is not so simple, because THG exhibits different properties depending upon the experimental conditions under which it is tested, and as a consequence, experimental findings are both confusing and contradictory. The protein has been reported to act as an inhibitor [eg[30,34,87-91]] and a promoter.[29,34,92-93] The picture is further complicated by the fact that conflicting findings were obtained in the only studies in which the effect of THG was tested in undiluted urine: Hallson and Rose[29] and Rose and Sulaiman[92] found that THG enhanced the deposition of CaOx crystals from urine concentrated by evaporation to high osmolalities, whereas Ryall et al[30] and Grover et al[91] found that the protein was a potent inhibitor of CaOx crystal aggregation, although having no effect on CaOx deposition. An explanation for these opposing findings is to be found in a study by Grover et al[34] who tested the effect of THG in the experimental systems used by the two research groups and found that while THG undoubtedly promotes CaOx precipitation under conditions of high osmolality, where it also links CaOx crystals together into large, loosely connected agglomerates, it is a very effective inhibitor of crystal aggregation at more usual urinary concentrations. It is also apparent that THG inhibits crystal aggregation by steric hindrance, rather than by binding to the crystal surfaces[30] - binding may therefore not be a prerequisite for inhibitory potency after all! Moreover, however potent an inhibitor it may be, it cannot account for the total inhibitory effect of urinary macromolecules on CaOx crystal aggregation in centrifuged and filtered urine,[94] because it is removed from urine by centrifugation and filtration. Thus there can be no doubt that other urinary macromolecules contribute to this inhibition.[30]

A reflection of the disagreement surrounding the role of THG as a promoter or inhibitor of CaOx crystallization is to be found in similar conflict relating to its urinary excretion: if indeed THG does play a directive role in stone formation, we might expect that its excretion would be different in stone formers and normal subjects. But it is not.[95-98] However, it may be that stone formation is related more to the type of THG excreted than to the quantity. Hess et al[99] reported that the ability of THG from 6 patients with severe nephrocalcinosis to inhibit CaOx monohydrate crystal aggregation is reduced in comparison with that from healthy subjects, and that this is a result of a molecular abnormality of the protein in these patients. For a more detailed account of the role of THG in urolithiasis readers are referred to two recent reviews.[99,100]

It would be fair to say that we have not reaped the bounty we might have expected of the many years of study invested in THG. We know that the protein can act both as a promoter and an inhibitor of CaOx crystal processes in experimental crystallization systems, but we still cannot say with certainty whether it actually plays a key role in the formation of stones. However, of one thing we can be reasonably confident; whatever its role in stone pathogenesis, it is not the only urinary macromolecule likely to be involved. Further studies are still required to elucidate its real contribution, if any, to urolithiasis, and its interaction, if any, with its urinary companions. And it is to those companions that we will now direct our attention.

B. NEPHROCALCIN (NC)

Second only to THG, nephrocalcin (NC) has been the most widely studied protein reported in the stone literature. It was first described by Nakagawa and his colleagues in 1978 simply as an unidentified acidic polypeptide[32] and then for a number of years as a glycoprotein inhibitor of CaOx crystal growth.[101-106] In 1987, nine years after the first report of its isolation from urine, the protein was named nephrocalcin,[62] by analogy with the bone protein osteocalcin. NC has assumed a prominent position in urolithiasis research, having been claimed to be the principal inhibitor of CaOx crystallization in urine;[102] its activity reportedly accounting for approximately 90% of urine's total inhibitory effect on CaOx crystallization.[101-103]

The molecular weight of NC varies widely depending upon the state of aggregation of the protein, with the molecular weights of the monomer, dimer, trimer and tetramer being reported as 14-15, 23-30, 45-48 and 60-68 kDa, respectively.[101,104,107,108] The protein has been shown in immunohistochemical studies to be located in the epithelium of the proximal tubules and thick ascending limb of the loops of Henlé, in both human and mouse kidneys.[109] NC was originally isolated from human urine, but has also been purified from rat kidney and urine,[103] tissue culture medium of human kidney cell lines,[101,110] kidney stones,[107] and the urine of patients with renal stones[104,105,107] and renal cell carcinoma.[111] The finding that its levels were elevated in patients with advanced renal cell carcinoma prompted suggestions that it may be a useful tumour marker.

A glycoprotein, NC has been reported to occur in urine at concentrations ranging from 5 mg/L[102] to 16 mg/L[112] and to contain 2-3 residues of γ-carboxyglutamic acid (Gla) in its primary structure.[32,101,102,104-106] The Gla component has been assumed to confer the protein's potent ability to inhibit CaOx crystallization; the protein isolated from the urine of stone formers was reportedly deficient in this amino acid,[107] and the urine from these individuals had reduced inhibitory activity.[113] A lack of Gla in NC isolated from kidney stones was suggested as the reason why the stones had formed.[107] However, it is worth noting that Colette et al[114] found no difference in the protein-bound urinary Gla content of stone formers compared with normal individuals. NC has also been shown to inhibit the proliferation of BSC-1 monkey kidney epithelial cells induced by calcium oxalate monohydrate crystals.[115]

Despite its comparatively long history, and the fact that NC has come to be recognised as an important urinary protein inhibitor of CaOx crystallization *in vitro*, and by inference, as a significant determinant of CaOx stone formation, there has been no report in the literature of its primary amino acid sequence, and the exact nature of the protein remains unknown. It was recently claimed by Desbois et al[116] that the N-terminus of rat NC is almost identical to that of rat osteocalcin and that the two proteins have indistinguishable SDS-PAGE migration patterns. These authors provided no corroborative evidence for this claim, which was based on an earlier paper by Deyl et al;[117] however, NC is not even mentioned in the Deyl paper. In fact, as recently as 1990, the original authors[108] reported that, despite intensive efforts, they had been unable to obtain a primary amino acid sequence for the protein. Recently, our laboratory isolated from human urine a protein we deduced to be NC on the basis of SDS-PAGE, inhibitory, and gel filtration properties, and subjected it to amino acid sequence analysis.[118] Its molecular weight, as well as the amino acid sequences of two of its peptides, suggested its identity with fragment HI-14 of the light chain (bikunin) of inter-α-trypsin inhibitor (ITI).[119] We have tentatively concluded that NC represents a portion of the bikunin chain of ITI, and await with interest the results of amino acid sequence analysis of authentic NC prepared by the original authors, particularly since ITI contains no Gla. If our findings are correct, then it would appear that NC is closely related to uronic acid-rich protein (UAP), a newly described urinary inhibitor of CaOx crystallization which is discussed in detail below. However, irrespective of the true molecular nature of NC, more recent findings with other urinary proteins demand a reassessment of experiental data upon which its presumed role in stone formation has been based. Its description as the principal inhibitor of CaOx crystallization in urine,[102] accounting for around 90% of the total inhibitory effect of urine on CaOx crystallization[101-103] should now perhaps be reappraised, particularly since a recent paper by Worcester et al[120] reassessed its contribution to be no more than 16%. Moreover, all existing estimates of the protein's inhibitory potency have been obtained from crystallization systems based on inorganic metastable solutions: its activity has never been tested in urine. Although NC is undoubtedly a potent inhibitor of CaOx deposition in an inorganic metastable solution, an observation we[118] and others[121] have confirmed using material tentatively identified as NC, it is becoming increasingly apparent that this potency is shared with a number of other urinary proteins, such as uropontin, urinary prothrombin fragment 1 and UAP.

C. UROPONTIN (OSTEOPONTIN)

Recently, Shiraga et al[74] reported the isolation from human urine of a protein which they called uropontin (UP). UP was isolated by immuno-affinity chromatography using a monoclonal rat antibody that had been raised against a group of urinary proteins separated by DEAE cellulose, and had exhibited maximal inhibition of CaOx crystal growth in an inorganic metastable solution:[122] its effect on crystal aggregation has not been determined. The protein was purified by reversed-phase HPLC (RPHPLC) and subjected to N-terminal amino acid sequence analysis,[74] which revealed complete identity with the N-termini of osteopontin (OP)[123] and lactopontin.[124] Total amino acid analysis of UP,[74] which revealed a remarkably high proportion of aspartic acid residues, showed striking similarities with these two proteins. Molecular weight estimations of UP, which has an apparent Mr of 50 kDa in 16% SDS-PAGE gels and 72 kDa in 5-18% gradient gels were also similar to OP.[123] These similarities, together with identical nucleotide sequences of cDNAs encoding OP from human kidney[74] and bone[125] indicate that UP is not a distinct protein, but rather a urinary form of OP. As usually occurs when a protein is first described, OP has been referred to by a variety of names since its discovery,[126] including secreted phosphoprotein,[126] 44 kDa bone phosphoprotein,[123] 2ar[127] and now, uropontin. Although the identity of UP with OP is acknowledged by Hoyer,[128] he advocates continued use of the name UP on the basis of its source and role within the urinary tract, a practice not, however, adopted by Kohri and colleagues.[129,130] These authors conducted *in situ* hybridization studies[130] which showed that osteopontin mRNA in the kidneys of stone-forming rats was increased in proportion to the dosage and duration of the drugs used to induce lithogenesis, thereby reinforcing the possibility that OP may play a role in stone formation, although the increase could also have been secondary to epithelial cell injury caused by precipitation of CaOx crystals. However, in that, and another study in which they cloned and sequenced the cDNA encoding the same urinary stone protein,[129] the same authors specifically referred to the protein as osteopontin. Worcester et al,[131] who isolated the same protein from cultured mouse kidney cortical cells, also preferred use of this name. It is to be hoped that in the course of time, concensus will be reached about the name by which UP/urinary OP is to be referred in the stone literature.

Osteopontin is a protein important in bone mineralization, where it is thought to anchor osteoblasts to bone.[132] Originally isolated from rat bone matrix as a 44 kDa phosphorylated protein, it is rich in serine, aspartic acid and glutamic acid - acidic amino acids commonly found in proteins involved in biomineralization.[123] It has an amino acid sequence that serves as a recognition signal for interacting with cell surface receptor molecules known as integrins, which are involved in cell adhesion,[133] and is a member of a family of proteins rich in aspartic acid that have been shown *in vitro* to have stereospecific activity at the surface of crystals.[134]

The distribution of OP in humans was recently reported by Brown et al.[135] They detected the protein widely distributed on the luminal surfaces of specific epithelial cells in the gastrointestinal tract, gall bladder, pancreas, urinary and reproductive tracts, lung, breast, salivary glands and sweat glands. They also reported that the two principal sites of OP gene transcription detected by Northern analysis were the gallbladder and the kidney. In the latter, OP was specifically found in the cytoplasm of many epithelial cells of the distal tubules and collecting ducts. Others have reported OP to be present in mice, but only in the thick ascending limbs of the loop of Henlé and the distal convoluted tubules in a subset of nephrons.[136] It might perhaps be argued that the widespread distribution of OP in humans militates against its having a specific function in stone formation, but its potent effect on CaOx crystal growth would suggest that it may influence the course of the disease. Certainly, it is present in kidney stones, with quantities in those composed principally of CaOx dihydrate being considerably less than in calculi comprising mainly CaOx monohydrate,[128] where its abundance is substantially greater than that reported for NC.[107] UP is present in normal adult urine at a mean concentration of approximately 6×10^{-8} molar.[128] Assuming a molecular weight of 50 kDa, this value converts to approximately 3 mg/L, which is of the same order as that of NC, whose concentration has been reported as 5 mg/L[102] and 16 mg/L.[112] The proportional contribution of UP to the inhibitory activity of urine has not been assessed,[128] but the fact that it is present in stones in greater quantities than NC, despite its lower concentration in urine, might suggest that it binds more avidly to the CaOx crystal surface, and may consequently be a more potent inhibitor. However, like NC, its inhibitory effect on CaOx crystallization has not been tested in urine, so it is not presently possible to assess its likely effects on CaOx crystallization *in vivo*. Thus, like all proteins currently under investigation for their possible roles in stone formation, significantly more information must be obtained before it will be possible to state with certainty that the presence of UP in urine is related specifically to its ability to inhibit CaOx crystallization, and thereby, stone pathogenesis.

D. URINARY PROTHROMBIN FRAGMENT 1 (UPTF1)

Urinary prothrombin fragment 1 (UPTF1) differs from its fellow urinary proteins by virtue of the circumstances of its discovery. Unlike its peers, UPTF1 was not isolated directly from urine, but from CaOx crystals freshly precipitated from urine. Doyle et al[16] reasoned that the study of *crystals* enabled the study of urinary proteins directly involved in the crucial crystal nucleation phase of stone formation, free of any macromolecular contaminants that might be introduced into a *stone* by cellular injury. Adopting an approach previously used by Morse and Resnick,[67,137] they subjected the organix matrix remaining after EDTA demineralization of CaOx crystals to SDS-PAGE analysis and found that, despite the enormous array of proteins present in urine, the crystals precipitated from it contained relatively few proteins. In particular, one protein, with an apparent Mr of 31 kDa and staining characteristics of a glycoprotein, was selectively incorporated into the crystals in quantities far exceeding those of any other, and in amounts disproportionately greater than its concentration in the urine from which the crystals had been derived. Unable to identify the protein by Western blotting using commercial antibodies to likely candidates, such as α1-microglobulin, they named the protein crystal matrix protein, or CMP. However, it has since been shown, both immunologically and by amino acid sequence analysis that the protein is related to human prothrombin.[138,139] This finding, which established for the first time a link between urolithiasis and blood clotting, was recently confirmed when its close identity with the F1 activation peptide of human prothrombin was demonstrated.[140] To avoid confusion, the name crystal matrix protein has been abandoned and the protein is now known as urinary prothrombin fragment 1 (UPTF1).

The known characteristics of UPTF1 indicate that it may fulfil some function in stone pathogenesis. It is present in calcium stones,[141] having been detected in 9 of 10 calcium stones and in all (8/8) whose principal constituent was CaOx. Most notably, it was not found in two struvite stones, indicating that its presence in CaOx stones is a consequence of direct inclusion into the crystalline architecture, rather than a secondary product of tissue injury. Analysis of calcium phosphate crystal matrix reveals that UPTF1 is a major component, whereas in urate crystals it is only a very minor constituent, which reflects the known relationship between this protein and calcium ions.[141] Immunohistochemical studies have mapped its location in the human kidney specifically to the epithelial cells of the thick ascending limb of the loops of Henlé and the distal convoluted tubules, including the maculae densae of involved nephrons.[142] The protein was not detected in any other human tissues, with the exception of the cytoplasm of hepatocytes. Demonstration of its location here, by immunolabelling using a polyclonal antibody, most likely reflected the shared epitopes of UPTF1 and its parent molecule, prothrombin, which is exclusively synthesised in the liver. Limited data also demonstrated that the amount of UPTF1 in the kidneys of stone formers is significantly greater than in those from healthy subjects,[142] a finding which, though currently unexplained, raises a number of tantalising questions which future research must address.

Until recently, evidence that UPTF1 inhibits CaOx crystallization was only indirect. UPTF1, as stated above, is the most prominent protein in the organic matrix of CaOx crystals precipitated from fresh human urine.[16] This, combined with the observation that the organic matrix is the most potent macromolecular inhibitor of CaOx crystallization induced in human urine that has yet been described,[27] led to the presumption that this inhibitory activity was likely to be attributable to its component UPTF1. This presumption was largely justified, since UPTF1 purified by RPHPLC is now known to be a potent inhibitor of CaOx crystal aggregation in undiluted urine,[143] a feature that currently distinguishes it from its peers, whose inhibitory activities, with the exception of THG, have only ever been tested in inorganic solutions. There seems little doubt that the potent inhibitory effect of UPTF1 on CaOx crystallization can be ascribed to the Gla domain of the peptide. Derived from its parent prothrombin, this region of the protein's primary structure contains 10 Gla residues, which are known to be crucial to the function of the molecule during blood clotting,[144] being responsible for its binding to phospholipid membranes - a process essential to blood coagulation.

Along with that of its peers, UP and UAP, the study of UPTF1 is still in its early stages. Certainly, preliminary data would indicate that it possesses all the features expected of a significant macromolecular urinary inhibitor, including potent activity in undiluted urine. Nonetheless, like **all** other proteins presumed to fulfill some function in urolithiasis, the true role of UPTF1 must remain speculative until such time as a cause and effect relationship between the protein and stone pathogenesis can be unequivocally demonstrated.

E. URONIC-ACID-RICH PROTEIN (UAP)

The latest arrival on the stone scene, uronic-acid-rich protein was first described as recently as 1993 by Atmani et al.[121] At the present time there is relatively little published information about UAP, but like other proteins possessing the ability to inhibit CaOx crystallization, it will undoubtedly feature prominently in forthcoming stone literature. UAP earned its name because of its high content of uronic acid: D-glucuronic and L-iduronic acids are major constituents of GAGs. The protein was isolated from human urine by three gel filtration chromatographic steps, and in the reduced form, had an estimated molecular weight of 35 kDa on an 8 - 20% SDS gradient gel. A glycoprotein, carbohydrate accounts for 8.5% of its weight, and amino acid analysis reveals a protein rich in glutamic and aspartic acids, glycine and valine; it contains no Gla. N-terminal amino acid sequence analysis has demonstrated homology with the inter-α-trypsin inhibitor (ITI) family.[145] The inhibitory activity of UAP was determined in an inorganic CaOx crystallization system, where it strongly retarded CaOx crystal growth. More recently, it was reported that this activity is reduced in stone formers compared with normal controls.[146] The protein has also been isolated from rat urine[147] and shown to possess very similar properties to the human urinary protein, including identical molecular weights and total amino acid analyses. Like human UAP, rat UAP immunoreacts with an ITI antibody. This, together with its N-terminal amino acid identity to ITI strongly suggests that UAP is a relative of ITI, and this is supported by data from independent laboratories. We have also recently isolated a 35 kDa protein from urine[118] and, on the basis of its Mr and N-terminal amino acid sequence, have tentatively identified it as the light chain of ITI. These findings are in keeping with those of Atmani et al,[121,145] as well as those of Sørenson et al,[73] who reported the isolation from human urine of a protein with an Mr of approximately 40 kDa and a strong inhibitory effect on CaOx crystal growth. Although Sørenson et al[73] described the protein as "unidentified" their amino acid sequencing data revealed N-terminal homology with ITI.

ITI belongs to the Künitz-type protease inhibitor superfamily,[68,148] and in the intact form, has an Mr of 220,000, comprising two heavy chains (H1 and H2) and one light chain known by a confusing selection of names (eg, HI-30, urinary trypsin inhibitor, acid-resistant trypsin inhibitor, bikunin), linked covalently by a chondroitin sulphate chain.[68,148] Only the light chain, now more commonly known as bikunin, with which the anti-protease activity is associated, is excreted in human urine. It is difficult to escape the conclusion that the proteins described by Sørenson et al,[73] Atmani and his colleagues,[121,145] and ourselves,[118] are one and the same: furthermore the N-terminal amino acid homology with ITI in all three cases is compelling evidence that they are a form of bikunin. This supposition is further vindicated by the fact that the urinary derivative of ITI is also known to be a GAG-adduct[68] and has a relative Mr of 35 kDa on SDS-PAGE under reducing conditions.[119] The hypothesis of Atmani & Khan,[147] based on Ouchterlony immunodiffusion data, that UAP is not completely identical to ITI, warrants further detailed examination. The Ouchterlony method is notoriously susceptible to any factor affecting diffusion properties and has been largely superceded by the more sensitive and accurate Western blotting technique, using which Atmani & Khan[147] demonstrated immunoreactivity of both human and rat UAP with an anti-ITI antibody. The relationship between UAP, Sørensen's protein and the 35 kDa protein isolated by us, requires further clarification, as does the proven connection of all three with ITI. However, it would seem most improbable that urine would contain three distinct urinary proteins with similar molecular weights and N-terminal homology with ITI, and commonsense dictates that all three proteins are likely to be one form of the light chain of ITI, each isolated from urine by slightly different methods.

Despite having been the subject of investigation for many years, the true physiological function of ITI remains somewhat of a mystery, although it may play a role in cancer, adult respiratory distress syndrome and septic shock, and may be a useful therapeutic agent in such conditions.[148] If this is so, then it is possible that its clinical usefulness may also extend to the treatment of human kidney stones. Unfortunately, however, as with NC, and UP, the effects of purified bikunin, or of its putative relatives, UAP, the protein described by Sørenson et al,[73] and the 35 kDa protein isolated by us,[118] on CaOx crystallization in human urine have not yet been determined. It is therefore not yet possible to gauge whether or not the light chain of ITI is likely to retain its inhibitory potency under physiological conditions. What is clear, however, is that there is an urgent need to clarify the role of ITI in stone formation, especially in the light of recent evidence tentatively identifying NC as either the HI-14 derivative of bikunin, or a portion of HI14.[118]

V. AND FINALLY....

The study of stone matrix has come a long way in recent years, but the wealth of knowledge we have gained has been offset to a large extent by conflicting findings, some of which have simply served to deepen the mystery of the role of matrix in stone formation. New technology has enabled us to strip macromolecules of their anonymity so that now we can identify individual components of matrix, but in every case we cannot say with real certainty just why they are there - whether they are good, bad or indifferent. Perhaps the major reason for this is the use of inappropriate methodology for testing the effects of urine and matrix macromolecules on crystallization processes. As discussed earlier in this chapter, whatever possible effects, or combination of effects on crystallization we might like to imagine, we are almost certain to find at least one macromolecule that will fit the bill. And this will continue to be the case just as long as we persist in measuring inhibitory activity in synthetic aqueous media and force the resulting data to make the quantum leap to urine and physiological conditions. Much of the confusion and contradiction that abound in the literature concerning matrix macromolecules can be ascribed directly to the habit, long since rightly questioned, but nonetheless persistently adopted, of drawing conclusions about macromolecules' effects in stone formation from data derived from aqueous inorganic systems. Such systems do not reproduce the complex ionic milieu of urine and we cannot expect inhibitors to exhibit the same effects in splendid isolation at low ionic strength, as they would in the urinary soup. It is to be hoped that in the future, results derived from inorganic media will be regarded with an appropriate degree of caution and more information will be sought using crystallization systems based on urine. Of course, no experimental crystallization system will ever replace the surfaces, the fluid and concentration dynamics, the twists and turns of the environment of the human kidney; but urine itself is a good start.

Since we are only now equipped with the requisite tools to identify individual matrix constituents, to some extent we may regard the study of stone matrix as just beginning. So it is important at the outset to avoid a problem that has tended to plague the study of proteins. It is all too tempting to suppose that the protein *we* have isolated from urine, crystals or stone matrix, holds the key to stone formation and, moreover, that it is unique. The passage of time usually proves otherwise, and once a protein has been identified definitively by amino acid sequence analysis or specific immunoblotting, it would seem prudent to avoid potential confusion in the literature by using that protein's currently accepted name, rather than indulging our introspective tendencies by clinging tenaciously to the one we gave it ourselves. As each new component of matrix is added to the list it is becoming increasingly apparent that there is no single, magic ingredient that alone will carry the blame for the fact that some of us suffer from stones, or take the credit for the fact that the majority of us, happily, do not. Every component of matrix is potentially an active protagonist in stone pathogenesis until proven otherwise. Future researchers intent on identifying those macromolecules rightfully entitled to a place as participants in CaOx crystallization processes, should ensure that their effects are tested in urine, and not neglect the possible contribution of other urinary components, for this approach carries the promise of discovering their true role in stone pathogenesis.

REFERENCES

1. Boyce, W.H, Organic matrix of human urinary concretions, Am. J. Med., 45, 673, 1968.
2. Lowenstam, H.A. and Weiner, S., On Biomineralization, Oxford University Press, New York, 324, 1989.
3. Boyce, W. and Garvey, F.K., The amount and nature of the organic matrix in urinary calculi: a review, J. Urol., 76, 213, 1956.
4. Warpehoski, M.A., Buscemi, P.J., Osborn, D.C., Finlayson, B. and Goldberg, E.P., Distribution of organic matrix in calcium oxalate renal calculi, Calcif. Tissue Int., 33, 211, 1981.

5. Roberts, S.D. and Resnick, M.I., Urinary stone matrix, in Renal Tract Stone, Wickham, J.E.A. and Buck, A.C., Eds., Churchill Livbngstone, London, 1990, 59.

6. Morse, R.M. and Resnick, M.I., Urinary stone matrix, J. Urol., 139, 602, 1988.

7. Khan, S.R. and Hackett , R.L., Role of organic matrix in urinary stone formation: An ultrastructural study of crystal matrix interface of calcium oxalate monohydrate stones, J. Urol., 150, 239, 1993.

8. Carr, J.A., The pathology of urinary calculi: radial striation. Br. J. Urol., 25, 26, 1953.

9. Khan, S.R. and Hackett, R.L., Crystal-matrix relationships in experientelly induced calcium oxalate monohydrate crystals, and ultrastuctural study, Calcif. Tissue Int., 41, 157, 1987.

10. Khan, S.R., Finlayson, B. and Hackett, R.L., Stone matrix as proteins adsorbed on crystal surfaces: A microscopic study, Scanning Microscopy 1, 379, 1983.

11. Khan, S.R. and Hackett, R.L., Microstructure of decalcified human calcium oxalate urinary stones, Scanning Electron Microscopy, 2, 935, 1984.

12. Khan, S.R., Hackett, R.L., and Finlayson, B., Morphology of urinary stone particles resulting from ESWL treatment, J. Urol., 136, 1367, 1986.

13. Spector, A.R., Gray, A. and Prien, E.L. Jr., Kidney stone matrix. Differences in acidic protein composition, J. Urol., 13, 387, 1976.

14. Boyce, W.H., King, J. and Fielden, M., Total non-dialyzable solids (TNDS) in human urine XIII. Immunological detection of a component peculiar to renal calculous matrix and to urine of calculous patients, J. Clin. Invest., 41, 1180, 1962.

15. Iwata, H., Kamei, O., Abe, Y., Nishio, S., Wakatsuki, A., Ochi, K. and Takeuchi, M., The organic matrix of urinary uric acid crystals, J. Urol., 139, 607, 1988

16. Doyle, I.R., Ryall, R.L., and Marshall, V.R., Inclusion of proteins into calcium oxalate crystal precipitated from human urine: a highly selective phenomenon, Clin. Chem., 37, 1589, 1991.

17. Petersen, T.E., Thørgesen, I. and Petersen, S.E., Identification of hemoglobin and to serine proteases in acid extracts of calcium containing kidney stones, J. Urol., 142, 176, 1989.

18. Khan, S.R., Shevock, P.N. and Hackett, R.L., Urinary enzymes and calcium oxalate urolithiasis, J. Urol., 142, 846, 1989.

19. Kim, K.M., Mulberry particles formed by red blood cells in human weddelite stones, J. Urol., 129, 855, 1983.

20. Finlayson, B., Khan, S.R. and Hackett, R.L., Mechanisms of stone formation: an overview, Scanning Microscopy, 3, 1419, 1984.

21. Khan, S.R., Shevock, P.N. and Hackett, R.L., In vitro precipitation of calcium oxalate in the presence of whole matrix or lipid components of urinary stones, J. Urol., 139, 418, 1988 .

22. Khan, S.R., Shevock, P.N. and Hackett, R.L., Presence of lipids in urinary stones: results of preliminary studies, Calcif. Tissue Int., 42, 91, 1988.

23. King, J.S. and Boyce, W.H., Immunological studies of serum and urinary proteins in urolith matrix in man, Ann. N.Y. Acad. Sci., 104, 579, 1963.

24. Robertson, W.G., Peacock, M. and Nordin, B.E.C., Activity products in stone-forming and non-stone-forming urine, Clin. Sci., 34, 579, 1968.

25. Robertson, W.G. and Peacock, M., Calcium oxalate crystalluria and inhibitors of crystallisation in recurrent renal stone-formers, Clin. Sci., 43, 499, 1972.

26. Hallson, P.C. and Rose, G.A., Crystalluria in normal subjects and stone formers with and without thiazide and cellulose phosphate treatment, Br. J. Urol., 48, 515, 1976.

27. Doyle, I.R., Marshall, V.R., Dawson, C.J. and Ryall, R.L., Calcium oxalate crystal matrix extract: The most potent macromolecular inhibitor of calcium oxalate crystallization yet tested in indiluted urine in vitro, Urol. Res., (In press), 1995.

28. Drach, G.W., Thorson, S. and Randolph, A., Effects of urinary organic macromolecules on crystallization of calcium oxalate: enhancement of nucleation, J. Urol., 123, 519, 1980.

29. Hallson, P.C. and Rose, G.A., Uromucoids and urinary stone formation, Lancet I, 1000, 1979.

30. Ryall, R.L., Harnett, R.M., Hibberd, C.M., Edyvane, K.A. and Marshall, V.R. Effects of chondroitin sulphate, human serum albumin and Tamm-Horsfall mucoprotein on calcium oxalate crystallization in undiluted human urine, Urol. Res., 19, 181, 1991.

31. Resnick, M.I. Sorrel, M.E., Barclay, J.A. and Boyce, W.H., Inhibitory effects of urinary calcium binding substances on calcium oxalate crystallization, J. Urol., 127, 568, 1982.

32. Nakagawa, Y., Kaiser, E.T. and Coe, F.L., Isolation and characterization of calcium oxalate crystal growth inhibitors from human urine, Biochem. Biophys. Res. Commun., 84, 1038, 1978.

33. Koide, T., Takemoto, M., Itatani, H., Takaha, M. and Sonoda, T., Urinary macromolecular substances as natural inhibitors of calcium oxalate crystal aggregation, Invest. Urol., 18, 382, 1981.

34. Grover, P.K., Ryall, R.L. and Marshall, V.R., Does Tamm-Horsfall mucoprotein inhibit or promote calcium oxalate crystallization in human urine? Clin. Chim. Acta, 190, 223, 1990.

35. Campbell, A.A., Ebrahimpour, A., Perez, L., Smesko, S.A. and Nancollas, G.H., The dual role of polyelectrolytes and proteins as mineralization promoters and inhibitors of calcium oxalate monohydrate, Calcif. Tissue Int., 45, 122, 1989.

36. Utsunomiya, M., Koide, T., Yoshioka, T., Yamaguchi, S. and Okuyama, A., Influence of ionic strength on crystal adsorption and inhibitory activity of macromolecules, Br. J. Urol., 71, 516, 1993.

37. Edyvane, K.A., Ryall, R.L. and Marshall, V.R., The influence of serum and serum proteins on calcium oxalate crystal growth and aggregation, Clin. Chim. Acta, 157, 81, 1986.

38. Ryall, R.L., Harnett, R.M.and Marshall, V.R., The effect of urine, pyrophoshate, citrate, magnesium and glycosaminoglycans on the growth and aggregation of calcium oxalate crystals in vitro, Clin. Chim. Acta, 112, 349, 1981.

39. Sugimoto, T., Funae, Y., Rübben, H., Nishio, S., Hautmann, R. and Lutzeyer, W., Resolution of proteins in the kidney stone matrix using high-performance liquid chromatography, Eur. Urol., 11, 334, 1985.

40. Kim, K.M., Lipid matrix of dystrophic calcification and urinary stone, Scanning Electron Microscopy, 3, 1275, 1983.

41. Boskey, A.L., Current concepts of the physiology and biochemistry of calcification, Clin. Orthop. Rel. Res. 157, 225, 1981.

42. Khan, S.R., Shevock, P.N. and Hackett, R.L., Membrane-associated crystallization of calcium oxalate in vitro, Calcif. Tissue Int., 46,116, 1990.

43. Khan, S.R., Whalen, P.O. and Glenton, P.A., Heterogeneous nucleation of calcium oxalate crystals in the presence of membrane vesicles, J. Crystal Growth, 134, 211, 1993.

44. Nishio, S., Abe, Y., Wakatsuki, A., Iwata, H., Ochi, K., Takeuchi, M. and Matsumoto, A., Matrix glycosaminoglycans in urinary stones, J. Urol., 134, 503, 1985.

45. Wessler, E., The nature of the non-ultrafiltrable glycosaminoglycans of normal human urine, Biochem. J., 122, 373, 1971.

46. Roberts, S.D. and Resnick, M.I., Glycosaminoglycans content of stone matrix, J. Urol., 135, 1078, 1986.

47. Yamaguchi, S., Yoshioka, T., Utsunomiya, M., Koide, T., Osafune, M., Okuyama, A. and Sonoda, T. Heparan Sulfate in the stone matrix and its inhibitory effect on calcium oxalate crystallization, Urol. Res., 21, 187, 1993.

48. Suzuki, K., Mayne, K., Doyle, I.R. and Ryall, R.L., Urinary glycosaminoglycans are selectively incorporated into calcium oxalate crystals, Scanning Microscopy, 8, 523, 1995.

49. Angell, A.H. and Resnick, M.I., Surface interaction between glycosaminoglycans and calcium oxalate, J. Urol., 141, 1255, 1989.

50. Shum, D.K.Y. and Gohel, M.D.I., Separate effects of urinary chondroitin sulphate and heparan sulphate on the crystallization of urinary calcium oxalate: differences between stone formers and normal control subjects, Clin. Sci., 85, 33, 1993.

51. Robertson, W.G., Peacock, M., Heyburn, P.J., Marshall, D.H. and Clark, P.B., Risk factors in calcium stone disease of the urinary tract, Br. J. Urol., 50, 449, 1978.

52. Nesse, A., Garbossa, G., Romero, M.C., Bogardo, C.E. and Zanchetta, J.R., Glycosaminoglycans in urolithiasis, Nephron, 62, 36, 1992.

53. Samuell, C.T., A study of glycosaminoglycan excretion in normal and stone-forming subjects using a modified cetylpyridinium chloride technique, Clin. Chim. Acta, 117, 63, 1981.

54. Ryall, R.L., Darroch, J.N. and Marshall, V.R., The evaluation of risk factors in male stone formers attending a general hospital out patient clinic, Br. J. Urol., 56, 116, 1984.

55. Shiba, K.S., Kanamori, K., Cho, H., Furuhata, N., Harada ,T., Shiba, A. and Nakao, M., Re-evaluation of normal urinary proteins fractionated by one dimensional sodium-dodecyl sulphate polyacrylamide electrophoresis, Clin. Chim. Acta, 172, 77, 1988.

56. Keutel, H.J,. King, J.S. and Boyce, W.H., Further studies of uromucoid in normal and stone urine, Urol. Int., 17, 324, 1964.

57. King, J.S. and Boyce, W.H., Amino acid and carbohydrate composition of the mucoprotein matrix in various calculi, Proc. Soc. Exp. Biol. Med., 95, 183, 1957.

58. Maxfield, M., Urinary mucopolysaccharides and calculi, Ann. Rev. Med., 14, 99, 1963.

59. Melick, R.A., Quelch, K.J. and Rhodes, M., The demonstration of sialic acid in kidney stone matrix, Clin. Sci., 59, 401, 1980.

60. Moore, S. and Gowland, G., The immunological integrity of matrix substance A and its possible detection and quantitation in urine, Br. J. Urol., 47, 489, 1975.

61. Wasserman, R.H. and Taylor, A.N., Evidence for vitamin-D-induced calcium-binding proteins in primates, Proc. Soc. Exp. Biol. Med., 136, 25, 1978.

62. Nakagawa, Y., Ahmed, M.A., Hall, S.L., Deganello, S. and Coe, F.L., Isolation from human calcium oxalate renal stones of nephrocalcin, a glycoprotein inhibitor of calcium oxalate crystal growth, J. Clin. Invest., 79, 1782, 1987.

63. Fraij, B.M., Separation and identification of urinary proteins and stone-matrix proteins by mini-slab sodium dodecyl sulfate-polyacrylamide gel electrophoresis, Clin. Chem., 35, 652, 1989.

64. Lian, J.B., Prien, E.L. Jr, Glimcher, M.J. and Gallop, P.M., The presence of protein-bound γ-carboxyglutamic acid in calcium-containing renal calculi, J. Clin. Invest., 59, 1151, 1977.

65. Friezner Degen, S.J. and Davie, E.W., Nucleotide sequence of the gene for human prothrombin, Biochemistry, 26, 6165, 1987.

66. Bezeaud, A. and Guillin, M.C., Quantitation of prothrombin activation products in human urine, Br. J. Haematol., 58, 597, 1984.

67. Morse, R.M. and Resnick, M.I., A new approach to the study of urinary macromolecules as a participant in calcium oxalate crystallization, J. Urol., 139, 869, 1988.

68. Salier, J-P., Inter-α-trypsin inhibitor: emergence of a family within the Kunitz-type protease inhibitor superfamily, T.I.B.S., 15, 435, 1990.

69. Binette, J.P., and Binette, M.B., The matrix of urinary tract stones: protein composition, antigenicity, and ultrastructure, Scanning Microscopy, 5, 1029, 1991.

70. Binette, J.P. and Binette, M.B., A cationic protein from a urate-calcium oxalate stone: Isolation and purification of a shared protein, Scanning Microscopy, 7, 1107, 1993.

71. Binette, J.P. and Binette, M.B., Sequencing of proteins extracted from stones, Scanning Microscopy, 8, 233, 1994.

72. Umekawa, T., Kohri, K., Amasaki, N., Yamate, T., Yoshida, K., Yamamoto, K., Suzuki, Y., Sinohara, H. and Kurita, T., Sequencing of a urinary stone protein identical to α1-Antitrypsin, Biochem. Biophys. Res. Commun., 193, 1049, 1993.

73. Sørenson, S., Hansen, K., Bak, S. and Justesen, S.J., An unidentified macromolecular inhibitory constituent of calcium oxalate crystal growth in human urine, Urol. Res., 18, 373, 1990.

74. Shiraga, H., Min, W., VanDusen, W.J., Clayman, M.D., Miner, D., Terrell, C.H., Sherbotie, J.R., Foreman, J.W., Przysiecki, C., Neilson, E.G. and Hoyer, J.R., Inhibition of calcium oxalate crystal growth in vitro by uropontin: another member of the aspartic acid-rich protein superfamily, Proc. Nat. Acad. Sci. USA, 89, 426, 1992.

75. Dalsing, M.C., Grosfeld, G.L. and Schiffler, M.A., Superoxide dismutase: A cellular protective enzyme in bowel ischemia, J. Surg. Res., 34, 589, 1983.

76. Kumar, S. and Muchmore, A., Tamm-Horsfall protein-uromodulin (1950-1990). Kidney Int., 37, 1395, 1990.

77. Tamm, I. and Horsfall, F.L., Characterisation and separation of an inhibitor of viral hemagglutination present in urine, Proc. Soc. Exp. Biol. Med., 74, 108, 1950.

78. Dulawa, J., Jann, K., Thomsen, M., Rambausek, M. and Ritz, E., Tamm-Horsfall glycoprotein interferes with bacterial adherence to human kidney cells, Eur. J. Clin. Invest., 18, 87, 1988.

79. Wallace, A.C. and Nairn, R.C. Tamm-Horsfall protein in kidneys of human embryos and foreign species, Pathol., 3, 303, 1971.

80. Sikri, K.L., Foster, C.L., MacHugh, N. and Marshall, R.D., Localization of Tamm-Horsfall glycoprotein in the human kidney using immuno-fluorescence and immuno-electron microscopical techniques, J. Anatomy, 132, 597, 1981.

81. Peach, R.J., Day, W.A., Ellingson, P.J. and McGiven, A.R., Ultrastructural localisation of Tamm-Horsfall protein in human kidney using ummunogold electron microscopy, Histochem. J., 20, 156, 1988.

82. Hunt, J.S., McGiven, A.R., Grouvsky, A., Lynn, K.L. and Taylor, M.C., Affinity-purified antibodies of defined specificity for use in a solid-phase microplate radioimmunoassay of human Tamm-Horsfall glycoprotein in urine, Biochem. J., 227, 957, 1994.

83. Muchmore, A.V. and Decker, J.M., Uromodulin: a unique 85-kilodalton immunosuppressive glycoprotein isolated from urine of pregnant women, Science, 479, 1985.

84. Stevenson, F.K., Cleave, A.J. and Kent, P.W., The effect of ions on the viscometric and ultracentrifugal behaviour of Tamm-Horsfall glycoprotein, Biochem. Biophys. Acta, 236, 59, 1971.

85. Pesce, A.J., Kant, K.S., Clyne, D.H. and Pollak, V.E., A model of urinary cast formation, Clin. Chem., 23, 1146 (abstract), 1977.

86. Curtain, C.C., The action of urea on the urinary inhibitor of influenza virus hemagglutination, Aust. J. Exp. Biol., 31, 615, 1953.

87. Robertson, W.G., Scurr, D.S. and Bridge, C.M., Factors influencing the crystallization of calcium oxalate in urine - a critique, J. Crystal Growth, 53, 182, 1981.

88. Kitamura, T. and Pak, C.Y.C., Tamm and Horsfall glycoprotein does not promote spontaneous precipitation and crystal growth of calcium oxalate in vitro, J. Urol., 127, 1024, 1982.

89. Fellström, B., Danielson, B.G., Ljunghall, S. and Wikström, B., Crystal inhibition: the effect of polyanions on calcium oxalate crystal growth, Clin. Chim. Acta, 158, 229, 1986.

90. Scurr, D.S. and Robertson, W.G. Modifiers of calcium oxalate crystallization found in urine. II. Studies on their mode of action in an artificial urine, J. Urol., 136, 128, 1986.

91. Grover, P.K., Ryall, R.L. and Marshall, V.R., Tamm-Horsfall mucoprotein reduces promotion of calcium oxalate crystal aggregation induced by urate in human urine in vitro, Clin. Sci., 87, 137,1994.

92. Rose, G.A. and Sulaiman, S., Tamm-Horsfall mucoprotein promotes calcium oxalate crystal formation in urine: Quantitative studies, J. Urol., 127, 177, 1982.

93. Yoshioka, T., Koide, T., Utsunomiya, M., Itatani, H., Oka, T. and Sonoda, T., Possible role of Tamm-Horsfall glycoprotein in calcium oxalate crystallization, Br. J. Urol., 64, 463, 1989.

94. Edyvane, K.A., Hibberd, C.M., Harnett, R.M., Marshall, V.R. and Ryall, R.L., Macromolecules inhibit calcium oxalate crystal growth and aggregation in whole human urine, Clin. Chim. Acta, 167, 329, 1987.

95. Grant, A.M.S., Baker, L.R.I. and Neuberger, A., Urinary Tamm-Horsfall glycoprotein in certain kidney disease and its content in renal and bladder calculi, Clin. Sci., 44, 377, 1973.

96. Bichler, K.H., Kirchner, C. and Ideler, V., Uromucoid excretion of normal individuals and stone formers, Br. J. Urol., 47, 733, 1976.

97. Samuell, C.T., Uromucoid excretion in normal subjects, calcium stone formers and in patients with chronic renal failure, Urol. Res., 7, 5, 1979.

98. Thornley, C., Dawnay, A. and Cattell, W.R., Human Tamm-Horsfall glycoprotein: urinary and plasma levels in normal subjects and patients with renal disease determined by a fully validated radioimmunoassay, Clin. Sci., 68, 529, 1985.

99. Hess, B., The role of Tamm-Horsfall glycoprotein and nephrocalcin in calcium oxalate monohydrate crystallization processes, Scanning Microscopy, 5, 689, 1991.

100. Hess, B. Tamm-Horsfall glycoprotein - Inhibitor or promoter of calcium oxalate monohydrate crystallization processes? Urol. Res., 20, 83, 1992.

101. Nakagawa,Y., Margolis, H.C., Yokoyama, S., Kézdy, F.J,. Kaiser, E.T. and Coe, F.L., Purification and characterization of a calcium oxalate monohydrate crystals growth inhibitor from human kidney tissue culture medium, J. Biol. Chem., 256, 3936, 1981.

102. Nakagawa, Y., Abram, V., Kézdy, F.J., Kaiser, E.T. and Coe, F.L., Purification and characterization of the principal inhibitor of calcium oxalate monohydrate crystal growth in human urine, J. Biol. Chem., 258, 12594, 1983.

103. Nakagawa, Y., Abram, V. and Coe, F.L., Isolation of calcium oxalate crystal growth inhibitor from rat kidney and urine, Am. J. Physiol., 247, F765, 1984.

104. Nakagawa, Y., Parks, J.H., Kézdy, F.J. and Coe, F.L., Molecular abnormality of urinary glycoprotein crystal growth inhibitor in calcium nephrolithiasis, Ass. Am. Phys. Tran., 98, 281, 1985.

105. Nakagawa, Y., Abram, V., Parks, J.H., Lau, H.S-H., Kawooya, J.K. and Coe, F.L., Urine glycoprotein crystal growth inhibitors. Evidence for a molecular abnormality in calcium oxalate nephrolithiasis, J. Clin. Invest., 76, 1455, 1985.

106. Coe, F.L., Margolis, L.H., Deutsch, L.H. and Strauss, A.L., Urinary macromolecular crystal growth inhibitors in calcium nephrolithiasis, Miner. Electrolyte Metab., 3, 268, 1980.

107. Nakagawa, Y., Ahmed, M., Hall, S.L., Deganello, S. and Coe, F.L., Isolation from human calcium oxalate renal stones of nephrocalcin, a glycoprotein inhibitor of calcium oxalate crystal growth. Evidence that nephrocalcin from patients with calcium oxalate nephrolithiasis is deficient in γ-carboxyglutamic acid, J. Clin. Invest., 79, 1782, 1987.

108. Netzer, M., Nakagawa, Y. and Coe, F.L., Characterization of a new antibody to nephrocalcin (NC), a major urinary inhibitor of calcium oxalate monohydrate (COM) crystal growth, Kidney Int., 37, 474 (abstract), 1990.

109. Nakagawa, Y., Netzer, M., and Coe, F.L., Immunohistochemical localization of nephrocalcin (NC) to proximal tubule and thick ascending limb of Henle's loop (TALH) of human and mouse kidney: resolution of a conflict, Kidney Int., 37, 474 (abstract), 1990.

110. Nakagawa, Y., Sirivongs, D., Novy, M.B., Netzer, M., Michaels, E., Vogelzang, N.J. and Coe, F.L., Nephrocalcin: biosynthesis by human renal carcinoma cells in vitro and in vivo, Cancer Res. 52, 1573, 1992.

111. Nakagawa, Y., Netzer, M., Michaels, E.K., Suzuki, F. and Ito, H., Nephrocalcin in patients with renal cell carcinoma, J. Urol. 152, 29, 1994.

112. Kaiser, E.T. and Bock, S.C., Protein inhibitors of crystal growth, J. Urol., 141, 750, 1989.

113. Hess, B., Nakagawa, Y. and Coe, F.L., Nephrocalcin isolated from human kidney stones is a defective calcium-oxalate-monohydrate crystal-aggregation inhibitor, in Urolithiasis, Walker, V.R., Sutton, R.A.L., Cameron, E.C., Pak, C.Y.C. and Robertson, W.G. Eds., Plenum, New York and London, 1989, 137.

114. Colette. C., Benmbarek, A., Boniface, H., Astre, C., Pares-Herbute, N., Monnier, L. and Guitter, J., Determination of protein-bound urinary gammacarboxyglutamic acid in calcium nephrolithiasis, Clin. Chim. Acta, 204, 43, 1991.

115. Lieske, J.C., Walsh-Reitz, M.M. and Toback, F.G., Calcium oxalate monohydrate crystals are endocytosed by renal epithelial cells and induce proliferation, Am. J. Physiol., 262, F622, 1992.

116. Desbois, C., Hogue, D.A and Kasenty G., The mouse osteocalcin gene cluster contains three genes with two separate spatial and temporal patterns of expression. J. Biol. Chem., 269, 1183, 1994.

117. Deyl, Z., Vancikova, O. and Macek, K., γ-carboxyglutamic acid-containing protein of rat kidney cortex. Changes with high fat diet and molecular parameters, Hoppe-Seyler's Z. Physiol. Chem., 361, 1767, 1980.

118. Tang, Y., Grover, P.K., Moritz, R.L., Simpson, R.J. and Ryall, R.L., Is nephrocalcin related to the urinary derivative (bikunin) of inter-α-trypsin inhibitor? Br. J. Urol., In submission, 1995.

119. Hochstrasser, K., Reisinger, P., Albrecht, G., Wachter, E. and Schönberger, Ö.L., Isolation of acid-resistant urinary trypsin inhibitors by high performance liquid chromatography and their characterization by N-terminal amino-acid sequence determination, Hoppe-Seyler's Z. Physiol. Chem., 365, 1123, 1984.

120. Worcester, E.M., Sebastian, J.L., Hiatt, J.G., Beshensky, A.M. and Sadowski, J.A., The effect of warfarin on urine calcium oxalate crystal growth inhibition and urinary excretion of calcium and nephrocalcin, Calcif. Tissue Int., 53, 242, 1993.

121. Atmani, F., Lacour, B., Drüeke, T. and Daudon, M., Isolation and purification of a new glycoprotein from human urine inhibiting calcium oxalate crystallization, Urol. Res., 21, 61, 1993.

122. Shiraga, H., Clayman, M.D., Neilson, E.G. and Hoyer, J.R., Affinity purification of urinary crystal growth inhibitor (CGI), Kidney Int., 35, 363 (abstract), 1989.

123. Prince, C.W., Oosawa, T., Butler, W.T., Tomann, M., Bhown, A.S., Bhown, M. and Schrohenloher, R.E., Isolation, characterization and biosynthesis of a phosphorylated glycoprotein from rat bone, J. Biol. Chem., 262, 2900, 1987.

124. Senger, D.R., Perruzzi, C.A., Papadopoulos, A. and Tenen, D.G., Purification of a human milk protein closely similar to tumor-secreted phosphoproteins and osteopontin, Biochem. Biophys. Acta, 996, 43, 1989 .

125. Kiefer, M.C., Bauer, D.M. and Barr, P.J., The cDNA and derived amino acid sequence for human osteopontin, Nucl. Acids Res., 17, 3306, 1989.

126. Franzén, A. and Heinegard, D., Isolation and characterization of two sialoproteins present only in stone calcified matrix, Biochem. J., 232, 4473, 1985.

127. Smith, J.H. and Denhardt, D.T., Molecular cloning of a tumor promotor-inducible mRNA found in JB6 epidermal cells: induction is stable at high, but not low, cell densities, J. Cell Biochem., 34, 13, 1987.

128. Hoyer, J.R., Uropontin in urinary calcium stone formation, Miner. Electrolyte Metab., In press, 1995.

129. Kohri, K., Suzuki, Y., Yoshida, K., Yamamoto, K., Amasaki, N., Yamate, T., Umekawa, T., Iguchi, M., Sinohara, H. and Kurita, T., Molecular cloning and sequencing of cDNA encoding urinary stone protein, which is identical to osteopontin, Biochem. Biophys. Res. Commun., 184, 859 1992.

130. Kohri, K., Nomura, S., Kitamura, Y., Nagata, T., Yoshioka, K., Iguchi, M., Yamate, T., Umekawa,T., Suzuki, Y., Sinohara, H. and Kurita, T., Structure and expression of the mRNA encoding urinary stone protein (osteopontin), J. Biol. Chem., 268, 15180, 1993.

131. Worcester, E.M., Blumenthal, S.S. and Beshensky, A., The calcium oxalate crystal growth inhibitor produced by mouse kidney cortical cells in culture is osteopontin, J. Bone Miner. Res., 7, 1029, 1992.

132. Reinholt, F.P., Hultenby, K., Oldberg, A. and Heingard, D., Osteopontin - a possible anchor of osteoblasts to bone, Proc. Nat. Acad. Sci. USA, 87, 4473, 1990.

133. Oldberg, A., Franzen, A. and Heinegard, D., Cloning and sequence analysis of rat bone sialoprotein (osteopontin) cDNA reveals an Arg-Gly-Asp cell-binding sequence, Proc. Natl. Acad. Sci. USA, 83, 8819, 1986.

134. Addadi, L. and Weiner, S., Interactions between acidic proteins and crystals: stereochemical requirements in biomineralization, Proc. Natl. Acad. Sci. USA, 82, 4110, 1985.

135. Brown, L.F., Berse, B., Van De Water, L., Papadopoulos-Sergiou, A., Perruzzi, C.A., Manseau, E.J., Dvorak, H.F. and Senger, D.R., Expression and distribution of osteopontin in human tissues: widespread association with luminal epithelial surfaces, Mol. Biol. Cell., 3, 1169, 1992.

136. Lopez, C.A., Hoyer, J.R., Wilson, P.D., Waterhouse, P. and Denhardt, D.T., Heterogeneity of osteopontin expression among nephrons in mouse kidneys and enhanced expression in sclerotic glomeruli, Lab. Invest., 69, 355, 1993.

137. Morse, R.M. and Resnick, M.I., A study of the incorporation of urinary macromolecules onto crystals of different mineral compositions, J.Urol., 141, 641, 1989.

138. Stapleton, A.M.F., Simpson, R.J. and Ryall, R.L., Crystal matrix protein is related to human prothrombin, Biochem. Biophys. Res. Comm., 195, 1199, 1993.

139. Suzuki, K., Moriyama, M., Nakajima, C., Kawamura, K., Miyazawa, K., Tsugawa, R., Kikuchi, N. and Nagata, K., Isolation and partial characterization of crystal matrix protein as a potent inhibitor of calcium oxalate crystal aggregation: evidence of activation peptide of human prothrombin, Urol. Res., 22, 45, 1994.

140. Stapleton, A.M.F. and Ryall, R.L., Blood coagulation proteins and urolithiasis are linked: Crystal matrix protein is the F1 activation peptide of human prothrombin, Br. J. Urol., In Press, 1995.

141. Stapleton, A.M.F. and Ryall, R.L., Crystal matrix protein - Getting blood out of a stone, Miner Electrolyte Metab., In press, 1995.

142. Stapleton, A.M.F., Seymour, A.E., Brennan, J.S., Doyle, I.R., Marshall, V.R. and Ryall, R.L., Immunohistochemical distribution and quantification of crystal matrix protein, Kidney Int., 44, 817, 1993.

143. Ryall, R.L., Grover, P.K., Stapleton, A.M.F., Barrell, D.K., Tang, Y., Moritz, R.L. and Simpson, R.J., The urinary F1 activation peptide of human prothrombin is a potent inhibitor of calcium oxalate crystallization in undiluted human urine in vitro, Clin. Sci., In submission, 1995.

144. Soriano-Garcia, M., Padmanabhan, K., De-Vos, A.M. and Tulinsky, A. The Ca^{2+} ion and membrane binding structure of the Gla domain of Ca-prothrombin fragment 1, Biochemistry, 31, 2554, 1992.

145.   Atmani, F., Lacour, B., Strecker, G., Parvy, P., Drüeke, T. and Daudon, M., Molecular characteristics of uronic-acid-rich protein, a strong inhibitor of calcium oxalate crystallization in vitro, <u>Biochem. Biophys. Res. Comm.</u>, 191, 1158, 1993.
146.   Atmani, F., Lacour, B., Jungers, P., Drüeke, T. and Daudon, M., Reduced inhibitory activity of uronic-acid-rich protein (UAP) in the urine of stone formers, <u>Urol. Res.</u>, 22, 257, 1994.
147.   Atmani, F. and Khan, S.R., Characterization of uronic-acid-rich inhibitor of calcium oxalate crystallization isolated from rat urine, <u>Urol. Res.</u>, In press, 1955.
148.   Michalski, C., Piva, F., Balduyck, M., Mizon, C., Burnouf, T., Huart, J-J. and Mizon, J., Preparation and properties of a therapeutic inter-alpha-trypsin inhibitor concentrate from human plasma, <u>Vox Sang.</u>, 67, 329, 1994.

Chapter 14

LIPID MATRIX OF URINARY CALCIUM OXALATE CRYSTALS AND STONES

Saeed R. Khan

I. INTRODUCTION

All products of biological mineralization in mammals consist of crystals and organic matrix and all such matrices contain lipids. Presence of lipids has been demonstrated, both histochemically and biochemically at both the physiological as well as pathological calcification sites.[1-13] Even though lipids are ubiquitous in mineralized substrates, they account for a relatively small proportion of the organic matrices: [10,12] 7-14% of bone, 2-6% of dentin, 12-22% of newly mineralized enamel, 9.6% of submandibular salivary gland calculi and 10.2% of supragingival calculi. Main reason for the presence of lipids at calcification sites is the involvement of cellular membranes in crystallization of calcific crystals. Lipids, particularly the acidic phospholipids of cellular membranes frequently act as a substrate for heterogeneous nucleation of calcium phosphate. According to the current concepts initial CaP deposition in a number of calcific diseases[14] occurs on cellular membranes which are present at the calcification site either as limiting membrane of the so-called matrix vesicles or as cellular degradation products. Even the calcification of bioprosthetic heart valves, the main cause of their failure, is shown to be membrane mediated.[15] Initial nuclei of calcification are associated with cellular membrane fragments derived from the pig cusp cells in case of porcine valves and from connective tissue cells in case of bovine pericardium. Calcification of intrauterine devices also appears to be initiated by cellular membranous material that is deposited on the devices during their exposure to the uterine fluid.[16] Dental plaque and calculus formation is yet another example of calcification initiated by cellular membrane.[1-5, 14] Membranes of microorganisms present in the dental plaque nucleate calcium hydroxyapatite and thus initiate calculus formation. *In vitro*, membranes, acidic phospholipids, lipid extracts from various calcified tissues, as well as lipid containing liposomes have been shown to initiate calcium phosphate formation from metastable solutions.[17-22]

There is a general consensus that nucleation of calcium oxalate (CaOx) crystals during urolithiasis is heterogeneous i.e. requires the presence of a solid phase.[23] Similar to the nucleation of calcium phosphate, cellular membranes and their constituent lipids may also act as substrates for

0-8493-7673-4/95/$0.00+$.50

nucleation of CaOx crystals. In this chapter we will review the results of studies investigating the following, 1. the presence of cellular membranes and lipids in urine and urinary stones using histochemical and biochemical techniques, 2. nucleation of CaOx crystals *in vitro* and *in vivo*, from metastable solutions in the presence of cellular membranes, 3. molecular interaction between membrane lipids and CaOx crystals.

II. URINARY LIPIDS

A. LIPID PROFILE OF STONE FORMERS URINE

Cellular degradation products are quite common in human urine[24] and small amounts of lipids appear in the urine under normal circumstances.[25]

Table 1

Significant Differences in Urinary Lipid Profiles of Normal and Calcium Oxalate Stone Formers

Lipid Species	Normal mg/24hr	Stone Former mg/24 hr	Normal/Calcific Stone Former P value
Phospholipids	0.323+.06 (8)	0.719+0.13 (14)	0.035
Cholesterol	1.978+0.28 (8)	3.811+0.55 (14)	0.018
Cholesterol Ester	1.176+0.2 (8)	2.58+0.45 (14)	0.007
Triglycerides	8.793+2.3 (8)	17.15+3.43 (13)	0.047
Gangliosides	1.39+0.47 (8)	3.4+0.95 (6)	0.064
Sulfatides	0.357+0.06 (8)	1.67+0.46 (6)	0.007
Digalacto-diglycerides	0.646+0.10 (7)	3.5+1.17 (11)	0.07

To determine the differences in urinary lipid profiles of normal controls and stone patients we isolated and identified various types of lipids present in the urine. Twenty four hour urine samples were collected from male and female CaOx stone formers and age and sex matched normal individuals with no history of urinary stones or other related disorders. Phospholipids, neutral lipids as well as glycolipids were isolated, identified and quantified. There were no significant differences in the urinary lipid profiles of male and females therefore the results were combined. All urine samples, males, females, normals or stone formers appeared to contain similar types of lipids but in different quantities. Table 1 shows the significant differences in the urinary lipid profile between stone forming patients and normal individuals. Other differences included the presence of lysophospholipids in the urine of stone forming patients and their absence from normal urine. A urine sample from uric acid stone former was also analyzed. Urine from the uric acid stone former contained very low amounts of phospholipids. The results indicate that urine of calcific stone formers has a different lipid profile than the urine of non stone formers.

B. CRYSTAL INDUCTION IN HUMAN URINE

Individual 24 hour urine specimens were collected from stone formers as well as healthy subjects with no known urological or nephrological problem. Samples were refrigerated during the collection. Calcium oxalate crystallization was induced by adding sodium oxalate to the urine.[26] Crystals were isolated by centrifugation at 10,000 g for 30 minutes in a J2-21 centrifuge. They were identified using scanning electron microscopy and X-ray diffraction. Crystal matrix was obtained by demineralization using 0.25M ethylenediamine tetra-acetic acid (EDTA) at pH 8 for 24 hours at 4°C. Samples were dialyzed, lyophilized and subjected to 10% sodium dodecyl sulfate polyacrylamide gel electrophoresis (SDS-PAGE) and transfer blotted for immunological identification. Lipids were extracted from the crystal matrix.[27] Crystal matrix was also examined by transmission electron microscopy.[28]

Calcium oxalate crystals were generally a mixture of monoclinic monohydrate and pyramidal dihydrate. Transmission electron microscopy of demineralized CaOx crystals revealed that crystal ghosts were intimately associated with cellular degradation products. Crystal matrix contained a mixture of phospholipids, glycolipids as well as neutral lipids. Crystals induced in stone formers urine contained significantly more phospholipids than the crystals made in urine from normal individuals. Matrix of CaOx crystals induced in urine from calcium oxalate stone formers contained more phospholipids and total lipids than the crystals induced in urine from normal non-stone forming individuals.

III. MEMBRANES AND LIPIDS IN STONE MATRIX

A. ULTRASTRUCTURAL STUDIES

Boyce demonstrated[29] the presence of an osmiophilic material in

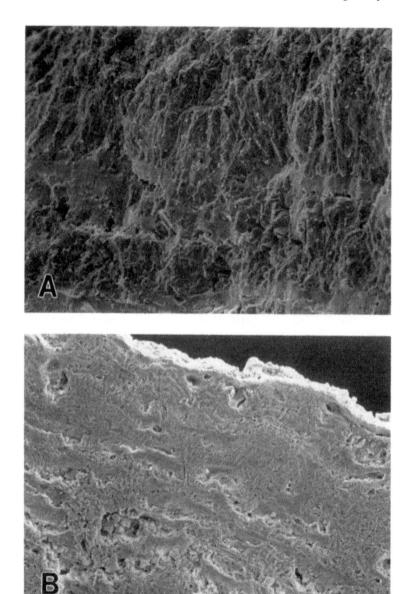

Figure 1. Fractured surface of calcium oxalate stone displaying concentrically laminated and radially striated appearance. **A.** Native X170. **B.** Demineralized, X45. Despite a loss of the mineral, architectural integrity of the stone is maintained because of the pervasive nature of the matrix.

organic matrices of urinary stones but interpreted the material as glycoprotein. Kim and Johnson suggested that the positive staining of urinary stones by osmium tetroxide was caused by lipids.[30] Kim later isolated membranous vesicles from the organic matrix of calcific urinary stones and proposed a role for such vesicles in the formation of urinary stones.[31]

We developed a technique for the examination of CaOx stone matrix[32] in which stone fragments were first embedded in agar and then simultaneously demineralized and fixed by treatment with a mixture of EDTA, glutaraldehyde and formaldehyde. Such a technique resulted in maintenance of stone shape even after the loss of crystalline components and for the first time revealed the pervasive nature and intimate involvement of organic matrix in the stone architecture (Figs.1 A,B). Crystal- matrix association is so intimate that dipyramidal habit of CaOx dihydrate, monoclinic habit of CaOx monohydrate and spherulitic habit of calcium phosphate stayed intact even after demineralization.[32-34] Crystals appeared as ghosts and EDTA insoluble organic matrix was found both on the crystal surfaces as well "inside" the crystals. Associated with the crystal ghosts were cellular degradation products; vesicles, mitochondria, nuclei, even the entire cells.[34] Membranous vesicles were frequently seen tightly bound to CaOx crystals ghosts.

For cytochemical localization of lipids in stone matrix, the demineralized stones were stained with sudan black B for light microscopic examination[27] and with malachite green[35] for transmission electron microscopy. Both stains showed positive staining of crystal-associated matrix (Figs. 2A, B) indicating the presence of phospholipids therein. Pretreatment of the paraffin sections with chloroform-methanol resulted in total loss of sudanophilia. Lipid extraction from the demineralized stones before malachite green staining reduced intensity of the stain.

B. BIOCHEMICAL EXTRACTION OF LIPIDS

Lipids of CaOx stones were extracted, isolated, identified and quantified.[27] An average of 3% of stone was organic matrix. They contained 3.8 ± 0.5 mg total lipid, and 0.42 ± 0.12 mg phospholipid/g of stone. Phospholipids isolated from the stone matrices included sphingomyelin, phosphatidyl ethanolamine, phosphatidyl choline, phosphatidyl inositol, phosphatidyl serine, phosphatidyl glycerol, and cardiolipin. Neutral lipids included cholesterol, cholesterol ester, free fatty acids, and mono, di and triglycerides. Small amounts of glycolipids including sulfatides, galactosyl-di-glycerides, I, II galactocerebroside and glucocerebroside were also present.

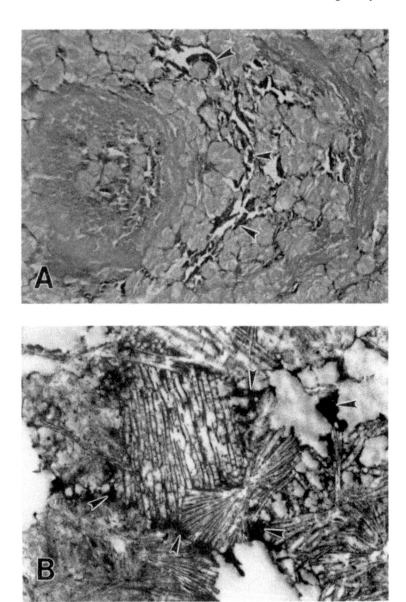

Figure 2. A. Paraffin section of a demineralized kidney stone stained with Sudan black B. **B.** Transmission electron micrograph of a malachite green stained demineralized kidney stone. In both cases crystal-associated organic matrix stained positive (arrowheads).

IV. NUCLEATION POTENTIAL OF CELLULAR MEMBRANES

We investigated the nucleation potential of the renal proximal tubular brush border membrane (BBM) both *in vitro* and *in vivo*. Proximal tubular BBM is, in many respects, similar to the membrane of matrix vesicles which are widely regarded as the initial site of both physiological and pathological calcification. Moreover, proximal tubular BBM is the single largest membrane pool in the mammalian kidney, and proximal tubular epithelial cells are highly susceptible to injury resulting in membrane sloughing. Membrane of the renal proximal tubular brush border can provide a suitable surface for crystal nucleation in the urinary environment and is a prime candidate as crystal nucleator. For *in vitro* studies BBM vesicles were obtained from rat kidney. For *in vivo* studies we induced membrane shedding from the proximal tubular brush border into the urine in the presence of acute or chronic hyperoxaluria.

A. CALCIUM OXALATE CRYSTALLIZATION *In Vitro*

Brush border membrane vesicles isolated from rat kidneys, were incubated, 0.1 mg/ml in a metastable low supersaturation solutions of CaOx at 37°C and 6.5 pH for 24, 48, 72 or 96 hours. Solutions were filtered through 0.2µm nucleopore filters, and depletion of calcium and oxalate in the filtrate was studied.[36] There was a gradual increase in the depletion of calcium and oxalate from the metastable solution with concomitant decrease in CaOx relative supersaturation. By 72 hours with 49.8% decrease in calcium, 43.6% in oxalate and 60.4% in CaOx relative supersaturation, crystals of CaOx were seen associated with the substrate (Fig.3A). There was no change in the calcium or oxalate level of the metastable solution of CaOx incubated without the brush border substrate. We conclude that renal epithelial brush border membranes can induce CaOx crystallization from a metastable solution which would otherwise not support it.

In constant composition crystallization experiments calcium and oxalate consumption started much earlier when BBM vesicles were present in the solution, within 258, 32, or 8 minutes of incubation in a metastable CaOx solution of 6, 10 or 12 relative supersaturation respectively.[37] Control solutions without the added BBM vesicles stayed free of crystallization for the duration. The plate-like crystals of calcium oxalate were formed in association with the membrane vesicles. Since crystallization started within 8 minutes of the incubation of brush border membrane in a CaOx solution of a low relative supersaturation of 12, there is a distinct possibility of membrane associated crystallization of CaOx in the urine of stone formers which is known to have a much higher CaOx relative supersaturation.

The effect of CaOx urinary stone matrix and its lipid contents on *in vitro* crystallization of CaOx was also studied.[38] Incubation of the whole

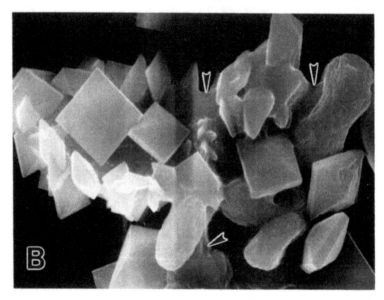

Figure 3. A. Calcium oxalate crystals produced *in vitro* by incubating brush border membrane in a metastable solution of calcium oxalate. X5,000. **B.** Calcium oxalate crystals produced *in vivo*, isolated from the urine of a gentamicin treated hyperoxaluric male rat. X4,250. In both examples crystals are associated with an organic substrate (arrowheads).

matrix or its lipids in a metastable solution of CaOx resulted in the depletion of calcium and oxalate, decrease of the CaOx relative supersaturation and the formation of CaOx crystals. No calcium or oxalate depletion and crystallization was detected in the metastable solution without the substrate.

B. CALCIUM OXALATE CRYSTALLIZATION *In Vivo*

In *in vivo* study of the role of membranes in CaOx crystallization both membranuria and hyperoxaluria were induced in male rats. Membranuria was provoked by administration of gentamicin sulphate, which specifically affects the proximal tubular epithelial cells. Low doses of ethylene glycol were utilized to avoid deposition of CaOx crystals in the kidneys. Male Sprague-Dawley rats were given 0.25% ethylene glycol in drinking water and daily subcutaneous injections of gentamicin sulphate, 100 mg/kg rat body weight.[39] One group of rats was given ethylene glycol only, another gentamicin sulphate only, and the third group of rats received both simultaneously. Rat urine were collected daily, and examined by light microscopy with both bright field and polarized optics. Urinary pH was measured. Urinary Ca^{++}, Na^+, K^+, Mg^{++}, ammonium, phosphate, sulphate, citrate and oxalate were determined. With these data urinary CaOx supersaturation were calculated. Kidneys as well as urinary sediment were examined by light, scanning and transmission electron microscopy.

With ethylene glycol only, no morphological changes were found in the kidneys but urinary oxalate levels were elevated. Gentamicin administration alone resulted in progressive renal tubular damage and an increase in membranous cellular degradation products in the urine. Neither ethylene glycol nor gentamicin alone induced CaOx crystals. However simultaneous administration of the two resulted in CaOx crystalluria in all animals by day 4 without overt tubular necrosis and crystal deposition in the kidneys. In addition there was an increase in both urinary excretion of calcium and oxalate as well as CaOx supersaturation. Supersaturation was however not high enough for the homogeneous nucleation of crystals. Scanning and transmission electron microscopic examination of urinary crystals demonstrated that they were intimately associated with cellular degradation products (Fig. 3B).[33, 39] It was concluded that membranous cellular degradation products were responsible for crystallization of CaOx in this model.

In separate experiments we induced acute hyperoxaluria in male rats by intraperitoneal administration of sodium oxalate. Calcium oxalate crystals formed in renal tubular lumina in close contact with membranous vesicles of mainly the brush border origin.[40]

V. MOLECULAR INTERACTION BETWEEN MEMBRANE LIPIDS AND CALCIUM OXALATE CRYSTALS

Once crystals are formed in the renal tubules they can either be excreted with the urine as crystalluria particles or retained in the kidneys and grow into kidney stones.[41] Almost all experimental data obtained from animal model[33,36,40] and cell culture[42-47] studies indicate the development of a complex relationship between CaOx crystals and cellular surfaces, which ultimately results in crystal retention within the renal tubules. This subject is reviewed by Hackett and Shevock in greater detail elsewhere in this volume. First step appears to be the attachment of luminal crystals to microvilli and apical membrane of the epithelial cells. Mandel's cell culture studies[45-47] have shown that crystals bind only to the cells which have altered membrane composition and/or structure and crystal-membrane phospholipid interactions play major role in crystal attachment. Primary cultures of inner medullary late collecting duct (IMCD) cells grown in monolayers were exposed to crystals of CaOx monohydrate, hydroxyapatite and uric acid. Crystals attached preferentially to cells present in aggregates and demonstrated saturable binding patterns. Transmission electron microscopic studies demonstrated that many of the cells in the aggregates had lost tight junctions between neighboring cells. Crystals preferentially adhered to the cells that had lost their polarity. Loss of tight junctions results in loss of polarity, migration of basolateral membrane components to the apical surface and major realignment of membrane lipids and other macromolecules.

Recently Mandel altered the phospholipid composition of IMCD cells and then exposed them to CaOx crystals.[48] Enrichment with phosphatidyl-serine (PS) increased crystal attachment. In addition crystals now attached to cells in monolayer that as we mentioned earlier normally do not bind crystals. Enrichment with phosphatidyl choline (PC) on the other hand, significantly reduced the crystal attachment. To test the possibility of crystal attachment caused by changes the membrane fluidity, cholesterol contents of IMCD cell membrane were altered. Increased cholesterol contents resulted in decreased fluidity and crystal attachment. Decrease in membrane fluidity hinders the lateral movement of lipids thereby impeding the development of long range structural repeats necessary for crystal membrane interaction and crystal attachment.

Mandel also examined epitaxial matches with less than 15% misfit between headgroups of phospholipids; PS, PC, phosphatidyl ethanolamine (PE) and phosphatidyl inositol (PI) and crystals of CaOx monohydrate and dihydrate, uric acid and hydroxyapatite.[48] Calcium oxalate monohydrate has the largest number of acceptable matches: 29 with PC, 10 with PE , 10 with PS and 3 with PI. CaOx dihydrate has only 3 matches with PC, none with PE, 1 with PS and 4 with PI. Thus there is a great likelihood of interaction between luminal CaOx monohydrate crystals and renal epithelial cell surfaces because PC is the most abundant phospholipid in outer leaflets of the cellular membranes.

VI. CONCLUDING REMARKS

Phospholipids of cellular membranes have both electrostatic, and calcium and hydrogen binding properties. Acidic phospholipids interacting with calcium and inorganic phosphate produce complexed phospholipids, the so-called acidic phospholipid-calcium phosphate complexes which in turn nucleate calcium phosphate.[11-13] Similarly, acidic phospholipids of membranes of cellular degradation products (CDP) can interact with calcium and oxalate to nucleate calcium oxalate crystals. These crystals can then adhere to the renal epithelial cells by interacting with other phospholipids of the membranes and thus be retained inside the renal tubules. Phospholipids of the CDP membranes present in the urine may interact with many crystals and help in linking them together to form crystal aggregates. Aggregation plays a critical role in crystal retention within the renal tubules, the formation of stone nidus and development of stone disease.[49] Formation of kidney stone and development of stone disease involves a cascade of events including crystal nucleation, growth, aggregation, and their retention within the kidneys. Membrane phospholipids are intimately involved in all these processes. Thus the presence of lipids in matrices of kidney stones is not just a consequence of adventitious incorporation but result of an active participation.

ACKNOWLEDGEMENTS

This work was supported in part by NIH grants # PO1 DK 20586 and RO1 DK 41434.

REFERENCES

1. **Irving, J.T.,** Histochemical changes in the early stages of calcification. *Clin. Orthop.,* 17, 93, 1959.
2. **Wuthier, R.E.,** Lipids of mineralizing epiphyseal tissues in the bovine fetus. *J. Lipid Research* 9, 68, 1968.
3. **Irving, J.T. and Wuthier, R.E.,** Histochemistry and biochemistry of calcification with special reference to the role of lipids. *Clin. Orthop.,* 56, 237, 1968.
4. **Odutuga, A.A. and Prout, R.E.S.,** Lipid analysis of human enamel and dentine. *Archs. oral. Biol.* 19, 729, 1971.

5. **Wuthier, R.E.,** The role of phospholipids in biological calcification. *Clin. Orthop.,* 90, 191, 1973.

6. **Vogel, J.J. and Boyan-Salyers, B.D.,** Acidic lipids associated with the local mechanisms of calcification. *Clin. Orthop. Rel. Res.,* 1118, 230, 1976.

7. **Boskey, A.L., Boyan-Salyers, B.D., Burstein, L.S. and Mandel, I.D.,** Lipids associated with mineralization of human submandibular gland sialoliths. Archs. Oral Biol. 26, 779, 1981.

8. **Slomiany B.L., Murty V.L.N., Aono M., Sarosiek J., Slomiany A., and Mandel I.D.,** Lipid composition of the matrix of human submandibular salivary gland stones. *Archs. oral Biol.,* 27, 673, 1982.

9. **Boskey, A.L., Burstein L.S. and Mandel I.D.,** Phospholipids associated with human parotid gland sialoliths. *Archs. oral Biol.,* 28, 655, 1983.

10. **Boskey A.L.** Current concepts of physiology and biochemistry of calcification. *Clin. Orthop.,* 157, 225, 1981.

11. **Boskey, A.L., Bullough, P.G., Vigorita, V. and di Carlo, E.,** Calcium-acidic phospholipid-phosphate compexes in human hydroxyapatite- containing pathologic deposits. *Am. J. Pathol.,* 133, 22, 1988.

12. **Boskey, A.L.,** Phospholipids and calcification, in: *Calcified Tissue,* Hukins, D.W.L. (Editor), CRC Press, Boca Raton, Florida, 215-243, 1989.

13. **Boyan, B.D., Schwartz, Z., Swain, L.D. and Khare, A.:** Role of lipids in calcification of cartilage. *Anat. Record,* 224, 211, 1989.

14. **Anderson H.C.,** Calcific diseases. *Arch. Pathol. Lab. Med.,* 107, 341, 1983.

15. **Schoen F.J., Harasaki H., Kim, K.M., Anderson, H.C., Levy, R.J.,** Biomaterial-assisted calcification: pathology, mechanisms, and strategies for prevention. *J. Biomed. Mater. Res.* 22(A1), 11, 1988.

16. **Khan, S.R. and Wilkinson, E.J.** Scanning electron microscopy, x-ray diffraction, and electron microprobe analysis of calcific deposits on intrauterine contraceptive devices. *Human Pathol.,* 16,732, 1985.

17. **Ennever, J., Vogel, J.J. and Benson, L.A.,** Lipid and calculus matrix calcification *in vitro. J. Dent. Res.,* 52, 1056, 1973.

18. **Odutuga, A.A., Prout, R.E.S. and Hoare ,R.J.** Hydroxyapatite precipitation *in vitro* by lipids extracted from mammalian hard and soft tissues. *Archs. Oral Biol.,* 20, 311, 1975.

19. **Vogel, J.J.,** Calcium Phosphate solid phase induction by dioleolphosphatidate liposomes. *J. Colloid and Interface Science,* 111, 152, 1986.

20 **Eanes, E.D. and Hailer, A.W.** Calcium phosphate precipitation in aqueous suspensions of phosphatidyl serine-containing anionic liposomes. *Calcif. Tissue Intl.,* 40, 3, 1987.

21. **Skrtic, D. and Eanes, E.D.,** Effect of different phospholipid-cholesterol membrane compositions on liposome-mediated formation of calcium phosphate. *Calcif. Tissue Intl.,* 50, 253, 1992.

22. **Skrtic, D. and Eanes, E.D.,** Membrane-mediated precipitation of calcium phosphate in model liposomes with matrix vesicle-like lipid composition. *Bone and Mineral,* 16, 109, 1992.

23. National Institute of Health Consensus Development Conference Statement Prevention and Treatment of Kidney Stones, March 28-30, 1988.

24. **Prescott, L.F.,** The normal urinary excretion rates of renal tubular cells, leucocytes, and red blood cells. *Clin. Sci.,* 31, 425, 1966.

25. **Martin, R.S. and Small, D.M.,** Physicochemical characterization of the urinary lipids from humans with nephrotic syndrome. *J. Lab. Clin. Med.,* 103, 798, 1984.

26. **Stapleton, A.M.F., Simpson, R.J., Ryall, R.L.,** Crystal matrix protein is related to human prothrombin. *Biochem. Biophys. Res. Commun.,* 195, 1199, 1993.

27. **Khan, S.R., Shevock, P.N. and Hackett, R.L.,** Presence of lipids in urinary stones: results of preliminary studies. *Calcif. Tissue Intl.,* 42, 91, 1988.

28. **Khan, S.R., Finlayson, B. and Hackett, R.L.,** Stone matrix as proteins adsorbed on crystal surfaces: a microscopic study. *Scan. Electr. Microsc.,* I, 379, 1983.

29. **Boyce, W.H.,** Some observations on the ultrastructure of "idiopathic" human renal calculi, in: *Urolithiasis: physical aspects,* Finlayson B., Hench L.L., Smith L.H. (eds), National Academy of Sciences, Washington, DC, 97-130, 1972.

30. **Kim, K.M. and Johnson, F.B.,** Calcium oxalate crystal growth in human urinary stones, *Scann. Electr. Microsc.,* III, 146, 1981.

31. **Kim, K.M.,** Lipid matrix of dystrophic calcification and urinary stone. *Scann. Electr. Microsc.,* 3, 1275, 1983.

32. **Khan, S.R., Finlayson, B. and Hackett, R.L.,** Agar-embedded urinary stones: a technique useful for studying microscopic architecture, *J. Urol.,* 130, 992, 1983.

33. **Khan, S.R., and Hackett, R.L.,** Crystal-matrix relationships in experimentally induced urinary calcium oxalate monohydrate crystals, an ultrastructural study. *Calcif. Tissue Intl.,* 41, 157, 1987.

34. **Khan, S.R., and Hackett R.L.,** Role of organic matrix in urinary stone formation: An ultrastructural study of crystal matrix interface of calcium oxalate monohydrate stones, *J. Urol.,* 150, 239, 1993.

35. **Goldberg, M., Lecolle, S., Ruch, J.V., Staub,i A., Septier, D.,** Lipid detection by malachite green-aldehyde in the dental basement membrane in the rat incisor. *Cell Tissue Res.,* 253, 685, 1988.

36. **Khan, S.R., Shevock, P.N., and Hackett R.L.**, Membrane-associated crystallization of calcium oxalate *in vitro. Calcif. Tissue Intl.*, 46, 116, 1990.
37. **Khan, S.R., Whalen, P.O., and Glenton, P.A.** Heterogeneous nucleation of calcium oxalate crystals in the presence of membrane vesicles. *J. Crystal Growth*, 134, 211, 1993.
38. **Khan, S.R., Shevock, P.N. and Hackett, R.L.**, *In vitro* precipitation of calcium oxalate in the presence of whole matrix or lipid components of the urinary stones. *J. Urol.* 139, 418, 1988.
39. **Hackett, R.L., Shevock, P.N., Khan, S.R.** Cell injury associated calcium oxalate crystalluria, *J. Urol.*, 144, 1535, 1990.
40. **Khan, S.R., Finlayson, B. and Hackett, R.L.**, Experimental calcium oxalate nephrolithiasis in rat, role of renal papilla, *Am. J. Pathol.* 107, 59, 1982.
41. **Finlayson, B. and Reid, F.M.**, The expectation of free and fixed particles in urinary stone disease. *Invest. Urol.* 15, 442, 1978.
42. **Hackett, R.L., Shevock, P.N., and Khan, S.R.**, Madin-Darby canine kidney cells are injured by exposure to oxalate and calcium oxalate crystals, *Urol. Res.* 22, 197, 1994.
43. **Lieske, J.C., Walsh-Reitz, M.M., and Toback, F.G.**, calcium oxalate monohydrate crystals are endocytosed by renal epithelial cells and induce proliferation, *Am. J. Physiol.* 262,F622, 1992.
44. **Lieske, J.C., Swift, H., Martin, T., Patterson, B., and Toback, F.G.**, Renal epithelial cells rapidly bind and internalize calcium oxalate monohydrate crystals, *P.N.A.S.*, 91, 6987, 1994.
45. **Riese, R.J., Riese, J.W., Kleinman, J.G., Wiessner, J.H., Mandel, G.S., and Mandel, N.S.**, Specificity in calcium oxalate adherence to papillary epithelial cells in culture, *Am. J. Physiol.*, 255, F1025, 1988.
46. **Mandel, N.S., and Riese, R.J.**, Crystal-cell interactions: crystal binding to rat renal papillary tip collecting cells in culture. *Am. J. Kid. Dis.*, 17, 402, 1991.
47. **Riese, R.J., Wiessner, J.H., Mandel, G.S., Mandel, N.S., and Kleinman, J.G.**, Cell polarity and calcium oxalate crystal adherence to cultured collecting duct cells, *Am. J. Physiol.*, 262, F177, 1992.
48. **Mandel, N.S.**, Crystal-membrane interaction in kidney stone disease, *J. Am. Soc. Nephrol.*, 5, S37, 1994.
49. **Kok, D.J., and Khan, S.R.**, Calcium oxalate nephrolithiasis, a free or fixed particle disease, *Kid. Intl.*, 46, 847, 1994.

Chapter 15

URATE AND CALCIUM OXALATE STONES: A NEW LOOK AT AN OLD CONTROVERSY

Phulwinder K. Grover and Rosemary L. Ryall
Flinders Medical Centre, Bedford Park, SA
Australia

I. INTRODUCTION

Despite the fact that the relationship between hyperuricosuria and calcium oxalate (CaOx) urolithiasis has been investigated for many years, the subject is still characterised by conflicting experimental and clinical findings, confusion, disagreement and general scepticism. The aim of this chapter is to clarify the topic by appraising the evidence that urate excretion influences CaOx stone formation. We will begin by critically analysing findings, both real and purported, that have led to the general recognition of a distinct clinical entity often referred to as "hyperuricosuric CaOx nephrolithiasis" and "hyperuricosuric CaOx stone formation". For the sake of simplicity, the terms urate and uric acid will be used interchangeably, unless otherwise indicated.

II. EVIDENCE LINKING URINARY URATE EXCRETION TO CALCIUM OXALATE STONE FORMATION

A. CALCIUM STONE DISEASE IN GOUTY PATIENTS

Though the association of gout with urinary calculi has been known since the 17th century [Greenhill 1844, cited in 1], the origin of the notion that patients excreting high levels of urinary urate experience an extraordinarily high incidence of CaOx stone formation is difficult to trace. To early workers, a logical starting point was to analyse the incidence of stones in patients suffering from gout, since their abnormal renal handling of urate leads to the urinary excretion of large quantities of urate. Reason dictated that if hyperuricosuria **does** predispose to stone formation, then these patients surely must have a higher incidence of stones than the population at large. Unfortunately, however, an unquestioning faith in logic simply served to perpetuate an unsubstantiated belief that gouty patients suffer more frequently from stones than do their normouricosuric colleagues. Eager to support the hypothesis that hyperuricosuria increases the risk of stone pathogenesis, countless papers over the years have unquestioningly declared that gouty individuals are prone to kidney stones, citing two works as supportive evidence, namely Prien & Prien[2] and Gutman & Yu.[3] Yet careful examination of these papers reveals that while

Prien and Prien[2] simply mentioned the existence of a common belief that calcium stones were often seen in association with gout, they did not actually analyse calculi from these patients. Only Yu and Gutman[4] reported the analysis of stones formed in gouty patients, and they later presented these results in 1968.[3] Their work is worthy of particular attention, for as far as we can ascertain, it alone carries the early burden of proof (however misplaced) implicating urate's involvement in CaOx stone formation. An analysis of relevant data in that paper reveals that these workers studied 1994 gouty patients, and simple calculation demonstrates that only 1.89% of these would have been expected to suffer from CaOx calculi. This figure is no higher, and is probably lower, than might be expected in the general population.[5-6] Thus the evidence commonly cited to support the notion that gouty patients are particularly at risk of CaOx stone formation is, at best, tenuous, at worst, the result of misinterpretation of published data.

B. "HYPERURICOSURIA" AND CALCIUM STONES

Another reason for relating uric acid excretion to CaOx stone formation was the empirical observation that hyperuricaemia and "hyperuricosuria" seemed to be more common in stone formers than in normal subjects.[7] Fortunately, this observation appeared mercifully easy to prove, and a large number of workers put it to the test. However, the findings, instead of resolving the issue, simply served to fuel the debate. While some workers reported stone formers to excrete more uric acid than normals,[7-18] others were unable to find any such difference.[19-25] Thus, as so frequently happens in stone research, what appeared to be an uncomplicated comparison of stone formers and healthy controls designed to yield conclusive findings, simply complicated the issue further. With the benefit of hindsight, these inconsistent findings can be attributed to several deficiencies in the design of virtually all of these studies. For instance, uric acid excretion is known to depend upon age, sex, body weight, diet, socioeconomic status and geographical location[22, 26] and fluctuates widely in any individual.[20, 27] Ideally therefore, all these factors should have been taken into account in studies which compared the daily urinary excretion of urate in normals and stone formers - obviously an impossible task. Furthermore, the definition of normal urate excretion, upon which the designation of "hyperuricosuria" depends, was highly questionable in many cases. The recognised dependence of urate excretion upon the various factors mentioned above dictates that a valid comparison should at least be based on the definition of a contemporary normal range established in individuals from the same geographical location, consuming a similar diet. However, while the normal values established almost 30 years ago by Gutman and Yu[3], served as the basis of comparison by a number of workers, others did not compare their stone formers with a control group at all, or even provide numerical definitions of their normal limits, opting instead to use such uninformative terms as "conventional criteria", "usual criteria" or "arbitrarily defined".[13, 14, 22] It is surely no wonder that, rather than resolve the matter, the findings of these studies simply compounded the confusion.

To date, no study has fully complied with what are, from a practical perspective, the impossible requirements that a truly valid comparison of urate excretion in stone formers and normals would entail. The unequivocal existence of a distinct clinical entity known as "hyperuricosuric CaOx nephrolithiasis"[14] therefore remains unproved. Nonetheless, the fact that true "hyperuricosuria" is not a well documented feature of CaOx urolithiasis, in no way diminishes the possibility that urate may still be an important factor in CaOx stone formation, and other evidence certainly supports this. For instance, recently Gambaro et al[28] reported an abnormal urate self-exchange in 30% of CaOx renal stone formers in whom urinary urate was frequently increased. Furthermore, the anomaly was associated with an increased frequency of hyperuricosuria and more intense disease activity.

C. ALLOPURINOL AND CALCIUM STONES

Perhaps the most compelling argument linking urate excretion to calcium stone formation can be found in the apparent success of allopurinol (a drug that lowers urinary output of urate) in reducing the rate of CaOx stone recurrences in patients excreting high levels of urinary urate.[11, 13, 29-33] However, with the benefit of hindsight, it is apparent that deficiencies in the design of these studies have lessened the impact of their findings: none were based on a double-blind design, nor were the subjects randomized between experimental and placebo groups.[34] These deficiencies were rectified in 1986, when Ettinger et al[35] conducted the first randomized, double-blind study and reported that allopurinol slightly reduced CaOx stone recurrences in "hyperuricosuric" subjects. However, while this trial corrected most of the shortcomings of its predecessors, one still remained. Hyperuricosuria was defined on the basis of Gutman and Yu's[3] normal range - despite the fact that in the intervening twenty years the upper normal limit of uric acid excretion must have risen in response to the well recognised increase in consumption of animal protein[36] during the same period. For this reason, the validity of even this, the most reliable study to date, has also been questioned.[37] But does the definition of hyperuricosuria really matter? Should the credibility of the trial be questioned because of the reference range used?

Allopurinol has been noted to reduce stone recurrences, even in normouricosuric subjects,[31, 38-40] an observation that has given rise to the notion that perhaps its good effect is attributable not to its reduction of urate output, but rather that of some other urinary metabolite which affects CaOx crystallization. This lateral thinking spawned several studies investigating the effect of allopurinol on excretion of compounds involved in, or already known to affect CaOx crystallization. As might be expected, the first on the list was the traditional villain, calcium. To be sure, urinary calcium was shown to be reduced during allopurinol therapy,[9, 29, 41-43] but the significance of the finding was overshadowed somewhat by other reports to the contrary.[30,44-49]

The excretion of oxalate was also investigated. A possible association between purine metabolism and oxalate excretion was suggested by Hodgkinson[20] as long ago as 1976, and this seemed to be confirmed by reports that allopurinol administration lowered oxalate excretion,[50-52] possibly by inhibiting absorption of purines from gut.[53-54] However, as so often transpires with investigations of CaOx stone disease, this observation was overwhelmingly swamped by reports to the contrary.[30, 43-45, 47, 49, 55-57] If not calcium and oxalate, why not urinary inhibitors of CaOx crystallization? But studies examining the effect of allopurinol on the excretion of such inhibitors were destined to suffer a similar fate. Reports that allopurinol increased the excretion of magnesium[41], pyrophosphate,[45] citrate and glycosaminoglycans were not substantiated by others.[30, 43-44, 47, 49, 58] Yet one other possibility remained: perhaps allopurinol directly affected CaOx? But this, too, was effectively discounted by several studies unable to demonstrate a direct effect of allopurinol or its metabolic derivatives[59-61] on CaOx crystallization *in vitro*; only one investigation Pak[45] reported the drug to inhibit precipitation of the mineral.

Thus it would be true to say that though possible explanations for the apparent beneficial effect of allopurinol in the treatment of stone recurrences are controversial, at least there is general concensus that it **is** beneficial in some patients. In the absence of experimental clinical or experimental evidence demonstrating conclusively that allopurinol's success lies elsewhere, we can then reasonably conclude that it can be attributable simply to its ability to reduce the urinary excretion of urate. But how?

III. THEORIES LINKING URATE EXCRETION TO CALCIUM OXALATE STONE FORMATION

A. EPITAXY

Epitaxy is the growth of crystals of one type upon the crystalline substrate of another, with a near geometrical fit between the networks in contact.[62] This process was first suggested by Modlin[63] to explain the simultaneous occurrence of minerals of different types in kidney stones. One year later, using x-ray crystallography, Lonsdale[64] demonstrated several crystal lattice fits for anhydrous uric acid, uric acid dihydrate, whewellite (CaOx monohydrate) and weddellite (CaOx dihydrate), thereby confirming the theoretical possibility that deposition of CaOx could occur upon the surface of uric acid crystals. In an attempt to explain the alternating layers of different minerals around a nucleus - commonly seen in urinary calculi - she proposed epitaxy as a mechanism of stone formation. This phenomenon was also subsequently invoked to explain CaOx crystallization in very acid urine.[65]

Human urine, particularly during morning hours, is commonly supersaturated with respect to urate,[30, 66-68] especially in patients excreting high levels of this salt. This favours the formation of urate particles which could, at least theoretically, act as a base for the deposition of CaOx. Experimental verification of this phenomenon was obtained 7 years later when Coe et al[69] and Pak and Arnold[70] demonstrated that seed crystals of monosodium urate (the stable phase of uric acid at physiological pH) and, to a significantly lesser extent, of uric acid, caused precipitation of CaOx in inorganic solutions *in vitro* . This is illustrated in Figure 1 where urate crystals in urine are depicted as promoting deposition of CaOx upon their surfaces. However, though sodium urate was an effective inducer of CaOx precipitation, uric acid proved to be relatively inefficient, although the effect could be overcome by the presence of small amounts of glutamic acid.[71] Since the latter is a natural component of urine, it was presumed that seed crystals of uric acid would also induce precipitation of CaOx in conditions likely to prevail in human urine *in vivo*. This reinforced the relevance of epitaxy as a potential mechanism of stone formation in general, and in patients excreting high levels of urinary urate, in particular.

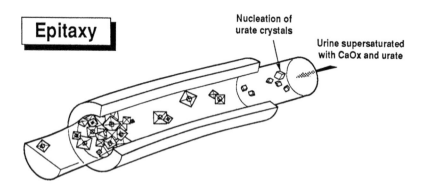

Fig 1. Diagrammatic representation of the epitaxy theory showing deposition of CaOx onto urate crystals (reprinted with permission from Mineral and Electrolyte Metabolism).

A basic requirement for epitaxy is the existence of a good three-dimensional lattice fit between the salts involved. However, though Lonsdale's[64] x-ray data confirmed the existence of a good two-dimensional lattice fit between anhydrous uric acid, uric acid dihydrate, and CaOx, it did not demonstrate a good three dimensional match. In fact, her data were limited only to two prominent crystal growth faces in uric acid as the substrate, and did not calculate the percentage of misfit in contact faces for any other common match. Furthermore, studies supporting this theory did not provide any direct evidence that the deposition of CaOx was actually caused by epitaxy. They were based simply on the observation that addition of urate seeds caused either a consistent decrease in soluble calcium and/or oxalate or a consistent increase in turbidity and/or particle number. One can reasonably argue that deposition of CaOx in this manner may be the result of the disturbance of ionic or thermodynamic equilibria by urate. This process is known as heterogeneous nucleation and does not necessarily require precipitation of one mineral onto the surface of another: as any undergraduate student of chemistry will testify, it can often be easily induced simply by scratching the walls of a glass vessel containing a supersaturated solution of a salt.

A further drawback is that most experiments supporting this theory were carried out either in inorganic synthetic media or in diluted human urine, despite the fact that results from such studies may bear no resemblance to events which would occur in whole urine.[72] Recently therefore, we tested the theory in undiluted human urine by studying the deposition of CaOx onto seed crystals of uric acid and sodium urate.[73] Analysis of data revealed that uric acid seeds increased the precipitation of CaOx from urine loaded with oxalate by 1.4% ($P<0.05$), sodium urate seeds by 5.2% ($P<0.01$) and CaOx by 54% ($P<0.001$). It is noteworthy that the duration of these experiments was 120min. However, the transit time of urine in renal tubules is of the order of 3-5min.[69, 74-75] Any promotion of CaOx deposition caused by urate or uric acid seed crystals during the time taken for urine to pass from the glomerulus to the renal pelvis would therefore be infinitesimally small. These data suggest that seeds of sodium urate and uric acid would not promote CaOx deposition to a significant degree under physiological conditions. Furthermore, scanning electron microscopic examination of the material precipitated at the end of our experiments showed that large aggregates of CaOx were deposited in the presence of CaOx seeds, which themselves were not visible, presumably because they had been enveloped by deposition of CaOx upon them. In contrast, urate seeds were clearly visible, being scattered like barnacles upon the surface of the much larger CaOx crystals. Many had been engulfed by the CaOx growth front, and were evident as smooth protuberances upon the surface of CaOx crystals (Figure 2). This was perhaps not demonstrative of epitaxial deposition of CaOx onto urate seeds, for in that case we might have expected the urate seeds to be invisible - completely contained within the CaOx deposited upon them. Of course, the possibility cannot be discounted that induction of CaOx crystal nucleation was triggered by just a few of the urate seeds, the remainder of which subsequently became attached the the rapidly growing, and considerably larger, CaOx crystals. What the study strongly suggests is that binding of sodium urate and uric acid crystals to, and their subsequent enclosure ("endocrystallosis") within actively growing CaOx crystals could occur in vivo, thereby increasing the overall volume of crystalline material precipitated and explaining the occurrence of mixed urate/oxalate stones. Certainly, the results provide no direct support for the significant promotion of CaOx crystal nucleation in urine by seed crystals of uric acid. In any event, if epitaxy were a common cause of CaOx stones then urate would be a frequent component of stones composed predominantly of CaOx. Such a combination is rarely observed in urinary calculi,[76-80] particularly in stones from the upper urinary tract.[37]

Fig 2. Scanning electron micrograph of CaOx crystals deposited from urine in the presence of sodium urate seeds. Note the presence of seed crystals of sodium urate lying on the surface of the CaOx crystals and on the membrane, and embedded inside the CaOx crystals - revealed as protuberances on the crystal surfaces (reprinted with permission from Mineral and Electrolyte Metabolism).

As if the foregoing were not sufficient to seriously question the role of epitaxy in CaOx stone formation, the theory is further undermined by the observation that crystallization of sodium urate requires the excretion of very large, physiological improbable, quantities of sodium and urate ions.[81] This may also explain the rare occurrence of crystalline urate in urine[14, 82] - an absolute prerequisite for the operation of this theory *in vivo* . Thus, although epitaxy has been the most commonly cited theory linking urate excretion to CaOx stone formation, and therefore, justifying the administration of allopurinol to recalcitrant CaOx stone formers, the majority of experimental and empirical evidence does not support a significant role for the process *in vivo*.

B. INTERACTION OF URATE WITH URINARY GLYCOSAMINOGLYCANS

Perhaps the most damning deficiency of the epitaxy theory is the virtual lack of crystalline urate in fresh warm human urine. To surmount this shortcoming, Robertson et al[83] proposed an alternative theory, the so-called glycosaminoglycan (GAG) depletion theory, which is depicted in Figure 3. This theory, like epitaxy, requires that urine be supersaturated with both CaOx and urate; however, in this case urate is proposed as occurring in urine as colloidal, rather than crystalline, particles. Its tenet is that GAGs are the principal inhibitors of CaOx crystallization and that binding of colloidal particles of urate to GAGs attenuates their inhibitory potency. This, in consequence, facilitates the nucleation, growth and aggregation of CaOx crystals.

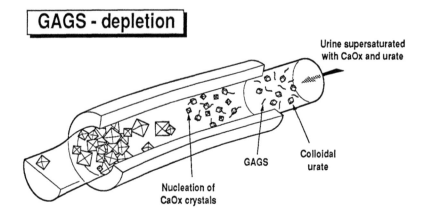

Fig 3. Diagrammatic representation of the GAGs depletion theory showing binding of GAGs to colloidal urate particles and subsequent crystallization of CaOx (reprinted with permission from Mineral and Electrolyte Metabolism).

There is some support for this theory: there is evidence to suggest that colloidal urate is stabilized by binding of GAGs, and about 30-50% of urate (at least in the urine of Dalmatian dogs) is associated with the GAGs fraction.[84]   In addition, the binding of GAGs to urate has been demonstrated, both indirectly[85-88] and directly,[89-92] and incubation with urate has been reported to decrease the inhibitory activity of heparin,[92-94] diluted macromolecular fractions of urine[95-96] and dilute unfractionated urine.[94, 97]   However, the urine data should be regarded with circumspection, since it is likely that a number of macromolecular urinary inhibitors other than GAGs, proteins for instance, will bind to urate crystals under such circumstances.

This theory is founded on a requirement for the presence of colloidal urate in human urine, but while some workers have reported its presence,[18, 30, 98] others have been unable to substantiate the observation.[81, 99] The theory also requires that GAGs stabilize urate in a colloidal form and prevent precipitation of the latter. If this is so, then adding urate to urines from which intact GAGs have been removed by ultrafiltration should encourage precipitation of the urate salt. If the sample were then to be re-ultrafiltered, the precipitated urate should be visible on the membrane and the concentration of urate in solution should fall. However, this was not what has been observed,[99] indicating that urate does not exist in a colloidal or particulate form, at least in human urine. It is remarkable that those studies reporting the presence of colloidal urate in urine based their conclusion simply on the observation that urinary uric acid was not completely ultrafiltrable. A critical look at their data reveals that the degree of partition they observed was almost identical to the surface effects associated with the ultrafiltration membrane itself, reported by McCulloch et al.[81] It is therefore conceivable that a similar effect may have been responsible for the results of Pak et al[30] and Baggio et al[18, 98] as well, especially since Bowyer et al[100] were unable to demonstrate that sodium urate, the stable form of uric acid at physiological pH, exists in colloidal form in aqueous buffered solutions. Furthermore, even although Baggio et al[18, 98] reported the presence of non-ultrafiltrable, presumably colloidal, urate in human urine, its level in stone formers was no higher than that found in individuals suffering from other pathological conditions such as diabetes and glomerulonephritis,[18] but not from kidney stones. Moreover, the concentraton of urate did not correlate with the endogenous GAGs content - something which might be expected in the presence of significant binding of GAGs to particulate urate.[18, 98]

Evidence in support of the GAGs-depletion theory is based largely on experimental evidence obtained using the model GAG heparin, which is virtually absent from human urine. The predominant GAGs in human urine are the chondroitin sulphates and non-sulphated chondroitin,[101-102] and results of studies testing their binding to urate have been contradictory.[81, 87, 90-91, 98, 103-104] Moreover, preincubation of chondroitin sulphate with sodium urate does not alter its inhibitory potency towards CaOx precipitation,[92] nor is the inhibitory activity of undiluted urine altered by reductions in the endogenous urate concentration[82, 105] or by preincubation with crystalline monosodium urate.[106-107] In any event, it has been shown that though chondroitin sulphate binds to calcium,[87, 90, 104] it is not a significant inhibitor of CaOx crystallization in undiluted urine[108] - an assumption upon which the theory's credibility totally depends. Finally, although formulation of the theory was based on the empirical observation that urinary inhibitory activity correlated more strongly with the ratio of GAGs to urate concentration in urine than to the concentrations of GAGs alone,[83] a later study using similar methodology failed to confirm this.[24]

At the present time, it would be fair to say that there is no experimental evidence unequivocally supporting GAGs depletion as a significant contributing mechanism to stone formation. What is perhaps most puzzling, is that during the last 25 years or so, a great deal of scientific energy has been been expended in attempts to obtain corroborative evidence for theories requiring the existence of solid or colloidal urate in urine, when by 1970, experimental data had already been reported that demonstrated that *dissolved* urate could directly precipitate CaOx crystals from urine.

C. SALTING-OUT

In 1963 at a meeting of the American Chemical Society in New York, Stern et al[109] reported that uric acid interferes with the solubility of CaOx. Five years later they published these findings in full[8] and shortly after, in 1970 and 1971, Kallistratos and and co-workers[110-111] reported that addition of dissolved urate to inorganic solutions containing calcium and oxalate ions, and to human urine, caused precipitation of CaOx. These workers attributed the effect to the principle of "salting-out". If this conclusion was correct, their findings were worthy of serious consideration and further experimental studies were warranted. Recently, we repeated and extended the studies of Kallistratos and his colleagues.[112] Using urine samples obtained from 20 normal men, we increased the concentration of urate by approximately 3-4 mmol/l by the addition of a filtered saturated solution of sodium urate dissolved in 1 mol/l sodium hydroxide. Two types of urines were observed. The more concentrated urines, designated "type A" spontaneously deposited CaOx dihydrate crystals upon the addition of urate alone. The dilute samples (type B) on the other hand, did not precipitate any crystalline material and in these, CaOx crystallization was induced by the oxalate load technique.[113] In these experiments,[112] addition of dissolved urate effectively halved the median metastable limit, trebled the amount of crystalline material deposited and markedly increased the median size of the precipitated particles, although the size of the *individual* crystals was substantially reduced - an observation consistent with urate's having enhanced their rate of nucleation. In addition to confirming the findings of Kallistratos et al[110] and Kallistratos and Timmermann,[111] our studies demonstrated that the volume and size of the particles deposited in urines enriched with urate were greater than in the corresponding controls. Both these phenomena have enormous ramifications for stone pathogenesis, as they increase the likelihood of renal tubule blockage and hence, of stone formation.

However, there was a potential flaw in these studies. The urate solution had been filtered through a 0.22 μm membrane prior to its addition to the urine: it was therefore possible that particles of undissolved sodium urate smaller than this size may have traversed the membrane and been added to the urine samples. Further, even had all the urate been completely in solution, we could not discount the possibility that small particles of the salt may have nucleated in the urine samples themselves *after* addition of the urate solution. Thus, although urate may indeed have induced CaOx precipitation by a salting-out mechanism, as proposed by Kallistratos et al,[110] it was still possible that small particles of it might also have contributed, either directly by epitaxial nucleation,[69-70] or indirectly by binding to and inactivating urinary GAGs.[83] Had this been the case then we would have expected a reduction in the urinary urate concentration following the induction of CaOx crystallization; but this did not occur.[114] This observation was consolidated by our failure to detect the presence of urate inside the CaOx crystals, by infra-red spectroscopy, powder x-ray diffraction analysis or wet chemical analysis using ultraviolet absorption spectroscopy: ultraviolet analysis can detect urate in mixtures of CaOx at concentrations as low as 0.055%. Studies with urines lacking intact GAGs, gave virtually identical results to that obtained in spun and Millipore filtered urine.which contain full complement of GAGs. Taken together, these results indicate that the promotory effect of dissolved urate cannot be attributed either to the epitaxial deposition of CaOx on to urate particles or their inactivation of urinary GAGs.

The enhanced deposition of CaOx by dissolved urate is consistent with the "salting out" of CaOx from solution as suggested previously by Kallistratos et al.[110] and Kallistratos and Timmermann.[111] "Salting out" is a process that occurs when a solid/liquid phase is formed from a liquid phase as a result of a reduction in its dissolving capacity, and may occur within a defined pH range when a less soluble salt or compound precipitates after addition of a more soluble substance. Over the pH range 5.7 to 6.5, CaOx is less soluble than uric acid.[110-111] Therefore, if uric acid is added to inorganic solutions containing calcium and oxalate ions, or to human urine, the solubility of the less soluble CaOx is considerably decreased and it precipitates from solution as a result of the "salting-out" effect, which is illustrated in Figure 4. Therefore, the excretion of large amounts of uric acid in the urine causes a reduction in the solubility of CaOx and favours the formation of CaOx crystals over the pH range of 5.7 to 6.5. Thus, the phenomenon of salting out provides a plausible scientific explanation for the empirical clinical observation that CaOx urolithiasis tends to be more commmon in people excreting large quantities of urate - without the need to invoke the presence of particles of sodium urate or uric acid in the urine.

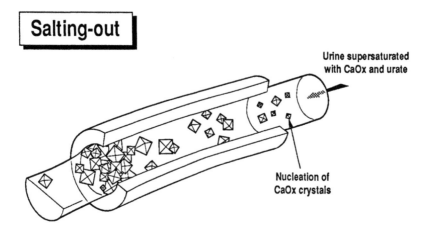

Fig 4.   Diagrammatic representation of the "salting-out" effect showing precipitation of CaOx from urine supersaturated with CaOx and urate.

The probability that CaOx will precipitate from urine does not depend entirely upon the concentration of urate: obviously, the prevailing concentrations of both calcium and oxalate are also important. We have recently studied the effect of ambient urate concentration [Ur] on the relationship between the concentration of calcium [Ca] and oxalate [Ox] at which spontaneous nucleation of CaOx occurs in undiluted urine of normal and stone former males. [Ox]total was calculated by summing the endogenous Ox concentration, [Ox]endo, and that required to reach the metastable limit. Plots relating [Ox]total to the Ca concentration [Ca] were normalised by logarithmic transformation: the slopes of the resulting lines were used as an index of the degree of dependency of [Ox]total on [Ca]. Comparison of normals and stone formers were made on the basis of whether the [Ur] was greater or less than 4 mmol/l. In normals, the slope of the line was 16.4% greater in urines where [Ur] was > 4 mM than in those in which [Ur] was < 4 mM, while in stone formers the corresponding slope was almost double. When data from normals and stone formers were combined, the slope at [Ur]>4mM was 48.4% greater than at [Ur] < 4mM. There was no difference between normals and stone formers with regard to [Ur] or [Ox]endo, but [Ox]total was significantly less (p<0.0001) in the stone formers, presumably because of their increased (p<0.05) [Ca]. This suggested that at urate concentrations > 4mM, the risk of spontaneous CaOx precipitation in urine is significantly raised in relation to increasing [Ca] within the physiological range: this risk is greater in stone formers than in normals subjects. These results re-emphasise the possible contributing role of urate in CaOx stone formation. But is this information of any practical consequence? Will it assist us in the treatment of stone patients - particularly those doomed to multiple episodes?

As far as treatment is concerned, we have the possibility of lowering urate, calcium and oxalate excretion, or finding some other agent which may attenuate urate's effects. For instance, pyrophosphate, magnesium and citrate have been reported to inhibit crystal growth in undiluted human urine, and magnesium has been shown to raise the urinary metastable limit with respect to CaOx.[115] The real potential benefit of such agents is that, unlike high molecular weight compounds, their concentration can be increased therapeutically. A detailed study of the effects of these low molecular weight inhibitors, especially citrate and magnesium, on the urate-induced promotion of CaOx crystallization is therefore warranted.

IV. SYNOPSIS

In this review we have discussed the theories that have been advanced to explain the apparent link between CaOx stone formation and urate excretion, and have critically analysed the experimental evidence supporting or refuting those theories. Several points deserve mention: (1) Although at the present time documentary proof that allopurinol reduces recurrent CaOx stone formation in individuals excreting high levels of urate is scarce, empirical evidence abounds and this should not be overlooked (2) To date, no study has been conducted to compare uric acid excretion in normals and stone formers, in which all factors affecting urate excretion have been adequately controlled. This must raise reasonable doubts about the unequivocal existence of "hyperuricosuric CaOx stone formation" as a distinct clinical entity. Nonetheless, urate excretion may still be an important factor in CaOx stone formation, as various studies have demonstrated that it can promote CaOx crystallization *in vitro*, and probably therefore, *in vivo*. (3) Seed crystals of sodium urate and uric acid do not induce epitaxial deposition of CaOx to a physiologically significant degree in urine: epitaxy as a likely mechanism of stone formation should therefore be discounted. (4) Likewise, the majority of experimental evidence militates against the credibility of the GAGs-depletion theory, which should therefore also be abandoned. (5) Currently, one explanation for the occurrence of CaOx urolithiasis in patients who excrete high levels of urate is that urate "salts out" CaOx from solution. Although this would also explain the apparent beneficial effect of allopurinol in recalcitrant CaOx stone formers, like other hypotheses before it, this too, awaits unambiguous experimental verification.

REFERENCES

1. Lavan, J.N., Neale, F.C., Posen, S., Urinary calculi: Clinical, biochemical and radiological studies in 619 patients, Med. J. Aust., 2, 1049, 1971.
2. Prien, E.L. and Prien, E.L. Jr., Composition and structure of urinary stone, Am. J. Med., 45, 654, 1968.
3. Gutman, A.B. and Yu, T.-F., Uric acid nephrolithiasis, Am. J. Med., 45, 756, 1968
4. Yu, T.-F. and Gutman, A.B., Uric acid nephrolithiasis in gout, Ann. Intern. Med., 67, 1133, 1967.
5. Ljunghall, S., Christensson, T. and Wengle, B., Prevalence and incidence of renal stone disease in a health screening programme, Scand. J. Urol. Nephrol. (Suppl.), 41, 39, 1977
6. Sierakowski, R., Finlayson, B., Landes, R.R., Finlayson, C.D. and Sierakowski, N., The frequncy of urolithiasis in hospital discharge diagnoses in the United States, Invest. Urol., 15, 438, 1978
7. Dent, C.E. and Sutor, D.J., Presence or absence of inhibitor of calcium oxalate crystal growth in urine of normals and stone-formers, Lancet, 2, 775, 1971.
8. Mayer, G.G., Chase, T., Farvar, B., Al Waidh, M., Longo, F., Karp, F. and Zinsser, H.H., Metabolic studies on the formation of calcium oxalate stones, with special emphasis on vitamin B6 and uric acid metabolism, Bull. N. Y. Acad. Med., 44, 28, 1968.
9. Smith, M.J.V., Hunt, L.D., King, J.S. Jr. and Boyce, W.H., Uricemia and urolithiasis, J. Urol., 101, 637, 1969.
10. Mugler, A., Lithiase oxalique et trouble du metabolisme de l'acide urique, Journal d' Urologie et Nephrologie, 76, 423, 1970.
11. Coe, F.L. and Raisen, L., Allopurinol treatment of uric-acid disorders in calcium-stone formers, Lancet, 1, 129, 1973.
12. Hartung, R. and Bergmann, M., Oxalatlithiasis and hyperurikurie, Helv. Chir. Acta, 41, 405, 1974.
13. Coe, F.L. and Kavalach, A.G., Hypercalciuria and hyperuricosuria in patients with calcium nephrolithiasis, N. Engl. J. Med., 291, 1344, 1974.
14. Coe, F.L., Hyperuricosuric calcium oxalate nephrolithiasis, Kidney Int., 13, 418, 1978.
15. Robertson, W.G., Peacock, M., Heyburn, P.J., Marshall, D.H. and Clark, P.B., Risk factors in calcium stone disease of the urinary tract, Br. J. Urol., 50, 449, 1978.
16. Broadus, A.E. and Thier, S.O., Metabolic basis of renal stone disease, N. Engl. J. Med., 300, 839, 1979.
17. Coe, F.L., Margolis, H.C., Deutsch, L.H. and Strauss, A.L., Urinary macromolecular crystal growth inhibitors in calcium nephrolithiasis, Miner. Electrolyte Metab., 3, 268, 1980.
18. Baggio, B., Gambaro, G., Marchi, A., Cicerello, E., Favaro, S. and Borsatti, A., The role of glycosaminoglycans and uric acid in idiopathic calcium nephrolithiasis, Contr. Nephrol., 37, 5, 1984.
19. Schwille, P.O., Samberger, N. and Wach, B., Fasting uric acid and phosphate in urine and plasma of renal calcium-stone formers, Nephron, 16, 116, 1976.
20. Hodgkinson, A., Uric acid disorders in patients with calcium stones, Br. J. Urol., 48, 1, 1976.
21. Pylypchuk, G., Ehrig, U. and Wilson, D.R., Idiopathic calcium nephrolithiasis. I. Differences in urine crystalloids, urine saturation with brushite and urine inhibitors of calcification between persons with and persons without recurrent kidney stone formation, Can. Med. Assoc. J., 120, 658, 1979.

22.   Fellstrom, B., Backman, U., Danielson, B.G., Johansson, G., Ljunghall, S. and Wikstrom, B., Urinary excretion of urate in renal calcium stone disease and in renal tubular acidification disturbances, J. Urol., 127, 589, 1982.

23.   Ryall, R.L. and Marshall, V.R., The value of 24-hour urine analysis in the assessment of stone-formers attending a general hospital outpatient clinic, Br. J. Urol., 55, 1, 1983.

24.   Ryall, R.L. and Marshall, V.R., The relationship between urinary inhibitory activity and endogenous concentrations of glycosaminoglycans and uric acid: comparison of urines from stone-formers and normal subjects, Clin. Chim. Acta, 141, 197, 1984.

25.   Ryall, R.L., Darroch, J.N. and Marshall, V.R., The evaluation of risk factors in male stone-formers attending a general hospital out-patient clinic, Br. J. Urol., 56, 116, 1984.

26.   Emmerson, B.T., How can one define urate over-production in man?, Adv. Exp. Biol. Med., 195 Part A, 287, 1986.

27.   Ryall, R.L. and Marshall, V.R., The investigation and management of idiopathic urolithiasis, in Renal Tract Stone: Metabolic Basis and Clinical Practice, Wickham, J.E.A. and Buck, A.C., Eds., Churchill Livingstone, 1990, 307.

28.   Gambaro, G., Vincenti, M., Marchini, F., D'Angelo, A. and Baggio, B., Abnormal urate transport in erythrocytes of patients with idiopathic calcium nephrolithiasis: a possible link with hyperuricosuria, Clin. Sci., 85, 41, 1993.

29.   Smith, M.J.V. and Boyce, W.H., Allopurinol and urolithiasis, J. Urol., 102, 750, 1969.

30.   Pak, C.Y.C., Barilla, D.E., Holt, K., Brinkley, L., Tolentino, R. and Zerwekh, J.E., Effect of oral purine load and allopurinol on the crystallization of calcium salts in urine of patients with hyperuricosuric calcium urolithiasis, Am. J. Med., 65, 593, 1978.

31.   Miano, L., Petta, S. and Gallucci, M., Allopurinol in the prevention of calcium oxalate renal stones, Eur. Urol., 5, 229, 1979.

32.   Maschio, G., Tessitore, N., D'Angelo, A., Fabris, A., Pagano, F., Tasca, A., Graziani, G., Aroldi, A., Surian, M., Colussi, G., Mandressi, A., Trinchieri, A., Rocco, F., Ponticelli, G. and Minetti, L., Prevention of calcium nephrolithiasis with low dose thiazide, amiloride and allopurinol, Amer. J. Med., 71, 623, 1981.

33.   Smith, M.J.V., Placebo versus allopurinol for recurrent urinary calculi, Proc. EDTA., 20, 422, 1983.

34.   Anonymous, Recurrence of renal stones, Lancet, 1, 187, 1980.

35.   Ettinger, B., Tang, A., Citron, J.T., Livermore, B. and Williams, T., Randomized trial of allopurinol in the prevention of calcium oxalate calculi, N. Engl. J. Med., 315, 1386, 1986.

36.   Robertson, W.G., Diet and calcium stones, Miner. Electrolyte Metab., 13, 228, 1987.

37.   Anonymous, Allopurinol for calcium oxalate stones?, Lancet, 1, 258, 1987.

38.   Berthoux, F.C., Juge, J., Genin, C., Sabatier, J.C. and Assenat, H., Double-blind trial of allopurinol as a preventative treatment in calcium urolithiasis, Miner. Electrolyte Metab., 2, 207, 1979

39.   Brien, G. and Bick, C., Allopurinol in the recurrence prevention of calcium oxalate lithiasis, Eur. Urol., 3, 35, 1977.

40.   Coe, F.L., Treated and untreated recurrent calcium nephrolithiasis in patients with idiopathic hypercalciuria, hyperuricosuria or no metbolic disorder, Ann. Intern. Med., 87, 404, 1977.

41.   Vabusek, M., Prevention of urinary calculi in hyperuricemia and gout; in Urolithiasis Research, Fleisch, H., Robertson, W.G., Smith, L.H. and Vahlensieck, W. Eds., Plenum Press, New York and London 1976, 565.

42.   Petit, C., Lithiase calcique: l'acide urique en question, Nephrologie, 5, 192, 1984.

43.   Marangella, M., Tricerri, A., Ronzani, M., Martini, C., Petrarulo, M., Daniele, P.G., Torrengo, S. and Linari, F., The relationship between clinical outcome and urine biochemistry during various forms of therapy for idiopathic calcium stone disease, in Urolithiasis and Related Clinical Research, Schwille, P.O., Smith, L.H., Robertson, W.G. and Vahlensieck, W. Eds., Plenum Press, New York and London, 1985, 561.

44.   Tiselius, H.-G. and Larsson, L., Urine composition in patients with renal stone disease during treatment with allopurinol, Scand. J. Urol. Nephrol., 14, 65, 1980.

45.   Pak, C.Y.C., The effects of allopurinol in calcium oxalate stone disease, in Urinary Calculus, Brockis, J.G. and Finlayson, B. Eds., PSG Publishing Company Inc., Littleton and Massachusetts, 1981, 469.

46.   Tomlinson, B., Al-Khader, A., Cohen, S.L., Kasidas, G.P., Krywawych, S., Edgar, L. and Rose, G.A., Mechanism of allopurinol action in calcium oxalate stone formers, Proc. EDTA., 18, 556, 1981.

47.   Baggio, B., Gambaro, G., Paleari, C., Cicerello, E., Marchi, A., Bragantini, L., Favaro, S. and Borsatti, A., Hydrochlorothiazide and allopurinol vs. placebo on urinary excretion of stone promoters and inhibitors, Curr. Ther. Res., 34, 145, 1983.

48.   Fellstrom, B., Backman, U., Danielson, B.G., Holmgren, K., Johansson, G., Lindsjo, M., Ljunghall, S. and Wikstrom, B., Allopurinol treatment of renal calcium stone disease, Br. J. Urol., 57, 375, 1985.

49.   Urivetzky, M., Braverman, S., Motola, J.A. and Smith, A.D., Absence of effect of allopurinol on oxalate excretion by stone patients on random and controlled diets, J. Urol., 144, 97, 1990.

50.   Scott, R., Paterson, P.J., Mathieson, A. and Smith, M., The effect of allopurinol on urinary oxalate excretion in stone formers, Br. J. Urol., 50, 455, 1978.

51.   Simmonds, H.A., Van Acker, K.J., Potter, C.F., Webster, D.R., Kasidas, G.P. and Rose, G.A., Influence of purine content of diet and allopurinol on uric acid and oxalate excretion levels, in Urolithiasis: Clinical and Basic Research, Smith, L.H., Robertson, W.G. and Finlayson, B. Eds., Plenum Press, New York and London 1981, 363.

52.   Tomlinson, B., Cohen, S.L., Al-Khader, A., Kasidas, G.P. and Rose, G.A., Further reduction of oxalate excretion by allopurinol in stone formers on low purine diet, in Urolithiasis and Related Clinical Research, Schwille, P.O., Smith, L.H., Robertson, W.G. and Vahlensieck, W. Eds., Plenum Press, New York and London 1985, 513.

53.   Shaw, M.I. and Parsons, D.S., Absorption and metabolism of allopurinol and oxypurinol by rat jejunum in vitro: effects on uric acid transport, Clin. Sci., 66, 257, 1984.

54.   Simmonds, H.A., Rising, T.J., Cadenhead, A., Hatfield, P.J., Jones, A.S. and Cameron, J.S., Radioisotope studies of purine metabolism during administration of guanine and allopurinol in the pig, Biochem. Pharmacol., 22, 2553, 1973.

55.   Tiselius, H.G., Larsson, L. and Hellgren, E., Clinical results of allopurinol treatment in prevention of calcium oxalate stone formation, J. Urol., 136, 50, 1986.

56.   Morris, G.S., Simmonds, H.A., Toseland, P.A., Van Acker, K.J., Davies, P.M. and Stutchbury, J.H., Urinary oxalate levels are not affected by dietary purine intake or allopurinol, Br. J. Urol., 60, 292, 1987.

57.   Pena, J.C., Monforte, M.F. and Briceno, A., The role of oxalate and calcium oxalate activity and formation product ratio in patients with renal stones before and during treatment, J. Urol., 138, 1137, 1987.

58.   Tiselius, H.-G., Inhibition of calcium oxalate crystal growth in urine during treatment with allopurinol, Br. J. Urol., 52, 189, 1980.

59.   Finlayson, B. and Reid, F., The effect of allopurinol on calcium oxalate (whewellite) precipitation, Invest. Urol., 15, 489, 1978.

60.   Finlayson, B., Burns, J., Smith, A. and Du Bois, L., Effect of oxipurinol and allopurinol riboside on whewellite crystallization. In vitro and in vivo observations, Invest. Urol., 17, 227, 1979.

61.   Goldwasser, B., Sarig, S., Azoury, R., Wax, Y., Hirsch, D., Perlberg, S. and Many, M., Changes in inhibitory potential in urine of hyperuricosuric calcium oxalate stone formers effected by allopurinol and orthophosphates, J. Urol., 132, 1008, 1984.

62.   Royer, L., Experimental research on parallel growth on mutual orientation of crystals of different species, Bulletinn de la Societe Francaise de mineralogie, 51, 7, 1928.

63.   Modlin, M., The aetiology of renal stones: A new concept arising from studies on a stone-free population, Ann. R. Coll. Surg. Engl. (London), 40, 155, 1967.

64.   Lonsdale, K., Epitaxy as a growth factor in urinary calculi and gallstones, Nature, 217, 56, 1968.

65.   Robertson, W.G., Peacock, M. and Nordin, B.E.C., Calcium oxalate crystalluria and stone saturation in recurrent renal stone formers, Clin. Sci., 40, 365, 1971.

66.   Pak, C.Y.C., Waters, O., Arnold, L., Holt, K., Cox, C. and Barilla, D., Mechanism for calcium urolithiasis among patients with hyperuricosuria. Supersaturation of urine with respect to monosodium urate, J. Clin. Invest., 59, 426, 1977.

67.   Coe, F.L., Strauss, A.L., Tembe, V. and Dun, S.L., Uric acid saturation in calcium nephrolithiasis, Kidney Int., 17, 662, 1980.

68.   Tiselius, H-G. and Larsson, L., Urinary excretion of urate in patients with calcium oxalate stone disease, Urol. Res., 11, 279, 1983.

69. Coe, F.L., Lawton, R.L., Goldstein, R.B. and Tembe, V., Sodium urate accelerates precipitation of calcium oxalate *in vitro*, Proc. Soc. Exp. Biol. Med., 149, 926, 1975.

70. Pak, C.Y.C. and Arnold, L.H., Heterogeneous nucleation of calcium oxalate by seeds of monosodium urate, Proc. Soc. Exp. Biol. Med., 149, 930, 1975.

71. Sarig, S., Hirsch, D., Garti, N. and Goldwasser, B., An extension of the concept of epitaxial growth, J. Crystal Growth, 69, 91, 1984.

72. Fleisch, H., Inhibitors and promotors of stone formation, Kidney Int., 13, 361, 1978.

73. Ryall, R.L. and Grover, P.K., The relationship between urate seeds and calcium oxalate crystallization: epitaxy or endocrystallosis?, Urol. Res., 21, 150, 1993.

74. Finlayson, B. and Reid, F., The expectation of free and fixed particles in urinary stone disease, Invest. Urol., 15, 442, 1978.

75. Finlayson, B., Khan, S.R. and Hackett, R.L., Mechanisms of stone formation - an overview, Scan. Electron Microsc., 3, 1419, 1984.

76. Prien, E.L. and Frondel, C., Studies in urolithiasis. I. The composition of urinary calculi, J. Urol., 57, 949, 1947.

77. Herring, L.C., Observations on the analysis of ten thousand urinary calculi, J. Urol., 88, 545, 1962.

78. Prien, E.L., Crystallographic analysis of urinary calculi: A 23-year survey study, J. Urol., 80, 917, 1963.

79. Cifuentes-Delatte, L., Bellanato, J., Santos, M. and Rodriguez-Minon, L., Monosodium urate in urinary calculi, Eur. Urol., 4, 441, 1978.

80. Rapado, A., Castrillo, J.M., Diaz-Curiel, M., Traba, M.L., Santos, M. and Cifuentes-Delatte, L., Uric acid/calcium oxalate nephrolithiasis. Clinical and biochemical findings in 86 patients, Adv. Exp. Med. Biol., 122A, 121, 1980.

81. McCulloch, R.K., Bowyer, R.C. and Brockis, J.G., The possible role of urate in calcium urolithiasis, in Urinary Calculus, Brockis, J.G. and Finlayson, B., Eds., PSG Publishing Company Inc., Littleton and Massachusetts 1981, 347.

82. Hallson, P.C., Rose, G.A. and Sulaiman, S., Urate does not influence the formation of calcium oxalate crystals in whole human urine at pH 5.3, Clin. Sci., 62, 421, 1982.

83. Robertson, W.G., Knowles, F. and Peacock, M., Urinary acid mucopolysaccharide inhibitors of calcium oxalate crystallization, in Urolithisis Research, Fleisch, H., Robertson, W.G., Smith, L.H., Vahlensieck, W., Eds., Plenum Press, New York and London 1976, 331.

84. Porter, P., Colloidal properties of urates in relation to calculus formation, Res. Vet. Sci., 7, 128, 1966.

85. Katz, W.A., Effect of glycosaminoglycans in urate precipitation in gout, Clin. Res., 21, 973, 1973.

86. Katz, W.A. andSchubert, M., The interaction of monosodium urate with connective tissue components, J. Clin. Invest., 49, 1783, 1970.

87. Hesse, A., Wurzel, H., Krampitz, G. and Vahlensieck, W., Experimental determination of the kinetics of calcium-binding with chondroitin sulphate and the effects of uric acid on this process, Urol. Res., 15, 93, 1987.

88. Pak, C.Y.C., Holt, K., Britton, F., Peterson, R., Crowther, C. and Ward, D., Assessment of pathogenetic roles of uric acid, monopotassium urate, monoammonium urate and monosodium urate in hyperuricosuric calcium oxalate nephrolithiasis, Miner. Electrolyte Metab., 4, 130, 1980.

89. Finlayson, B., Du Bois, L., Adsorption of heparin on sodium acid urate, Clin. Chim. Acta, 84, 203, 1978.

90. Baggio, G., Gambaro, G., De Nardo, L. and Borsatti, A., The role played by acid mucopolysaccharides in calcium oxalate nephrolithiasis, Kidney Int., 22, 95, 1982.

91. Fellstrom, B., Lindsjo, M., Danielson, B.G., Ljunghall, S and Wikstrom, B., Binding of glycosaminoglycans to sodium urate and uric acid crystals, Clin. Sci., 71, 61, 1986.

92. Pak, C.Y.C., Holt, K. and Zerwekh, J.E., Attenuation by monosodium urate of the inhibitory effect of glycosaminoglycans on calcium oxalate nucleation, Invest. Urol., 17, 138, 1979.

93. Pak, C.Y.C., Holt, K., Britton, F., Peterson, R., Crowther, C. and Ward, D., Assessment of pathogenetic roles of uric acid, monopotassium urate, monoammonium urate and monosodium urate in hyperuricosuric calcium oxalate nephrolithiasis, Miner. Electrolyte Metab., 4, 130, 1980.

94. Ryall, R.L., Harnett, R.M. and Marshall, V.R., The effect of monosodium urate on the capacity of urine, chondroitin sulphate and heparin to inhibit calcium oxalate crystal growth and aggregation, J. Urol., 135, 174, 1986.

95.   Fellstrom, B., Backman, U., Danielson, B.G., Holmgren, K., Ljunghall, S. and Wikstrom, B., Inhibitory activity of human urine on calcium oxalate crystal growth: effects of sodium urate and uric acid, Clin. Sci., 62, 509, 1982.

96.   Zerwekh, J.E., Holt, K. and Pak, C.Y.C., Natural urinary macromolecular inhibitors: attenuation of inhibitory activity by urate salts, Kidney Int., 23, 838, 1983.

97.   Tiselius, H-G., Effects of sodium urate and uric acid crystals on the crystllization of calcium oxalate, Urol. Res., 12, 11, 1984.

98.   Baggio, B., Gambaro, G., Cicerello, E., Marchini, F. and Borsatti, A., Further studies on the possible lithogenetic role of uric acid in calcium oxalate stone disease, in Urolithiasis and Related Clinical Research, Schwille, P.O., Smith, L.H., Robertson, W.G. and Vahlensieck, W., Eds., Plenum Press, New York and London 1985, 863.

99.   Grover, P.K., Ryall, R.L. and Marshall, V.R., Calcium oxalate crystallization in urine: role of urate and glycosaminoglycans, Kidney Int., 41, 149, 1992.

100.   Bowyer, R.C., McCulloch, R.K., Brockis, J.G. and Ryan, G.D., Factors affecting the solubility of ammonium acid urate, Clin. Chem. Acta, 95, 17, 1979.

101.   Wessler, E., The nature of the non-ultrafiltrable glycosaminoglycans of normal human urine, Biochem. J., 122, 373, 1971.

102.   Varadi, D.P., Cifonelli, J.A., Dorfman, A., The acid mucopolysaccharides in normal urine, Biochem. Biophys. Acta, 141, 103, 1967.

103.   Laurent, T.C., Solubility of sodium urate in the presence of chondroitin-4-sulphate, Nature, 202, 1334, 1964.

104.   Grases, F., Costa-Bauza, A., March, J.G. and Masarove, L., Glycosaminoglycans, uric acid and calcium oxalate urolithiasis, Urol. Res., 19, 375, 1991.

105.   Grover, P.K., Ryall, R.L. and Marshall, V.R., The effect of decreasing the concentration of urinary urate on the crystallization of calcium oxalate in undiluted human urine, J. Urol., 143, 1057, 1990.

106.   Goldwasser, B., Sarig, S., Azoury, R., Wax, Y. and Many, M., Hyperuricosuria and calcium oxalate stone formation, in Urolithiasis and Related Clinical Reseach, Schwille, P.O., Smith, L.H., Robertson, W.G. and Vahlensieck, W., Eds., Plenum Press, New York and London 1985, 859.

107.   Ryall, R.L., Hibberd, C.M. and Marshall, V.R., The effect of crystalline monosodium urate on the crystallization of calcium oxalate in whole human urine, Urol. Res., 14, 63, 1986.

108.   Ryall, R.L., Harnett, R.M., Hibberd, C.M., Edyvane, K.A. and Marshall, V.R., The effects of chondroitin sulphate, human serum albumin and Tamm-Horsfall mucoprotein on calcium oxalate crystallization in undiluted urine, Urol. Res., 19, 181, 1991.

109.   Stern, F., Mayer, G.G., Light, I.S. and Zinsser, H.H., Interference by uric acid in the analysis of urine for oxalic acid. Presented at the meeting of the American Chemical Society, New York, NY, September 1963.

110.   Kallistratos, G., Timmermann, A. and Fenner, O., Zum Einflub des Aussalzeffektes auf die Bildung von Calciumoxalat-Kristallen im menschlichen Harn, Naturwissenschaften, 57, 198, 1970.

111.   Kallistratos, G. and Timmermann, A., The "salting-out" effect as a possible causative factor for the formation of calcium oxalate crystals in human urine. Paper presented at the 66th annual meeting of the American Urological Association in Chicago, Illinois, USA, May 17, 1971, Urological Research Forum.

112   Grover, P.K., Ryall, R.L. and Marshall, V.R., Effect of urate on calcium oxalate crystallization in human urine: evidence for a promotory role of hyperuricosuria in urolithiasis, Clin. Sci., 79, 9, 1990.

113.   Ryall, R.L., Hibberd, C.M. and Marshall, V.R., A method for studying inhibitory activity in whole uriine, Urol. Res., 13, 285, 1985.

114.   Grover, P.K., Ryall, R.L. and Marshall, V.R., Dissolved urate promotes calcium oxalate crystallization: epitaxy is not the cause, Clin. Sci., 85, 303, 1993.

115.   Ryall, R.L., Grover, P.K., Harnett, R.M., Hibberd, C.M. and Marshall, V.R., Small molecular weight inhibitors, in Urolithiasis, Walker, V.R., Sutton, R.A.L., Cameron, E.C.B., Pak, C.Y.C. and Robertson, W.G., Eds., Plenum Press, New York and London, 1989, 91.

Chapter 16

CRYSTAL-CELL INTERACTION: ITS ROLE IN THE DEVELOPMENT OF STONE DISEASE

Raymond L. Hackett, M.D. and Paula N. Shevock, B.A.

I. HUMAN STUDIES

A. OVERVIEW

Formation of a kidney stone represents the culmination of a series of complex physical, chemical and biological events most probably occurring as the result of synchronous and metachronous steps. Some of the stages involved in this process include heterogenous nucleation, aggregation of crystalline complexes and retention within the urinary tract especially the renal medulla and papilla. Of these processes, crystal retention plays a central role, and will be the major focus of this chapter. No precise, generally acceptable definition of stone exists. In our laboratory, stone is defined as the crystalline formation product precipitated from the urine whether the crystal be of microscopic or macroscopic size. When a stone obstructs urinary flow or causes associated irreversible morphologic changes in renal structure, the term stone disease is employed. Precedence for this approach to nomenclature exists. For example, Anderson and McDonald [1] defined a calculus as "a size sufficient to be seen easily under the lower-power microscope lens and measuring at least five to six times the size of tubular cells." In our opinion, stone disease is a phenomenon intimately involved with cell injury and crystal/cell interactions, resulting in retention, concepts supported by three broad lines of evidence involving human studies, animal experiments, and cell culture investigations. Animal research is reviewed elsewhere in this volume [2]. In this chapter, we will review briefly some of the evidence derived from the other two approaches.

Although Huggins, in 1933, [3] suggested a connection between tissue damage and stones as indicated by medullary ossification in the presence of large uroliths, it remained for Randall to bring this relationship into sharp focus [4,5,6]. Of his numerous writings on the topic of stones, and although he subsequently expanded his series of observations, Randall's 1937 [4] paper remains one of the most seminal contributions to the field. After reviewing the current hypotheses for stone formation and reporting on relatively unsuccessful attempts to produce stones experimentally, he described results from examining human kidneys at necropsy. Randall meticulously dissected 429 pairs of kidneys utilizing hand lens and in some

0-8493-7673-4/95/$0.00+$.50
© 1995 by CRC Press

cases light microscopy and found papillary alterations in 17% of kidneys. These consisted of interstitial calcium deposits which, in some instances, eroded the overlying papillary epithelium. In some kidneys, calculi attached to such foci were found, as illustrated in Fig. 7 of his 1937 article [4]. Chemical analysis revealed the portion of the stone attached to papillary plaque to be comprised of calcium carbonate and calcium phosphate, whereas the bulk of the stone surrounding the nidus was calcium oxalate (CaOx). Figs. 5 and 6 from that work [4] illustrate the microscopic concomitant of similar stones and show a 2 mm concretion firmly attached to a papillary tip containing subepithelial calcific deposits. Randall concluded that papillary calcium plaques were the end stage of renal tubular damage and, while a much more common event than stone, when stone disease did occur the calculi deposited upon and were attached to an eroded subepithelial calcific plaque. Whether such attachment sites represented retention of a preformed stone, or a nucleation site for stone formation was never clearly delineated.

Randall's work stimulated several succeeding reports and investigations into the relationship between papillary injury, as manifested by calcific deposits, and calculous attachment [7-15]. Rosenow [7] found calcareous papillary plaques in 22% of 239 necropsies and described morphologically identifiable bacteria in 24 cases. This latter finding, however, has not been confirmed by subsequent investigators. Posey [8] postulated that the papillary changes preceded stone disease and that the initial derangement was a vascular degeneration, whereas Vermooten [9] felt that the calcification involved primarily any of the collagen fibers located in the papilla. More recently, the argument that the process is primarily vascular has been revived with the finding by Swagel et al. [10] that calcific deposits involve the vasa recta in the papillary tip. Most investigators described calcium deposits in only a percentage of kidneys involved, but Anderson and McDonald [1] found deposits in every one of 168 surgically removed kidneys. Based on histologic examination, they felt that the deposition resulted from ingestion of calcium by phagocytes which degenerate and coalesce into deposits. Cooke [11] disputed this in an another necropsy study and determined that the predominant focus for calcification was in the walls of the long loop of Henle. In a confirmatory electron microscopic study, he examined three nephrectomy specimens in which calcium was localized to collagen bundles oriented chiefly around the loops of Henle [12]. Electron microscopic examination of autopsy kidneys by Haggitt and Pitcock [13] placed calcium deposits in the interstitium and basement membrane of collecting ducts. Burry et al. [14] identified calcification around loops of Henle in the papilla in an extensive autopsy study in which he categorized medullary classifications into several varieties of which Randall's plaques were one form, but not necessarily related to the other types. More recently Cifuentes

Delatte et al. [15,16], in elegant scanning electron microscopic examinations of spontaneously passed calculi, described papillary stones. They concluded that of the three types of these calculi, two were associated with one or another form of Randall's plaques and, in general, the deposition of CaOx was secondary to a papillary nidus of different chemical composition. Of considerable interest, was their finding [16] that of 500 spontaneously passed stones, 28% were true papillary stones, whereas 72% showed a variety of non-papillary patterns. Some of these latter calculi, but certainly not all, may contain Randall's type II plaque, which is related to intra-tubular stone formation as described by Prien [17] rather than an intra-papillary sub-epithelial deposit.

B. CRITIQUE

The results from these earlier studies substantiated that adhesion of stones to some portion of the papilla was essential for the development of stone disease. The hypothetical basis for this idea was provided by one of the great theorists in experimental urolithiasis: Finlayson with Reid, in 1978 [18], after calculating renal tubular dimensions, urinary flow rates, and supersaturation postulated that CaOx stones could not grow quickly enough to be retained by size alone and therefore, fixation to an anatomic structure was essential for retention and stone growth. Subsequently, Kok and Khan [19] addressed this same issue with more recently available information and essentially confirmed the hypothesis but added the premise that crystal aggregation provided a key role in retention. Data from animal experiments, described in greater detail [2], generally support the fixation and retention concept. While in early animal experiments, Dunn et al. [20] felt that obstruction by stones was not important, more recent data suggest that in animal models, as reported by Dykstra and Hackett [21], Rushton et al. [22], and Khan and Hackett [23], stone retention could be demonstrated as intratubular crystals entangled with cell surface structures in proximal tubules, or as shown by Khan et al. [24] were intimately attached as aggregates to more distal structures. Support for these observations came as well from studies of kidneys from patients suffering primary oxalosis. Morgenroth et al. [25] clearly demonstrated entanglement of oxalate crystals with tubular epithelial microvilli (Fig. 1 in [25]) and complex interdigitation with a fibrous network (Fig. 3 in [25]), possibly basement membrane or interstitium. Lieske et al. [26] described changes in a kidney transplanted into a patient with oxalosis and illustrated endocytosis of oxalate crystals by tubular epithelial cells and apparent crystal movement into the interstitium.

In considering the information available from human materials, it is evident that such studies are critical to an understanding of the pathogenesis of stone disease and offer a cogent explanation for papillary tip stones, but not necessarily for other forms. Nonetheless, examination of crystal-cell interactions and stone attachment and retention in whole organisms, be they human or animal, presents difficulties in timing and sampling complicated by heterogeneity of renal tubular anatomy. Because of these problems, a more recent series of investigations has utilized cells in suspension or culture.

II.   CELLULAR FACTORS IN RETENTION

A.   CRYSTAL-CELL ATTACHMENT

When confluent monolayers of the established renal epithelial cell lines, Madin-Darby Canine Kidney (MDCK) [27] or porcine kidney (LLC-PK$_1$) [28] cells are exposed to calcium oxalate monohydrate (COM) crystals suspended in the nutrient media, crystals settle onto the cell surface. Here, they become entangled with the specialized membrane structures of cilia and microvilli. The cell membrane surface is structurally altered as patches of membrane devoid of microvilli are found adjacent to the crystal entrapped on the surface (Fig. 1). The process is rapid. Lieske et al. [29], using African green monkey (BSC-1) [30] renal epithelial cells exposed to ^{14}C labelled COM crystals, demonstrated crystal binding within 15 seconds of exposure, and to MDCK cells within minutes, the adherence being faster the lower the crystal concentration. This interaction between crystal and cell apparently depends on specific binding sites under appropriate circumstances of cell isolation and culture. Wiessner et al. [31] and Riese et al. [32] demonstrated in primary cultures of rat papillary collecting tubule cells that crystal binding occurred preferentially to cells in clumps rather than to the monolayer. Clustered cells are structurally different in that they have lost tight junctions to adjacent cells with an associated loss of polarity, alterations which could be viewed as a form of cell injury. Mandel [33] postulated that during this process basolateral membrane components are translocated to the apical surface resulting in modified proportions of the various membrane lipids. By exposing human red blood cells and inner medullary late collecting duct (IMCD) cells to liposomes containing specific lipids, and exposing the cells to COM crystals, he was able to demonstrate increased red blood cell lysis or increased crystal attachment to IMCD cells especially when cells were enriched with phosphatidyl serine.

These observations give strong support to the idea that injury to cells plays a central role in crystal attachment, an idea that enjoys support from animal experiments. For example, injury to rat urinary bladder urothelium results in increased binding of CaOx crystals to the surface [34,35], and hyperoxaluric rats form more crystals at earlier time periods, when treated with the nephrotoxin gentamicin [36,37]. While the

Figure 1. MDCK culture treated with 1 mM KOx and 50μg/ml 1 μm COM for 120'. Crystals are enmeshed with surface microvilli and cilia.

implications of the above data are yet to be completely understood, they are key in emphasizing a critical role for the cell membrane, especially altered membrane, in urolithiasis.

B. FATE OF CRYSTAL-CELL ATTACHMENT

Subsequent to crystal-cell membrane attachment, many COM crystals disappear from the cell surface and appear to be internalized (Fig. 2). This phenomenon is seemingly common to all cultured renal epithelial cells since it has been shown to occur in the MDCK [38], BSC-1 [39], and LLC-PK$_1$ [40] cell lines. The uptake of crystals appears to increase with time and crystal concentration [29] and stimulates a series of complex reactions. Morphologically, crystal aggregates are contained with intracellular vesicles which appear to be membrane lined whereas smaller crystals can be found in lysosomes [29,41]. In addition, cells exocytose the internalized crystals and in MDCK cells, crystal-cytoplasm complexes can be observed leaving cells at the intercellular junctions [38].   The interaction is further complicated by the fact that whole cell-crystal units detach from the culture substrate after exposure to crystals [38]. Associated with these processes is an alteration in portions of the cytoskeletal network with concentration of F-actin filaments in the region of crystal uptake, as well as dispersal of

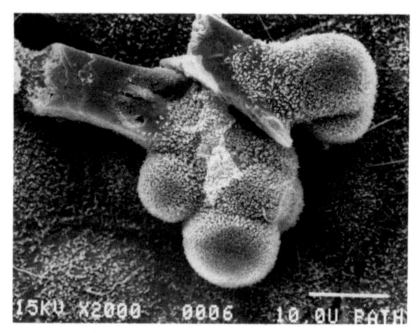

Figure 2. MDCK cells exposed to 1 mM KOx and 100 μg/ml 10 μm COM for 60'.
Crystals are shown in the process of interanalization.

a cytokeratin from its perinuclear concentration to a more diffuse
cytoplasmic distribution [29].

Additional effects of crystalline exposure include a mitogenic response in
both BSC-1 and MDCK cells. Using (^{3}H) thymidine to measure DNA
synthesis, Lieske [39] found that a maximal effect was noted in BSC-1 cells
24 hours after exposure to COM crystals and that such stimulation appeared
to be relatively cell type specific since little response was found in fibroblasts
exposed to COM crystals. In addition, internalized crystals are passed on
to daughter cells [29]. These observations suggest that COM endocytosis is
not necessarily lethal to renal epithelial cells. However, subsets of MDCK
cells in culture are definitely damaged by crystal exposure since a finite
number of cells are killed or injured as demonstrated by reduction of trypan
blue exclusion, and increased adenine, lysosomal, cytosolic and membrane
enzyme release [38]. In addition, as mentioned previously, whole cells are
lost from the culture when exposed to crystals. It appears that the act of
crystal ingestion results in a heterogenous cellular response during which
some, but not all, cells will perish from the experience.

C. MODULATION OF CRYSTAL-CELL ATTACHMENT

That the adhesion of crystal to cells is an active biologic process is
further supported by studies investigating mechanisms of attachment

inhibition. Ebisuno et al. [42] utilized confluent MDCK cells exposed to COM crystals in the absence or presence of various glycosaminoglycans (GAGs) including sodium pentosan polysulfate, chondroitin sulphate C, heparan sulphate, and hyaluronic acid. With increasing concentrations of all studied polyanions, a significant decrease in numbers of attached crystals was observed. Similarly, Lieske et al. [43] studied various polyaninons and their effect on crystal adhesion and internalization within cell cultures. They concluded that effect of adhesion and or uptake inhibition varied among urinary molecules and that anionic molecules act directly upon the crystal rather than the cell surface. Thus, the presence of GAGs was viewed as inhibitory to crystal attachment and supportive of older animal experiments [34,35] in which destruction of integrated cell membranes GAGs resulted in an increase in numbers of attached crystals. Lieske and Toback [44] also investigated the effects of a number of other diverse compounds on adhesion and endocytosis of COM crystals including Tamm-Horsfall glycoprotein [THP], arginine-glycine-aspartic acid-serine, fibronectin, transforming growth factor-β_2, and heparin. Whereas mitogenic signals such as epidermal growth factor increased endocytosis, the test compounds alone inhibited crystal uptake. The inhibitory effects occurred only when the agents were applied to the cultures and not when COM crystals were pre-coated with the substances implying that the surface of the cell and not the crystal was the "locus of inhibition of endocytosis by THP". Because these compounds interact with cells along different pathways, no single mechanism appears to be responsible for their activities except for the generalization that they interact with a variety of cell membrane receptors and most likely affect cell motility. All of the above studies indicate, however, that perturbations of cell membrane integrity play a key role in the adhesion process. Goswami et al. [41] studied another aspect of the adhesion process by growing renal epithelial ovine kidney (OK) [45] cells on collagen matrices. While growth collagen gel inhibited cell proliferation of untreated cultures, it increased crystal endocytosis in COM-treated cells. The reason for this is unclear, but the stimulation of interstitial cell growth demonstrated by the same authors may trigger collagen production with subsequent increase in endocytosis of COM crystals.

D. THE ROLE OF OXALATE IN CRYSTAL-CELL ATTACHMENT

Intuitively, CaOx crystals cannot form unless a medium contains a sufficient concentration of oxalate for crystals to nucleate. Since oxalate interacts with cells in a number of ways the question must be posed as to whether the presence of oxalate might or might not play a part in crystal-cell interactions. Some of the earliest work related to the oxalate question was by Baggio et al. [46]. They studied oxalate flux in the erythrocytes of human stone formers and established the existence of a defect in oxalate transport giving rise to the viewpoint that stone formation could be considered a systemic disorder. Later studies by Gambaro and Baggio [47] implicated reduced erythrocyte membrane GAGs, thus raising the level of anion exchangers resulting in increased oxalate exchange. Koul et al. [48] examined oxalate transporters in LLC-PK$_1$ cell cultures. In these cells maximal oxalate uptake occurred at a concentration of 500 μM and plateaued thereafter, suggesting transport-mediated cell uptake. Interestingly, transport sensitivity is polarized in that the apical surface is chloride sensitive, sulphate/bicarbonate insensitive whereas the reverse is true for the baso-lateral surface. Wandzilak et al. [49] reported similar findings not only in LLC-PK$_1$ cells, but also in MDCK and OK cells. That such transport mechanisms may play a role in experimental urolithiasis was demonstrated by Sigmon et al. [50]. In animals with stones produced by exposure to gentamicin plus oxalate, oxalate transport was significantly increased in isolated renal papillary cells. However, Verkoelen et al. [51] in another study, also utilizing LLC-PK$_1$ cells, could not demonstrate an oxalate transporter and suggested that the ion moved via a paracellular pathway. In addition to transport studies, data from different experiments, show that oxalate affects other cellular functions. Del Pizzo et al. [52] demonstrated that cellular uptake of calcium is increased in the presence of oxalate and when combined with gentamicin the calcium uptake is even greater. A mitogenic response due to oxalate resulted when LLC-PK$_1$ [53] cells were exposed to levels of oxalate from 0 to 0.4 mM with a resulting concentration dependent increase in DNA synthesis peaking at 0.3 to 0.4 mM. Of interest was a biphasic response in cell numbers: they were increased at concentrations of 0.08 to 0.32 mM, but significantly decreased at concentrations greater than 0.4 mM suggesting a toxic effect at higher levels of oxalate, a finding corroborated by work from our laboratory [40]. Supporting the concept that oxalate by itself may be an agent of cell injury, oxalate appears to be an oxidant and induces lipid peroxidation both in vitro and in experimental animals. Lipid peroxidation in isolated liver cells [54], exposed to oxalate, occurs at ranges of concentrations utilized in other cell culture experiments, and the oxidant effects can be prevented by exposure to antioxidants in oxalate treated rats [55,56].

While the contributions to crystal-cell interactions, described in previous paragraphs, characterize important elements of the mechanisms involved, they do not take into account the possible influence of oxalate upon the process. Since COM crystals exist in an oxalate environment, and since oxalate has cellular effects independent of crystals, it would appear germane to examine the combined effects. We tested the toxic effects of low levels of oxalate (0.25 mM, CaOx relative supersaturation 3.6) and of COM crystals, separately and combined, upon MDCK cells [38]. Each treatment by itself resulted in cell loss from the monolayer, a decrease in viability as demonstrated by trypan blue exclusion, increased tritiated adenine release, and excretion of cellular enzymes into the media. When cells were exposed to both agents, COM crystals in media containing 0.25 mM oxalate, the effects were potentiated in that all measures of cell injury occurred at earlier time periods than with either treatment separately. In this experimental system, when oxalate concentrations are raised to the upper limits of metastability for tissue culture media (1.0 mM, CaOx relative supersaturation 13.5), damage to MDCK and LLC-PK1 cells is evident at earlier time periods and greater with combined exposure to oxalate and crystals than to oxalate alone. Verkoelen et al. [57] also examined the effect of COM crystals on MDCK cells, using crystals suspended in media saturated with CaOx, and was unable to detect cell injury in the monolayer. The effect of combined oxalate and COM on the cells remaining on monolayer after treatment may very well be moot. While these monolayer cells demonstrate a stimulation in proliferation, it is the impact upon the cells released from the culture into the surrounding media that may be of greater import. There is no question that these cells are extensively and irreversibly damaged [40].

From all of the above studies it is apparent that exposure of cells to oxalate alone perturbs cell function, structure and viability, and that exposure to COM crystals alone has similar if not identical effects, and that many of these effects are magnified when treatments are combined, one question remaining to be determined is the reaction of cells exposed to continuously or sporadically elevated concentrations of oxalate as may occur in the hyperoxaluric stone former, and how such pre-conditioned cells affect crystal-cell interactions when limits of metastability are exceeded and COM crystals are formed.

E. CRITIQUE

Investigations into cell/oxalate crystal interactions are in their infancy and currently, such work suffers from a variety of problems: 1) The several investigators in the field use a variety of cell lines making it difficult to compare results; 2) COM crystal load varies from 50-500 μg/l and standardized crystal sizes, habits, or surface charges are not utilized; 3) incubation protocols, including media and cell substrate deviate widely; and 4) in most reports crystal data does not discount oxalate effects (Table 1). Nonetheless, when compared to our comprehension of the cell biology of urolithiasis a decade ago, such research offers great promise in understanding the fundamental biologic processes influencing urolithiasis.

IV. SUMMARY

In summarizing the available information, one can suggest that most stone retention results from either one or both of the following schemes:

A.   Two stage mechanism for stone attachment (predominantly papillary stones)

    1. Prior papillary injury

        a. Possible causes

            1. Ischemia

            2. Toxins

            3. Electrolyte exchange imbalance

            4. Other

        b. Results

            1. Deposition of calcium salts, chiefly phosphate

            2. Erosion of overlying epithelium

            3. Formation of encrustation platform

    2. Stone attachment

        a. CaOx relative supersaturation exceeds metastable limits

        b. Epitaxial nucleation of CaOx on encrustation platform

        c. Retention and stone growth leading to stone disease

B. One-stage mechanism for stone attachment (predominantly non- papillary stone)

    1. No prior papillary injury

        a. CaOx relative supersaturation exceeds metastable limits

            1.   CaOx crystals adhere to cell membrane specialized structures

                a. Specific receptors required

                 b. Potentiated by injury due to oxalate, crystals, other

                c.   Modified by re-arrangement of membrane lipids

            2. Endocytosis

                a. Mitogenesis; potentiated by oxalate, also a mitogen

                b. Alteration of interstitium by cytokines

                c. Interstitial fibrosis

            3. Stone attachment

                a.   To   basement   membrane   by transcellular transport

                b.   To exposed basement membrane, secondary to epithelial cell loss

                c. To altered interstitium

                d. Retention and stone growth leading to stone disease

TABLE 1

Year (Ref)	Cell Type	Cell Origin	CaOx Crystal Studies	Oxalate Studies	Conclusion
1986 (46)	Erythrocytes	Human Blood	-	^{14}C-Oxalate flux in patient RBC	Oxalate exchange is significantly greater in CaOx stone formers. This may represent an inheritable cellular defect in oxalate transport.
1986 (61)	Erythrocytes	Human Blood	COM, COD; 5 - 27.5 mg	-	COM are more membranolytic than COD at constant surface area. Altering COM morphology changes membranolytic potential.
1987 (31)	RPCT Primary Culture	Rat Kidney Papilla	COM, COD slurry; Crystal amount not stated	-	Crystals bind preferentially to cell clumps rather than the flat monolayer suggesting that cell injury potentiates crystal adhesion.
1987 (58)	Erythrocytes	Human Blood	COM; 10 - 1000 μg/ml	-	COM crystals are membranolytic.

1988 (32)	RPCT Primary Culture	Rat Kidney Papilla	^{14}C-COM, AP; 0.076 - 1.21 mg/cm^2	-	Crystal adherence is saturable in cell clumps suggesting that binding is specific. A mathematical model describes both cell and crystal contributions to binding.
1991 (59)	RCPT Primary Culture	Rat Kidney Papilla	^{14}C-COM, ^{45}-Ca-HA, ^{14}C-UA; 0 - 2 mg/cm^2	-	Clumped cells lose polarity. Common binding sites may occur for COM, HA, and UA crystals, but unique sites are present also.
1991 (49)	LLC-PK$_1$	Established cell line	-	^{14}C-Oxalate transport; 10-600μm	Oxalate transport is time, concentration and energy dependent and is inhibited by anion exchange inhibitors.
1992 (60)	IMCD Primary Culture	Rat Kidney Papilla	^{14}C-CaOx; 0 - 70 μg/cm^2	-	Lysis of intracellular junctions results in greatly increased CaOx crystal adherence. Experimental loss of polarity may mimic events in renal tubular injury.

Year (Ref)	Cell	Source	Substrate	Measurement	Comments
1992 (47)	Erythrocytes	Human Blood	-	^{14}C-Oxalate flux in patient RBC	Membrane GAGs are decreased, band 3 phosphorylation is increased resulting in increased oxalate flux. Administration of GAGs corrects the anomaly.
1993 (44)	BSC-1 MDCK LLC-PK$_1$	Kidney Cell (distal) Kidney Cell (distal) Kidney Cell (proximal)	^{14}C-COM, HA, Brushite, Latex Beads; 100 - 200 µg/ml	-	Endocytosis of crystals is inhibited by diverse molecules including THP, heparin, fibronectin, growth factor β_2.
1993 (51)	LLC-PK$_1$ MDCK	Kidney Cell (proximal) Kidney Cell (distal)	-	^{14}C-Oxalate transport; 50 µm	Oxalate is not transported by a specific transcellular mechanism; a paracellular pathway may be involved.
1993 (42)	MDCK	Kidney Cell (distal)	COM; 0.5 - 2.0 mg/ml	-	Pre-treatment of cell monolayers with various GAGs significantly decreases CaOx crystal adhesion.

Year (Ref)	Cell Line	Tissue/Cell Type	Substrate	Transport/Concentration	Observations
1994 (33)	IMCD Primary Culture Erythrocytes	Rat Kidney Human Blood	COM, COD, UA, HA	-	COM adherence is related to altered membrane composition; specific membrane phospholipids are critical for binding
1994 (29)	BSC-1 MDCK	Kidney Cell (distal) Kidney Cell (distal)	^{14}C-COM; 10 - 300 µg/ml	-	Crystals adhere to cell membrane within minutes and are endocytosed; crystals are present in intracellular vesicles; the cytoskeleton is reorganized
1994 (62)	BSC-1	Kidney Cell (distal)	COM; Crystal amount not stated	-	Specific genes are induced including c-myc, but other genes are unaltered or not induced including heat shock protein and c-fos
1994 (48)	LLC-PK$_1$	Kidney Cell (proximal)	-	^{14}C-Oxalate transport; 50 -100 µM	Oxalate is transported by unique transcellular systems; oxalate/chloride at the apical surface membrane, oxalate/sulfate at the basolateral surface.
1994 (53)	LLC-PK$_1$	Kidney Cell (proximal)	-	^{14}C-Oxalate transport; 0 -1600 µM	Oxalate stimulates DNA synthesis and induces c-myc. High concentrations of oxalate result in reduced number of cells.

1994 (52)	OK	Kidney Cell	-	^{45}Ca-transport; > 0.5 mM Ox	Oxalate increases cellular uptake of calcium. This uptake is increased when the nephrotoxin gentamicin is added.
1994 (38)	MDCK	Kidney Cell (distal)	COM; 500 μg/ml slurry	KOx; 0.25 mM	Oxalate, at low concentrations, and COM crystals are each toxic for cells; the effect is synergistic when the agents are combined.
1995 (40)	MDCK LLC-PK$_1$	Kidney Cell (distal) Kidney cell (proximal)	COM; 500 μg/ml slurry	KOx; 1.00 mM	Oxalate, at metastable limits for CaOx, when combined with COM crystals are toxic for the monolayer. Cells lost into the media exhibit significant injury.
1995 (41)	OK	Kidney Cell	^{45}Ca-COM; 2.5 - 20 μg/ml	-	COM crystal-cell interaction stimulates proliferation of interstitial cells. This may mimic the interstitial scarring without epithelial hyperplasia seen in stone patients.
1995 (57)	MDCK	Kidney Cell (distal)	COM; Crystal amount not stated	-	Cells endocytose CaOx crystals from a CaOx saturated media; no cell injury is detected.

Table abbreviations: CaOx: Calcium oxalate; COM: Calcium oxalate monohydrate; COD: Calcium oxalate dihydrate; HA: Hydroxyapatite; AP: apatite; UA: uric acid; BSC-1: Green Monkey Kidney Cell Line; MDCK: Madin Darby Canine Kidney Cell Line; LLC-PK$_1$: Porcine Kidney Cell Line; IMCD: Intermediate Collecting Duct; THP: Tamm-Horsfall Protein; GAGs: Glycosaminoglycans; RPCT: Rat inner papillary collecting tubule.

ACKNOWLEDGEMENTS

We would like to thank Shiro Fujito for his translation of the paper by Ebisuno, et al [42]. This work is supported by N.I.H. Grant 5PO1 DK 20586-18.

VI. REFERENCES

1. **Anderson, L. and J. R. McDonald,** The origin, frequency, and significance of microscopic calculi in the kidney. *Surgery Gynec. Obstet.* 82:275. 1946.

2. **Khan, S. R.**, Animal model of calcium oxalate nephrolithiasis. in *Calcium Oxalate in Biological Systems*, Khan, S. R. , Ed. C.R.C. Press, Inc., Boca Raton, FL. 1995.

3. **Huggins, C. B,** Bone and calculi in the collecting tubules of the kidney. *Arch. Surgery* 27:203. 1933.

4. **Randall, A,** The origin and growth of renal calculi. *Ann. Surgery* 105:1009. 1937.

5. **Randall, A,** The etiology of primary renal calculus. *Intern. Abstract Surgery* 71(3):209. 1940.

6. **Randall, A,** Papillary pathology as a precursor of primary renal calculus. *J. Urol.* 44:580. 1940.

7. **Rosenow, E. C,** Renal calculi: a study of papillary calcification. *J. Urol.* 49:19. 1940.

8. **Posey, L. C.**, Urinary concretions: II. A study of the primary calculous lesions. *J. Urol.* 48:300. 1942.

9. **Vermooten, V.,** The origin and development in the renal papilla of Randall's calcium plaques. *J. Urol.* 48:27. 1942.

10. **Swagel, E., Low, R. K., and Stoller, M. L.,** Histopathology of renal papillae suggesting a vascular related etiology for nephrolithiasis. *J. Urol.* 153(4):351A. 1995.

11. **Cooke, S. A. R.**, The site of calcification in the human renal papilla. *Brit. J. Surg.* 57(12):890. 1970.

12. **Weller, R. O., Nester, B., and Cooke, S. A. R.,** Calcification in the human renal papilla: an electron-microscope study. *J. Path.* 107:211. 1972.

13. **Haggitt, R. C. and Pitcock, J. A.**, Renal medullary calcifications: a light and electron microscopic study. *J. Urol.* 106:342. 1971.

14. **Burry, A. F., Axelsen, R. A., Trolove, P., and Saal, J. R.**, Calcification in the renal medulla: a classification based on a prospective study of 2261 necropsies. *Human Pathol.* 7(4):435. 1976.

15. **Cifuentes Delatte, L., Minon-Cifuentes, J. L. R., and Medina, J. A.**, Papillary stones: calcified renal tubules in randall's plaques. *J. Urol.* 133:490. 1985.

16. **Cifuentes Delatte, L., Minon-Cifuentes, J., and Medina, J. A.**, New studies on papillary calculi. *J. Urol.* 137:1024. 1987.

17. **Prien, E. L,** The riddle of Randall's plaques. *J. Urol.* 114:500. 1975.

18. **Finlayson, B. and Reid, F.**, The expectation of free and fixed particles in urinary stone disease. *Invest. Urol.* 15(6):442. 1978.

19. **Kok, D. J. and Khan, S. R.**, Calcium oxalate nephrolithiasis, a free or fixed particle disease. *Kidney Int.* 46:847. 1994.

20. **Dunn, J. S., Haworth, A., and Jones, N. A.**, The pathology of oxalate nephritis. *J. Pathol. Bacteriol.* 27:299. 1924.

21. **Dykstra, M. J. and Hackett, R. L.**, Ultrastructural events in early calcium oxalate crystal formation in rats. *Kidney Int.* 15:640. 1979.

22. **Rushton, H. G., Spector, M., Rodgers, A. L., Hughson, M., and Magura, C. E.**, Developmental aspects of calcium oxalate tubular deposits and calculi induced in rat kidneys. *Invest. Urol.* 19(1):52. 1981.

23. **Khan, S. R. and Hackett, R. L.**, Retention of calcium oxalate crystals in renal tubules. *Scanning Microsc.* 5(3):707. 1991.

24. **Khan, S. R., Finlayson, B., and Hackett, R. L.**, Experimental calcium oxalate nephrolithiasis in the rat, role of the renal papilla. *Am. J. Pathol.* 107(1):59. 1982.

25. **Morgenroth, K., Backmann, R., and Blaschke, R.**, On the forms of deposits of calcium oxalate in the human kidney in oxalosis. *Beitr. path. Anat.* 136:454. 1968.

26. **Lieske, J. C., Spargo, B. H., and Toback, F. G.**, Endocytosis of calcium oxalate crystals and proliferation of renal tubular epithelial cells in a patient with type 1 primary hyperoxaluria. *J. Urol.* 148:1517. 1992.

27. **Gaush, C. R., Hard, W. L., and Smith, T. F.**, Characterization of an established line of canine kidney cells (MDCK) (31293). *Proc. Soc. Exp. Biol. Med.* 122:931. 1966.

28. **Hull, R. N., Cherry, W. R., and Weaver, G. W.**, The origin and characteristics of a pig kidney cell strain, LLC-PK$_1$. *In Vitro* 12(10):670. 1976.

29. **Lieske, J. C., Swift, H., Martin, T., Patterson, B., and Toback, F. G.**, Renal epithelial cells rapidly bind and internalize calcium oxalate monohydrate crystals. *P. N. A. S.* 91:6987. 1994.

30. **Hopps, H. E., Bernheim, B. C., Nisalak, A., Hin Tjio, J., and Smadel, J. E.**, Biologic characteristics of a continuous kidney cell line derived from the African green monkey. *J. Immunol.* 91:416. 1963.

31. **Wiessner, J. H., Kleinman, J. C., Blumenthal, S. S., Garancis, J. C., Mandel, G. S., and Mandel, N. S.**, Calcium oxalate crystal interaction with rat renal inner papillary collecting tubule cells. *J. Urol.* 128:640. 1987.

32. **Riese, R. J., Riese, J. W., Kleinman, J. G., Wiessner, J. H., Mandel, G. S., and Mandel, N. S.**, Specificity in calcium oxalate adherence to papillary epithelial cells in culture. *Am. J. Physiol.* 255:F1025. 1988.

33. **Mandel, N,** Crystal membrane interaction in kidney stone disease. *J. Am. Soc. Nephrol.* 5:S37. 1994.

34. **Gill, W. B., Ruggiero, K., and Straus, F. H.**, Crystallization studies in a urothelial-lined living test tube (the catheterized female rat-bladder) I. Calcium oxalate crystal adhesion to the chemically injured rat bladder. *Invest. Urol.* 17(3):257. 1979.

35. **Khan, S. R., Cockrell, C. A., Finlayson, B., and Hackett, R. L.**, Crystal retention by injured urothelium of the rat urinary bladder. *J. Urol.* 132:153. 1984.

36. **Hackett, R. L., Shevock, P. N., and Khan, S. R.**, Cell injury associated calcium oxalate crystalluria. *J. Urol.* 144:1535. 1990.

37. **Kumar, S., Sigmon, D., Miller, T., Carpenter, B., Khan, S. R., Malhotra, R., Scheid, C., and Menon, M.**, A new model of nephrolithiasis involving tubular dysfunction/injury. *J. Urol.* 146:1384. 1991.

38. **Hackett, R. L., Shevock, P. N., and Khan, S. R.**, Madin-Darby canine kidney cells are injured by exposure to oxalate and calcium oxalate crystals. *Urol. Res.* 22:197. 1994.

39. **Lieske, J. C., Walsh-Reitz, M. M., and Toback, F. G.**, Calcium oxalate monohydrate crystals are endocytosed by renal epithelial cells and induce proliferation. *Am. J. Physiol.* 262:F622. 1992.

40. **Hackett, R. L., Shevock, P. N., and Khan, S. R.**, Alterations in MDCK and LLC-PK$_1$ cells exposed to oxalate and calcium oxalate monohydrate crystals. *Scanning Microsc.* (Submitted). 1995.

41. **Goswami, A., Singhal, P. C., Wagner, J. D., Urivetzky, M., Valderrama, E., and Smith, A. D.**, Matrix modulates uptake of calcium oxalate crystals and cell growth of renal epithelial cells. *J. Urol.* 153:206. 1995.

42. **Ebisuno, S., Yoshida, T., Kohjimoto, Y. and Ohkawa, T.**, Adhesion of calcium oxalate crystal to Madin-Darby canine kidney cells: quantitative determination and effects of glycosaminoglycans (GAG) and cell injuries on adhesion. *Jap. J. Urol.* 84:1980. 1993.

43. **Lieske, J. C., Leonard, R., and Toback, F. G.**, Adhesion of calcium oxalate monohydrate crystals to renal epithelial cells is inhibited by specific ions. *Am. J. Physiol.* (In Press). 1995.

44. **Lieske, J. C. and Toback, F. G.**, Regulation of renal epithelial cell endocytosis of calcium oxalate monohydrate crystals. *Am. J. Physiol.* 264:F800. 1993.

45. **Koyama, H., Goodpasture, C., Miller, M. M., Teplitz, R. L., and Riggs, A. D.**, Establishment and characterization of a cell line from the American Opossum (*Didelphys virginiana*). *In Vitro* 14(3):239. 1978.

46. **Baggio, B., Gambaro, G., Marchini, F., Cicerello, E., Tenconi, R., Clementi, M., and Borsatti, A.**, An inheritable anomaly of red-cell oxalate transport in "primary" calcium nephrolithiasis correctable with diuretics. *N. Engl. J. Med.* 314(10):599. 1986.

47. **Gambaro, G. and Baggio, B.**, Idiopathic calcium oxalate nephrolithiasis: a cellular disease. *Scanning Microsc.* 6(1):247. 1992.

48. **Koul, H., Ebisuno, S., Renzulli, L., Yanagawa, M., Menon, M., and Scheid, C.**, Polarized distribution of oxalate transport systems in LLC-PK$_1$ cells, a line of renal epithelial cells. *Am. J. Physiol.* 266:F266. 1994.

49. **Wandzilak, T. R., Calo, L., D'Andre, S., Borsatti, A., and Williams, H. E.**, Oxalate transport in cultured porcine epithelial cells. *Urol. Res.* 20:341. 1992.

50. **Sigmon, D., Kumar, S., Carpenter, B., Miller, T., Menon, M., and Scheid, C.**, Oxalate transport in renal tubular cells from normal and stone-forming animals. *Am. J. Kid. Dis.* XVII(4):376. 1991.

51. **Verkoelen, C. F., Romijn, J. C., de Bruijn, W. C., Boevé, E. R., Cao, L. C., and Schröder, F. H.**, Absence of a transcellular oxalate transport mechanism in LLC-PK$_1$ and MDCK cells cultures on porous supports. *Scanning Microsc.* 7(3):1031. 1993.

52. **Del Pizzo, J., Urivetzky, M., and Smith, A.**, Calcium transport by a renal tubular epithelial cells. The effect of calcium, oxalate and gentamicin. *J. Urol.* 151(5):424A. 1994.

53. **Koul, H., Kennnington, L., Nair, G., Honeyman, T., Menon, M., and Scheid, C.**, Oxalate-induced initiation of DNA synthesis in LLC-PK$_1$ cells, a line or renal epithelial cells. *Biochemical and Biophysical Research Communications* 205(3):1632. 1994.

54. **Ernster, L. and Nordenbrand, K.**, Microsomal lipid peroxidation. *Methods Enzymol.* X:574. 1967.

55. **Thamilselvan S.**, Lipid peroxidation in experimental urolithiasis: Role of vitamin E and mannitol on the retention of calcium oxalate in the kidney. Ph.D. Dissertation. University of Madras, India. 1992.

56. **Thamilselvan, S. and Selvam R.**, Antioxidant therapy prevents

calcium oxalate deposition in acute hyperoxaluria. *J. Urol.* (Submitted). 1995.

57. **Verkoelen, C. F., Romijn, J. C., de Bruijn, W. C., Boeve, E. R., Cao, L. C., and Schroder, F. H.,** Association of calcium oxalate monohydrate crystals with MDCK cells. *J. Urol.* 153:350A.   1995.
58. **Elferink, J. G. R,** The mechanism of calcium oxalate crystal-induced haemolysis of human erythrocytes. *Br. J. exp. Path.* 68:551. 1987.
59. **Mandel, N. and Riese R,** Crystal-cell interactions: crystal binding to rat renal papillary tip collecting duct cells in culture. *Am. J. Kid. Dis.* VXII(4):402. 1991.
60. **Riese, R. J., Mandel, N. S., Wiessner, J. H., Mandel, G. S., Becker, C. G., and Kleinman, J. C.,** Cell polarity and calcium oxalate crystal adherence to cultured collecting duct cells. *Am. J. Physiol.* 262:F177. 1992.
61. **Wiessner, J. H., Mandel, G. S., and Mandel, N. S.** Membrane interactions with calcium oxalate crystals: variation in hemolytic potentials with crystal morphology. *J. Urol.* 135:835. 1986.
62. **Hammes, M. S., Lieske, J. C., Pawar, S., Keeley, E., and Toback, F. G.,**   Calcium oxalate monohydrate (COM) crystals induce gene expression in kidney epithelial cells. *J. Am. Soc. Nephrol.* 5(3):864. 1994.

Chapter 17

ANIMAL MODEL OF CALCIUM OXALATE NEPHROLITHIASIS

Saeed R. Khan

I. INTRODUCTION

Crystallization of salts in the urine is a process common to many mammals including humans. The crystals are either harmlessly excreted as crystalluria particles or are retained inside the kidneys or other parts of the urinary tract causing stone disease or urolithiasis. Nephrolithiasis specifically refers to crystal deposition inside the renal tubules. Many *in vitro* and *in vivo* models have been developed to understand the mechanisms involved in the formation of human urinary stones and to determine the effects of various therapeutic agents and protocols on development and progression of the disease. Since calcium oxalate (CaOx) stones are most common, CaOx urolithiasis has been studied in greater details. Since rat is the most easily available and commonly used laboratory animal, it is the most frequently used animal in many such studies. Pathogenesis of urolithiasis involves crystal nucleation, growth, aggregation and retention within the urinary system. The final event in stone formation is evolution of retained crystals into urinary stones. A variety of modulators control these processes by promoting, inhibiting or modifying them.

II. CRYSTAL DEPOSITION IN THE KIDNEYS

A. HYPEROXALURIA

Like humans, there is no spontaneous production of urinary stones by normal rats. Induction of hyperoxaluria is essential.[1] Thus rat models are similar to human conditions because in humans too, hyperoxaluria is considered as one of the most common causes of idiopathic CaOx nephrolithiasis.[2] Hyperoxaluria can be induced by direct dispensation of oxalate in the form of potassium, sodium or ammonium oxalate or by administration of oxalate precursors like ethylene glycol (EG), hydroxy-l-proline or glycollic acid. Hyperoxaluria can also be induced by other dietary manipulations including pyridoxine deficient diets, low phosphate diets and high protein diets. Various methods of induction result in two basic types of hyperoxaluria,[3] 1. chronic, when a rat is challenged with multiple, generally small doses of lithogen for a long period of time, and 2. acute, when a single, generally large dose is administered. In male

0-8493-7673-4/95/$0.00+$.50
© 1995 by CRC Press

Sprague-Dawley rats with acute hyperoxaluria or moderate to high grade chronic hyperoxaluria, CaOx crystal deposition begins in proximal segments of the renal tubules. On the other hand in male rats with a low grade chronic hyperoxaluria deposition of crystals starts in the distal segments of the renal tubules.[4] Obviously in the first situation, urine becomes sufficiently supersaturated with respect to CaOx much earlier in the nephron, while in the second, urine attains appropriate level of supersaturation when it reaches the collecting duct system. The latter condition may be similar to what happens in humans, where renal papilla is the most common site of kidney stone formation. In both acute and chronic hyperoxaluria crystals are first seen in the tubular lumen (Fig. 1) and then in the interstitium indicating movement of crystals from lumen to the interstitium. Administration of ammonium chloride (AC) to reduce urinary pH expedites the CaOx nephrolithiasis in hyperoxaluric rats. We have studied nephrolithiasis in male Sprague-Dawley rats induced by both chronic and acute hyperoxaluria.

1. Chronic Hyperoxaluria

Chronic hyperoxaluria induced by 0.75% ethylene glycol (EG) alone or with 2% ammonium chloride (AC), first produced CaOx crystalluria and then nephrolithiasis.[3,4] Ethylene glycol administration with AC resulted in the development of persistent CaOx crystalluria in all rats by day 3 and nephrolithiasis by day 7. However on EG administration alone, it took about 12 days for all rats to show persistent crystalluria and 2 to 3 weeks for nephrolithiasis. In the beginning smaller dipyramidal CaOx dihydrate crystals were seen in the urine. Later most of the crystals were large aggregates of dumbbell shaped CaOx monohydrate crystals and twinned CaOx dihydrate crystals. Ministones ranging in size from $75\mu m$ to $200\mu m$ across were seen in the bladder aspirate.

Initially the crystals were distributed randomly in the renal medulla. Eventually renal papillary tips and calices at the papillary base were the preferred sites for crystal deposition. Most crystals were present as aggregates in the tubular lumen. Many of them were also seen in the intercellular spaces of the tubular epithelium as well as inside the epithelial cells and the interstitium. After only one week on the treatment, rats on EG+AC showed more and larger crystal aggregates in the kidneys and had small stones on their renal papillary surfaces. On the other hand rats on EG alone even after 8 weeks, had fewer crystal deposits in their kidneys and only a few of the rats had crystals at their renal papillary tips.

Urinary excretion of oxalate increases rapidly and significantly during chronic administration of ethylene glycol as 0.5%, 0.75% or 1% aqueous solution in drinking water reaching its peak by day 7 or 10 (Table 1). Urinary calcium oxalte supersaturation increases accordingly. Excretion of calcium, magnesium and citrate goes down.

Table 1

Urinary chemistry of male rats on 0.75% EG in drinking water

Day	Calcium mM	Oxalate mM	Magnesium mM	Citrate mM	CaOx Relative Supersaturation
0	6.94±0.7	1.12±0.1	17.6±0.9	29.5±2.2	7.98±1.1
7	2.70±0.3	3.83±0.3	14.8±1.7	18.0±2.8	18.03±2.5
14	1.95±0.5	4.86±0.5	12.7±1.3	15.4±2.8	14.68±3.1
21	2.16±0.8	4.20±0.6	14.1±1.8	11.2±2.4	16.09±2.0
28	2.04±0.4	4.15±0.6	13.1±1.2	9.9±2.6	18.53±2.8
35	2.06±0.4	4.11±0.6	13.3±1.7	9.7±2.1	19.68±3.01

2. Acute Hyperoxaluria

Administration of a sodium oxalate solution in saline by intraperitoneal injection resulted in almost an instant increase in urinary excretion of oxalate which reached its peak by about 6 hours and returned to normal after 48 hours after the injection.[3,5] Acute hyperoxaluria induced by 3mg/100g rat body weight dose of sodium oxalate was associated with the appearance of CaOx crystals in cortical tubules within 15 minutes, in both cortical and medullary tubules within 30 minutes, in medullary segments only within 1 hour. Kidneys were free of crystals within 3 hours. Such a rapid movement indicates that crystals were present in the tubular lumen and were moving with the urine. Morphological examination of the kidneys also demonstrated the crystals in tubular lumen. A similar pattern was seen at higher doses of sodium oxalate.[6,7] However higher doses resulted in larger crystals and crystal aggregates, much larger in size than even the diameter of tubular lumen and crystals remained in the kidneys for longer duration. Many of them could be seen weeks after the initial oxalate challenge and were localized in the renal interstitium. Crystals seen inside the tubules within 15 minutes of injection were mostly CaOx dihydrate, but at later times most of the crystals were CaOx monohydrate.

III. RENAL EPITHELIAL INJURY

A. MORPHOLOGICAL CHANGES

Crystal formation and deposition in the kidneys is almost always associated with damage to the renal tubular epithelium. Morphological studies of kidneys after acute hyperoxaluric challenge showed that severe

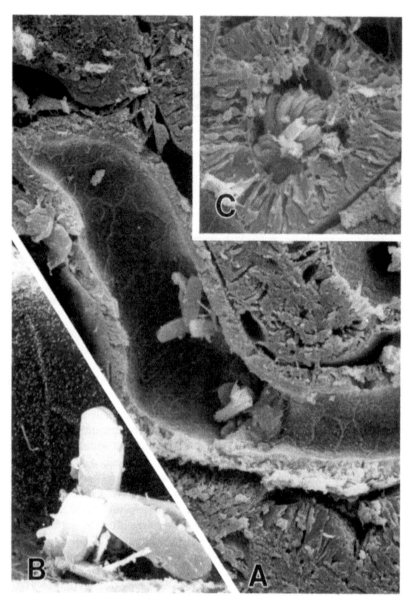

Figure 1. Plate-like crystals of calcium oxalate monohydrate are present in the lumen of renal tubules one hour after a sodium oxalate injcetion. (A) loop of Henle. X 1.5K (B) Higher magnification of crystals in A, showing crystal attachment to the epithelial surface. X 3.5K (C) Crystal aggregate blocking the lumen of a proximal tubule. X 2K (From Khan, S.R. and Hackett, R.L., *Scanning Microscopy* 5, 707-712, 1991. With permission)

damage was mostly restricted to the tubules containing the crystals.[1,7] Proximal tubular brush border was distorted by clubbing of the microvilli, formation of blebs and localized loss of microvilli. General cellular necrosis was seen in all parts of the nephron. Epithelial cells contained increased number of lysosomes some of which were autophagic vacuoles. Degenerative changes included swelling of mitochondria, dilatation of endoplasmic reticulum, cytoplasmic edema, and vacuolation. The intercellular spaces between the intact epithelial cells appeared enlarged. Luminal cell membranes of degenerative cells appeared to burst releasing their contents into the tubular lumen. Focally, cells appeared sheared from the tubular basement membrane, their contents contributing to the cellular degradation products present in the lumen of all parts of the nephron. Numerous dividing cells were found in the epithelial lining of the tubules. Surface of the renal papillary tip was badly damaged.[7] Its covering urothelium was lost exposing the basement membrane.

B.   ENZYMURIA

Both acute and chronic hyperoxaluria result in increased urinary excretion of enzymes.[6,8] Acute hyperoxaluria induced by administration of single intraperitoneal injection of sodium oxalate at a dose of 10mg/100 rat body weight was associated with increased urinary excretion of of alkaline phosphatase, leucine aminopeptidase and gamma glutamyl transpeptidase.[6] The enzymes increased within the first 24 hours and started to return back to the normal range by day 3. These are membrane marker enzymes and their increase in the urine indicates damage to the epithelial brush border membrane. After 3 days of the injection there was an increase in urinary excretion of lysosomal enzyme N-acetyl-β-glucosminidase indicating a latter involvement of cellular lysosomal system in the process.

In a separate experiment we demonstrated enhanced urinary excretion of enzymes by rats as a result of low grade chronic hyperoxaluria without CaOx crystal deposition in the kidneys.[8] Hyperoxaluria was induced by administration of EG, hydroxy-L-proline or ammonium oxalate. N-acetyl-β-glucosaminidase, the lysosomal enzyme whose irregular excretion in urine is regarded as an indicator of chronic renal disease, was the first to significantly increase in the urine. The increase was positively and significantly correlated with increased urinary excretion of oxalate. Membrane marker enzymes alkaline phosphatase and gamma glutamyl transpeptidase were also affected by the challenge but not as significantly and were elevated in later urinary collections. Display of enzymuria in the presence of mild hyperoxaluria is significant since human stone formers who generally have only a mild hyperoxaluria also demonstrate enzymuria of proximal tubular origin.[9]

Increased lipid peroxidation has also been described in the kidneys of stone forming rats.[10,11] Many types of crystals have deleterious effect on cells. Recent studies using cultured epithelial cells of renal origin have demonstrated that both oxalate ions and CaOx crystals are injurious to the cells.[12] Apparently, nephrolithiasis is associated with renal injury which is the result of a combination of two factors, increased oxalate load and crystal retention.

IV. SEX HORMONES AND NEPHROLITHIASIS

Idiopathic calcific stone disease is two to three times more common in men than women.[13] Kidney stones found in men are normally CaOx while those in females are usually calcium phosphate.[14] Reasons for these differences between sexes are still not clear. Lower rate of calcific nephrolithiasis in women is suggested by some to be a result of higher urinary excretion of citrate[15] and by others an effect of lower urinary excretion of calcium and oxalate.[16] Results of studies involving rat model of nephrolithiasis have provided some explanation for this interesting phenomenon.

A.   CALCIUM PHOSPHATE NEPHROLITHIASIS
Spontaneous deposition of calcium phosphate in the kidneys of rats on semipurified diets has been reported by many investigators and the incidence of lesion formation is consistently higher in females than males.[17,18] Calcification starts intraluminally at the cortico-medullary junction spreading into the medulla. Estrogen, in addition to the dietary levels of calcium, phosphorus and magnesium plays a significant role in the development of this disease. Ovariectomy resulted in cessation of calcification while replacement therapy with estrogen following gonadectomy resulted in calcium phosphate nephrolithiasis in both male and female rats.[17]

B.   CALCIUM OXALATE NEPHROLITHIASIS
Administration of 1% EG in drinking water for 4 weeks produced CaOx nephrolithiasis in 3/13 males but 0/12 females.[19] Urinary acidification increased the incidence to 5/6 males and 1/9 females. In an attempt to determine the role of testosterone, 0.5% EG was used to induce CaOx nephrolithiasis in male and female, normal and gonadectomized rats.[20] Low level CaOx crystal deposition was found in all rats but in normal males, 5/7 produced kidney stones and 4/7 had massive CaOx crystal deposition. Only 1/7 castrated males produced kidney stones and none of them had massive CaOx crystal deposition in their kidneys. None of the female rats normal or castrated produced any kidney stones or massive crystal deposition in their kidneys. It was concluded that testosterone plays a determinant role in the pathogenesis of urolithiasis.

Our studies of calcium phosphate nephrolithiasis induced by semipurified diet and CaOx nephrolithiasis induced by EG showed that all rats of both sexes receiving ethylene glycol had CaOx deposits in their kidneys but only male rats had massive crystal deposition and crystals at the renal papillary tips.[21] Only female rats produced calcium phosphate deposits on the semipurified diet. Calcium phosphate deposition was confined to the cortico-medullary junction of the kidneys and CaOx was mostly localized to renal papillae and fornices. But both male and female rats had similar relative supersaturations for both CaOx and calcium phosphate. Then why do male and female rats behave differently when it came to nephrolithiasis? Nephrolithiasis involves crystal formation and retention.[22] The later can be facilitated by crystal aggregation.[23] Perhaps there are some differences in urinary inhibitors of crystal aggregation between males and females.

V. CRYSTAL NUCLEATION

A. HETEROGENEOUS NUCLEATION

Both the studies of human urolithiasis and *in vivo* studies using rat models demonstrate that nephrolithiasis occur at relative supersaturation of approximately 30,[2,24,25] which is not high enough for homogeneous nucleation of CaOx crystals. Thus it is most likely that crystal nucleation in both rat and human urine is heterogeneous. It has been suggested that calcium phosphate crystals may act as nucleators of CaOx, since calcium phosphate is the most common crystal found in human urine and urinary stones[26,27] and can cause nucleation of CaOx *in vitro* from a metastable solution of CaOx.[28] Our extensive experimental studies of CaOx nephrolithiasis have failed to demonstrate any calcium phosphate in association with deposits of CaOx in the rat kidneys.

1. Calcium Phosphate as Nucleator

We tested the hypothesis that renal deposits of calcium phosphate can provoke precipitation of CaOx within the renal tubules of hyperoxaluric rats.[21] In one experiment both male and female rats were first given a diet that induces calcium phosphate deposits in the kidneys and then put on a hyperoxaluria-inducing protocol. Female rats responded to the calcium phosphate-inducing diet by generating calcium phosphate crystal deposits in renal tubules at the cortico-medullary junction. Male rats responded to the hyperoxaluric challenge by producing CaOx crystals in the collecting ducts of the renal papillae. Small deposits of CaOx were seen in kidneys of the female rats on hyperoxaluria-inducing protocol. Male rats on calcium phosphate-inducing diet did not produce any calcific deposits in their kidneys. In another experiment a calcium phosphate-inducing diet and hyperoxaluric challenge were simultaneously delivered to female rats. This resulted in calcium phosphate and CaOx crystalluria as well as nephrolithiasis, but CaOx crystals were not observed in association with calcium phosphate crystals. Results of these experiments suggest that calcium phosphate is not necessary for CaOx nephrolithiasis.

2. Cellular Membranes As Nucleator

Then what is the heterogeneous nucleator of CaOx crystals? Urinary as well as renal CaOx crystals induced in rats are almost always found associated with cellular degradation products.[1,4,7] Membranous cellular degradation products and their constituent lipids are an important part of the matrix of human urinary stones.[29] Both the lipids isolated from human urinary stones and the membrane vesicles obtained from renal proximal tubular brush border have been shown to induce crystallization of CaOx from a metastable solution.[30,31] As described earlier, both hyperoxaluria and CaOx crystals can be injurious to the tubular epithelial cells and cause cellular necrosis and degeneration. As a result, urine of hyperoxaluric rat contains an abundant supply of membranous cellular degradation products which can promote heterogeneous nucleation of CaOx crystals.

Mammalian urine contains both inhibitors and promoters of crystallization.[2,11,32] In normal individuals, inhibitors may have the upper hand, but in stone formers, urine is either less inhibitory or becomes less inhibitory during stone forming episodes. Alternatively, the promotory potential of urine may increase by virtue of an increase in heterogeneous nucleators. We tested this hypothesis by increasing the membranous cellular degradation products as heterogeneous nucleators in the urine.[33] Membranuria was induced by administration of gentamicin sulphate to male Sprague-Dawley rats. A low grade hyperoxaluria was produced by EG administration. Hyperoxaluria or gentamicin treatment alone did not cause the formation of crystals in the urine, but simultaneous production of membranuria and hyperoxaluria resulted in CaOx crystalluria. Crystals were formed in association with the membranous substances in the urine. These observations were later confirmed when a higher grade hyperoxaluria in association with gentamicin-induced membranuria resulted in 63% of rats with CaOx nephrolithiasis.[34] Hyperoxaluria alone induced nephrolithiasis in only 6% of the treated rats.

VI. CRYSTALLIZATION MODULATORS

There are two major categories of crystallization modulators present in human urine. One consists of substances like citrate, phosphocitrate, pyrophosphate and magnesium and the other of urinary macromolecules, glycosaminoglycans, Tamm-Horsfall protein (THP), osteopontin (OP), nephrocalcin (NC), uronic acid rich protein (UAP), and inter-α-trypsin inhibitor (ITI). Magnesium and citrate form soluble complexes with oxalate and calcium respectively.[11] Citrate also binds to the surface of CaOx crystals, interferes with their growth and modifies their morphology.[35,36]

Magnesium has been extensively studied in animal models of CaOx urolithiasis and it has been shown that magnesium administration in the form of alkalinizing salts causes a decrease in CaOx crystal deposition in the kidneys of hyperoxaluric rats.[25,37,38] Magnesium therapy reduces the amount of oxalate excreted in the urine, and increases the urinary excretion of citrate resulting in a reduced urinary CaOx relative supersaturation. Citrate, magnesium and pyrophosphate are discussed in detail in a separate chapter in this volume.

Much of the information about THP, OP (uropontin), NC, and UAP, including their normal renal distribution has been obtained from studies of rat kidneys by using immunohistological techniques and in situ hybridization. Macrmolecular modulators influence various aspects of crystallization ranging from crystal nucleation to growth and aggregation. Many of them are also suggested to be involved in crystal endocytosis by renal tubular epithelial cells. All these aspects are discussed in great detail in a separate chapter on the subject. We are currently investigating the role of THP and OP in experimental nephrolithiasis induced by EG administration to male rats. Both THP and OP are found associated with CaOx crystal deposits in the kidneys including the renal papilla.[39] Osteopontin appears to be more closely involved and seen even inside the large crystals and crystal deposits present at the papillary tips.[40] Many urinary macromolecules involved in crystallization inhibition are produced by renal tubular epithelial cells particularly the cells lining the proximal tubules and loops of Henle,[41] the same tubular segment which has been shown to be susceptible to damage by exposure to oxalate and CaOx crystals. Any damage to these cells may interfere with production of the macromolecules and influence the inhibitory potential of the urine. Urine of rats treated with gentamicin sulfate which is specifically toxic to proximal tubular epithelium has already been shown less inhibitory to CaOx crystallization *in vitro*.[42]

VII. CRYSTAL RETENTION

A.   CRYSTAL AGGREGATION
Crystal retention is indispensable to the development of kidney stones. Since size may play an important role in this process, any mechanism that can result in mass accretion by the crystal deposits, can regulate the process of crystal retention. Taking into account the rate of crystal growth, dimension of the renal tubules and time it takes the urine to pass through the nephron, Finlayson and Reid[22] concluded that crystals of CaOx can not grow fast enough to be retained within the renal tubules because of the size alone. They concluded that it is necessary for the crystals to attach to the renal tubular epithelium for retention within the kidneys. We recently revisited this issue, calculated the rate of mass accretion by crystal deposits

taking into account the process of crystal aggregation, and found that crystal aggregates can become large enough to block the renal tubules (Fig. 1) and thus be retained solely because of the aggregate size.[23] Studies using rat model of nephrolithiasis have demonstrated that as the disease progresses, more aggregated crystals are formed and the aggregates are generally large enough to be retained without attachment to the renal tubules.[4,43] Often the crystals deposit at sites where luminal diameter of the tubules narrows e.g. cortico-medullary junction where proximal tubules with larger luminal diameter meet narrow loops of Henle or at the papillary base where renal tubules bend and result in the development of kinks. Another site of crystal retention and development of urinary stone, the renal papillary tips, also display a transition from larger to narrower lumen. Of all renal tubular segments, ducts of Bellini have largest luminal diameter but their openings into the renal pelvis are very narrow and slit-like.[23]

B.   CRYSTAL FIXATION

Crystals have also been found retained inside the tubules by attachment to the renal epithelial cell surface (Fig. 1) as well as to the epithelial basement membrane.[4,7,43] As discussed earlier, hyperoxaluria and CaOx crystal deposition in the renal tubules is associated with injury to the epithelium which may cause a loss of the cell surface glycocalyx, damage the luminal plasma membrane, and the tight junctions between the epithelial cells, and result in total cellular necrosis. Loss of glycocalyx can induce crystal attachment to the cell surface. Damaged cell membrane can no longer maintain the homeostatic function; calcium, oxalate and other ions may freely move in and out, resulting in increased supersaturation within the confines of the cells, eventually causing the formation of crystals within the cells. Loss of tight junctions is suggested to result in relocation of macromolecules from the basal lamina side to the luminal side of the cells. These macromolecules may be involved in attachment of crystal to the luminal plasma membrane of the tubular epithelial cells.[43] Total cellular necrosis results in cell sloughing and exposure of the basement membrane. Calcium oxalate crystals have been seen attached to the basement membrane in both acute and chronic hyperoxaluria models of nephrolithiasis.[4,7]

C.   CRYSTAL FORMATION IN THE INTERSTITIUM

It has been argued that crystals actually form in the renal papillary interstitium,[44] since most of the oxalate, particularly that in the renal papilla, is present in the interstitium, and the papilla has the highest concentration of calcium and oxalate within the kidney.[45] If crystals have already formed in the interstitium then retention will not be an issue. However, all morphological studies of experimental nephrolithiasis have clearly shown that initial crystal formation starts in the tubular lumen. Interstitial crystals are seen much later during nephrolithiasis. In addition, once crystals move into the interstitium they do not seem to grow. They actually seem to disappear.[6] Moreover, interstitial calcific deposits are quite common in human kidneys and degree of of this kind of calcification increases with age.[46,47] In many studies almost all the kidneys examined contained calcific deposits. On the other hand nephrolithiasis, the formation of calcific deposits in renal tubules is not a common phenomenon and its incidence actually decreases with age.

VIII. DEVELOPMENT OF URINARY STONES

Crystal retention is essential for the development of stone disease, but how do the retained crystals develop into kidney stones? No studies, animal or otherwise, have been performed to specifically study the transition from retained crystals to kidney stones. However, a number of observations have been made that can assist us in developing a probable scenario. To produce stones at the renal papillary tip Vermeulen[48] challenged the rats with a high dose of lithogen or what he termed as a triggering dose. Soon after that the dose was reduced to a quarter, the maintenance dose. The treatment resulted in severe stone disease. Our studies of male rats with acute hyperoxaluria have shown that crystals are first seen in the tubular lumen and later in the interstitium as well as inside the cells indicating movement from the tubular lumen to the outside. More recent studies of chronic hyperoxaluria in male rats have similarly demonstrated crystal movement from the luminal to extraluminal location in the interstitium.[4,49] Both studies have also shown that the number of crystals retained in the interstitium appears to decrease with time. In addition, we have recently observed interstitial crystals surrounded by macrophages and other cells involved in the inflammatory processes (unpublished personal observations).

Tissue culture studies have also provided some interesting information in this regard.[12] When Madin-Darby canine kidney epithelial cells were exposed to CaOx crystals, many of the crystals were endocytosed. Some crystals were seen in the intercellular spaces. Still others were found underneath the cells. Thus even in tissue culture studies crystals appear to migrate from the luminal to the basolateral side of the renal epithelial cells. Crystals are retained at many sites in the kidneys, most common being the cortico-medullary junction[32] Human kidney stones, however, develop on papillary surfaces in the renal calyx and fornix. What are the reasons for stone development in association with the renal papillae. One reason may be that the renal papilla has the highest concentration of calcium and oxalate relative to the cortex and medulla.[44,45] Other reasons may be related to the anatomy and histology of the papilla which make it possible for the crystals retained in the papilla to grow into stones. Apparently the crystals retained in more central ducts move into the renal interstitium towards the interior of the kidney. Once in the interstitium, they are not exposed to the urine and to the urinary calcium and oxalate ions. Thus they are unable to grow. Moreover now they may be subjected to body's inflammatory processes.

On the other hand when crystals are retained in the peripheral collecting ducts of the renal papilla, ducts at the papillary tips or base, they can migrate to the papillary surface. Once crystal deposits move to the papillary or forniceal surface, they are continuously bathed in slow moving urine of renal pelvis and calices from where they receive their nourishment of calcium and oxalate ions. They can grow indefinitely even in the absence of high grade hyperoxaluria since normal urine is generally supersaturated with respect to CaOx. Thus it is clear, that for the development of kidney stones, not only must the crystals be retained in the kidneys but they should be located in the tubules from where they can move to the surface, be exposed to urine and be available for further growth.

IX. SUMMARY

Calcium oxalate nephrolithiasis starts with hyperoxaluria which results in increased urinary CaOx supersaturation and formation of CaOx crystals within renal tubular lumen (Figure 2). Both CaOx crystals and oxalate ions provoke a response from renal tubular epithelial cells. Cells release cytoplasmic, lysosomal and membrane associated enzymes. Epithelial cells often undergo necrosis and are sloughed exposing the basement membrane. Sloughed cellular membranes provide substrate for the nucleation of CaOx crystals at lower supersaturation and may also be involved in crystal aggregation. Damage to the cells of proximal tubules and loop of Henle may interfere with production of crystallization modulators. Crystal retention is accomplished by aggregation or by adherence to the cells and finally by attachment to the basement membrane. Crystals move from the luminal to the extraluminal location. In the process, crystal deposits present in the superficial collecting ducts of the renal papilla erode to the papillary surface where they are exposed to the slow moving pelvic and forniceal urine and form a nidus for stone formation.

X. ACKNOWLEDGMENTS

This work was supported in part by NIH grants # PO1 DK 20586 and RO1 DK 41434.

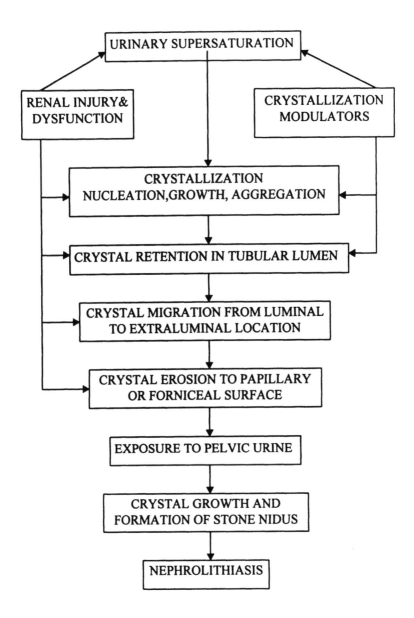

Figure. 2. Schematic diagram showing relationship between various factors involved in the formation of kidney stones.

REFERENCES

1. **Khan, S.R. and Hackett R.L.,** Calcium oxalate urolithiasis in the rat: Is it a model for human stone disease? A review of recent literature, *Scann. Electr. Microsc.*, II, 759, 1985.

2. **Finlayson, B.,** Physicochemical aspects of urolithiasis, *Kidney Intl.*, 13, 334, 1978.

3. **Khan, S.R.,** Pathogenesis of oxalate urolithiasis: Lessons from experimental studies with rats, *Am. J. Kid. Dis.*, 17, 398, 1991.

4. **Khan, S.R.,** Calcium oxalate crystal interaction with renal tubular epithelium, mechanism of crystal adhesion and its impact on stone development, *Urol. Res.*, in press, 1995.

5. **Khan, S.R., Finlayson, B. and Hackett, R.L.,** Histologic study of the early events in oxalate induced intranephronic calculosis, *Invest. Urol.*, 17,199, 1979.

6. **Khan, S.R., Shevock, P.N. and Hackett, R.L.,** Acute hyperoxaluria, renal injury and calcium oxalate urolithiasis, *J. Urol.*, 147, 226, 1992.

7. **Khan, S.R., Finlayson, B. and Hackett, R.L.,** Experimental calcium oxalate nephrolithiasis in the rat, role of renal papilla, *Am. J. Pathol.*, 107, 59, 1982.

8. **Khan, S.R., Shevock, P.N. and Hackett, R.L.,** Urinary enzymes and calcium oxalate urolithiasis, *J. Urol.*, 142, 846, 1989.

9. **Baggio, B., Gambaro, G., Ossi, E., Favaro, S. and Borsatti, A.,** Increased urinary excretion of renal enzymes in idiopathic calcium oxalate nephrolithiasis, *J. Urol.*, 129, 1161, 1983.

10. **Ravichandran, V. and Selvam, R.,** Increased lipid peroxidation in kidney of vitamin B-6 deficient rats, *Biochem. Intl.*, 21, 599, 1990.

11. **Rengaraju, M. and Selvam, R.,** Lipid changes in rat tissues in experimental urolithiasis, *Indian J. Exp. Biol.*, 27, 795, 1989.

12. **Hackett, R.L., Shevock, P.N. and Khan, S.R.,** Madine-Darby canine kidney cells are injured by exposure to oxalate and calcium oxalate crystals, *Urol. Res.*, 22, 197, 1994.

13. **Robertson, W.G. and Peacock, M.,** Pathogenesis of urolithiasis, in *Urolithiasis: etiology, diagnosis,* Schneider, H-J., (ed), Springer-Verlag, Berlin, 185-334, 1985.

14. **Otnes, B.,** Urinary stone analysis, methods, materials and value, *Scand. J. Urol. Nephrol.*, Suppl. 71, 1, 1983.

15. **Tiselius, T.G.,** Urinary excretion of citrate in normal subjects and in pateints with urolithiasis, in *Urolithiasis: Clinical and basic research,* Smith, L.H., Robertson W.G., Finlayson, B., (eds), Plenum Press, New York, 39-44, 1985.

16. **Robertson, W.G., Peacock, M., Heyburn, P.J., Marshall, D.H. and Clark, P.B.,** Risk factors in calcium stone disease of the urinary tract, *Br. J. Urol.*, 50, 449, 1978.

17. **Geary, C.P. and Cousins, F.B.,** An oestrogen-linked nephrocalcinosis in rats, *Br. J. Exp. Pathol.*, 50, 507, 1969.

18. **Nguyen, H.T. and Woodard, J.C.,** Intranephronic calculosis in rats, an ultrastructural study, *Am. J. Pathol.*, 100, 39, 1980.

19. **Lyon, E.S., Borden, T.A. and Vermeulen, C.W.,** Experimental oxalate lithiasis produced with EG, *Invest. Urol.*, 4, 143, 1966.

20. **Lee, Y.H., Huang, W.C., Chiang, H., Chen, M.T., Huang, J.K. and Chang, L.S.,** Determinant role of testerone in the pathogenesis of urolithiasis in rats, *J. Urol.*,147,1134, 1992.

21. **Khan, S.R. and Glenton, P.A.,** Deposition of calcium phosphate and calcium oxalate crystals in the kidneys, *J. Urol.*, In press, 1995.

22. **Finlayson, B. and Reid, F.,** The expectation of free and fixed particles in urinary stone disease, *Invest. Urol.*, 15, 442, 1978.

23. **Kok, D.J. and Khan, S.R.,** Calcium oxalate nephrolithiasis, a free or fixed particle disease, *Kidney Intl.*, 46, 847, 1994.

24. **Khan, S.R. and Hackett, R.L.,** Urolithogenesis of mixed foreign body stones, *J. Urol.*, 138, 1321, 1987.

25. **Khan, S.R., Shevock, P.N. and Hackett, R.L.,** Magnesium oxide administration and prevention of calcium oxalate nephrolithiasis, *J. Urol.*, 149, 412, 1992.

26. **Murphy, B.T. and Pyrah, L.N.,** The composition, structure and mechanisms of the formation of urinary calculi, *Br. J. Urol.*, 34, 129, 1962.

27. **Werness, P.G., Bergert, J.H. and Smith, L.H.,** Crystalluria, *J. Crystal Growth,* 30, 166, 1981.

28. **Meyer, J.L., Bergert, J.H. and Smith, L.H.,** Epitaxial relationships in urolithiasis; the calcium oxalate monohydrate-hydroxyapatite system, *Clin. Sci. Molec. Med.*, 52, 143, 1975.

29. **Khan, S.R., Shevock, P.N. and Hackett, R.L.,** Presence of lipids in urinary stones: results of preliminary studies, *Calcif. Tissue Intl.*,42, 91, 1988.

30. **Khan, S.R., Shevock, P.N. and Hackett, R.L.,** In vitro precipitation of calcium oxalate in the presence of whole matrix or lipid components of the urinary stones, *J. Urol.*, 139, 418, 1988.

31. **Khan, S.R., Whalen, P.O. and Glenton, P.A.,** Heterogeneous nucleation of calcium oxalate crystals in the presence of membrane vesicles, *J. Crystal Growth*, 134, 211, 1993.

32. **Khan, S.R.,** Structure and development of calcific urinary stones, in: *Calcification in Biological Systems*, Bonnucci, E., (ed), CRC Press, Boca Raton, Florida, 345-363, 1992.

33. **Hackett, R.L., Shevock, P.N. and Khan, S.R.,** Cell injury associated calcium oxalate crystalluria, *J. Urol.*, 144, 1535, 1990.

34. **Kumar, S., Sigmon, D., Miller, T., Carpenter, B., Khan, S., Malhotra, R., Scheid, C. and Menon, M.,** A new model of nephrolithiasis involving tubular dysfunction/injury. *J. Urol.,* 146, 1384, 1991.

35. **Antinozzi, P.A., Brown, C.M. and Purich, D.L.,** Calcium oxalate monohydrate crystallization: citrate inhibition of nucleation and growth steps, *J. Crystal Growth,* 125, 215, 1992.

36. **Shirane, Y. and Kagawa, S.,** Scanning electron microscopic study of the effect of citrate and pyrophosphate on calcium oxalate crystal morphology, *J. Urol.,* 150, 1980, 1993.

37. **Su, C-J., Shevock, P.N., Khan, S.R. and Hackett, R.L.,** Effect of magnesium on calcium oxalate urolithiasis, *J. Urol.,* 145, 1092, 1991.

38. **Ogawa, Y., Yamaguchi, K. and Morozumi, M.,** Effects of magnesium salts in preventing experimental oxalate urolithiasis in rats, *J. Urol.,* 144, 385, 1990.

39. **Gokhale, J.A., Glenton, P.A. and Khan, S.R.,** Localization of Tamm Horsfall protein and osteopontin in a rat nephrolithiasis model. J. Am. Soc. Nephrol., 5, 863, 1994.

40. **McKee, M.D., Nanci, A. and Khan, S.R.,** Ultrastructural immunodetection of osteopontin and osteocalcin as major matrix components of urinary calculi, *J. Bone Mineral Res.,* 9, S379, 1994.

41. **Coe, F.L., Nakagawa, Y. and Parks, J.H.,** Inhibitors within the nephron, *Am. J. Kid. Dis.,* 17, 407, 1991.

42. **Finlayson, B., Khan, S.R. and Hackett, R.L.,** Gentamicin accelerates calcium oxalate monohydrate (COM) nucleation, in: *Urolithiasis,* Walker, V.R., Sutton, R.A.L., Cameron, E.C.B., Pak, C.Y.C., (eds), Plenum Press, New York, 59-60, 1989.

43. **Riese, R.J., Riese, J.W., Kleinman, J.G., Wiessner, J.H., Mandel, G.S. and Mandel, N.S.,** Specificity in calcium oxalate adherence to papillary epithelial cells in culture, *Am. J. Physiol.,* 255, F1025, 1992.

44. **Hautmann, R. and Osswald, H.,** Concentration profiles of calcium and oxalate in urine, tubular fluid and renal tissue- some theoretical considerations, *J. Urol.,* 129, 433, 1983.

45. **Wright, R.J. and Hodgkinson, A.,** Oxalic acid, calcium, and phosphorus in the renal papilla of normal and stone forming rats, *Invest. Urol.,* 9, 369, 1972.

46. **Anderson, L. and McDonald, J.R.,** The origin, frequency, and significance of microscopic calculi in the kidney, *Surg. Gynecol. Obstet.,* 82: 275, 1946.

47. **Burry, A.F., Axelson, R.A., Trolove, P., and Saal, J.R.,** Calcification in the renal medulla, a classification based on a prospective study of 2261 necropsies, *Human Pathol.,* 7, 435, 1976.

48. **Vermeulen, C.W., Lyon, E.S. and, Borden, T.A.,** The renal papilla and the genesis of urinary calculi, *Trans. Am. Assoc. Gen. Urin. Surg.,* 58, 30, 1966.

49. **de Bruijn, W.C., Boeve, E.R., van Run, P.R.W.A., van Miert, P.P.M.C., Romijn, J.C., Verkoelen, C.F., Cao and L.C., Schroder, F.H.,** Etiology of experimental calcium oxalate monohydrate nephrolithiasis in rat kidneys, *Scanning Microsc.,* In press, 1995.

INDEX

A

Adsorption characteristics, inhibitors of crystallization, 25

Aerobic organisms, bacteria, oxalate-degrading, 143

Agaricus bisporus, fungi, calcium oxalate in, 73

Agaricus campestris, fungi, calcium oxalate in, 73

Agaricus carminescens, fungi, calcium oxalate in, 78

Age
  profile
    measurement, oxalate, 250
    plasma, measurement, oxalate, 252
  urolithiasis, 173

Agglomeration
  crystallization *in vitro*, 1
  inhibitors of crystallization, 27

Aggregate size, nephrolithiasis, animal model, 351

Aggregation of calcium oxalate, 37–52, 40, 45
  critical flocculation concentrations, 44
  floc formation, 44
  flocculation, 42, 46, 47
  heterocoagulation, 41
  hydroxyapatite, particle, 41
  insufficient polymer, 42
  kidney stone, 37, 38
  nephron, 37, 38, 39, 42
  opportunistic disease, 37
  particle aggregation in simple electrolytes, 40
  polyethyleneimine, 44
  Stern potential, 40
  stone formation, 37
  supersaturation, 37
  urine, 40

Alcalignes faecalis, bacteria, oxalate-degrading, 146

Allopurinol and calcium stones, urate excretion, 307

Ammonium chloride, nephrolithiasis, animal model, 344

Amniotic fluid, measurement, oxalate, 255

Anaerobic bacteria, bacteria, oxalate-degrading, 151

Anaerobic microbes, bacteria, oxalate-degrading, 140

Anaerobic organisms, bacteria, oxalate-degrading, 132

Angiosperms, plants, higher, oxalate formation, 53, 58

Animal models, bacteria, oxalate-degrading, 139

Anion
  exchange chromatography, measurement, oxalate, 246
  nephrolithiasis, cellular abnormalities, 209

Aphids, biosynthesis of calcium oxalate, 120

Ascorbic acid, plants, higher, oxalate formation, 54

Aspergillus, fungi, calcium oxalate in, 92

Atomic absorption spectrometry, measurement, oxalate, 241

B

Bacillus oxalophilus, bacteria, oxalate-degrading, 146

Bacteria, 131, 267
  infection, 274
  oxalate-degrading, 131–168, 133, 137, 138, 139, 140, 149, 150, 151
    aerobic organisms, 143
    Alcalignes faecalis, 146
    anaerobic bacteria, 151
    anaerobic microbes, 140
    anaerobic organisms, 132
    animal models, 139
    Bacillus oxalophilus, 146
    biosynthesis from oxalate, 153
    calcium oxalate, 132
      decarboxylation, 153
      degrading activities, 151
      degrading bacteria, 133, 137
      metabolism, 148
    catabolism, 148
    Clostridium, 135
    colon, 138
    colonic bacterial populations, 135
    colonization incidence in humans, 137
    Crohn's disease, 138

dietary oxalate, 156
earthworm intestine, 147
enteric hyperoxaluria, 138, 156
Eubacterium lentum, 137
gastrointestinal disorders, 138
gastrointestinal habitats, 147
gastrointestinal microbes, 132
gastrointestinal populations, 156
high oxalate diets, 132
hind gut, 133, 156
horses, 133
Hysterangium crassum, 144
inflammatory bowel disease, 138
intestinal microbes, 132
intestinal oxalate-degrading bacteria,
    155
isolations, 137, 146
    intestinal anaerobes, 133
jejunoileal bypass surgery, 138
laboratory rats, 139
large bowel, 156
microbes, in human feces, 135
Oxalobacter formigenes, 135
Oxalobacter vibrioformis, 141
Oxalophagus oxalicus, 141
phosphorous cycling, 144
Pseudomonas, 149
rhizomorphs, 144
rhizosphere soils, 143
rumen, 132, 148, 155
sediments, 140
significance of colonization by *O.*
    formigenes, 138
small intestine, 156
steatorrhea, 138
Streptomyces, 146
swine, 133
terrestrial ecosystems, 143
thermophilic habitats, 141
Thiobacillus novellus, 147
urolithiasis, 156
Xanthobacter, 147
Bacterial spores, fungi, calcium oxalate in,
    92
Basidiomycete species, fungi, calcium
    oxalate in, 78
Begonia, plants, higher, oxalate formation,
    59
Biochemical extraction of lipids, lipid
    matrix, 295
Biochemical prenatal diagnosis,
    hyperoxaluria type I, 198
Biomineralization, 53

plants, higher, oxalate formation,
    67
Bio-regulation of crystallization, inhibitors
    of crystallization, 31
Biosynthesis of calcium oxalate, 113–130,
    115
aphids, 120
brown-rot fungus, 116
calcium, 118
calcium oxalate, in fungi, 115
and calcium regulation, 118
cellulase, 122
chloride, 118
and defense mechanisms, 120
degrading enzymes in plants, 122
D-erythroascorbic acid, 116
formigenes, 120
functions of, 117, 121
hemicellulases, 122
in higher plants, 114
ion balance in plants, 117
L-ascorbic acid, 114, 115
Lathyrus sativus, 123
leafhoppers, 120
Lemna minor, 114
magnesium, 113, 118
mineral elements, 118
Pelargonium crispum, 115
photosynthesis, 119
phytopathogenic fungi, 121
polygalacturonase, 122
potassium, 113, 118
prickly-pear cactus, 121
red beets, 114
Sclerotium rolfsii, 116
sodium, 113, 118
spinach, 114
wood-rot, 122
xylem, 118, 119
Bladder calculi, urolithiasis, 169
Bladder stone
in children, urolithiasis, 169
urolithiasis, 169
Body weight, urate excretion, 306
Bone
lipid matrix, 291
specimens, measurement, oxalate,
    255
Brown-rot fungus, biosynthesis of calcium
    oxalate, 116
Brush border, lipid matrix, 299
Bulbillomyces farinosus, fungi, calcium
    oxalate in, 76

C

Calcific deposits
  crystal, cell, interaction of, 324
  nephrolithiasis, animal model, 353
Calcification
  nephrolithiasis, animal model, 348
  sites, lipid matrix, 291
Calcium
  biosynthesis of calcium oxalate, 118
  crystal, cell, interaction of, 324
  deposits, crystal, cell, interaction of, 324
  epithelial transport, 223
  nephrolithiasis, animal model, 349
  regulation, biosynthesis of calcium
      oxalate, 118
  restriction, epithelial transport, 217
  stone
    formation, urate excretion, 307
    in gouty patients, urate excretion, 305
  urate excretion, 308
  urolithiasis, 178
    nephrolithiasis, cellular abnormalities,
      207
Calcium oxalate, 297
  absorption and secretion, 217
  bacteria, oxalate-degrading, 132
    degrading activities, 151
    degrading bacteria, 133
  biosynthesis of calcium oxalate, in fungi,
      115
  crystal, cell, interaction of, flux, 329
  crystallization, crystallization in vitro, 1
  dihydrate
    crystallization in vitro, 8
    fungi, calcium oxalate in, 82
  epithelial transport, 229
    bioavailability, 224
    flux, 219
    and oxalate homeostasis, 222
    transport, 227
    in urine, 217, 224
  excretion, 218
  measurement, oxalate, 239, 243
    concentration, 245
    decarboxylase, 247
    determination in urine and blood, 241
    excretion, 252
    oxidase, 246
    related hyperacidurias, 254
    retention, 250
  monohydrate
    crystallization in vitro, 8

  fungi, calcium oxalate in, 82
  nephrolithiasis, nephrolithiasis, animal
      model, 348
  transport, epithelial, uptake, 219
  trihydrate, crystallization in vitro, 12
Calcium phosphate
  -inducing diet, nephrolithiasis, animal
      model, 349
  nephrolithiasis, nephrolithiasis, animal
      model, 348
  as nucleator, nephrolithiasis, animal
      model, 349
Callose, plants, higher, oxalate formation, 57
Canavalia leaves, plants, higher, oxalate
      formation, 59
Caseinkinase, nephrolithiasis, cellular
      abnormalities, 211
Cation transport, nephrolithiasis, cellular
      abnormalities, 209
Cell
  degradation products, lipid matrix, 292
  factors in retention, crystal, cell,
      interaction of, 326
  membranes as nucleator, nephrolithiasis,
      animal model, 350
  walls, plants, higher, oxalate formation,
      56
Cellulase, biosynthesis of calcium oxalate,
      122
Cellulose, plants, higher, oxalate formation,
      57
Charcoal
  adsorption, measurement, oxalate, 242
  treatment, measurement, oxalate, 247
Chemical techniques, measurement, oxalate,
      241
Chloride, biosynthesis of calcium oxalate,
      118
Chromatography
  assay, measurement, oxalate, 243
  measurement, oxalate, 243
  techniques, measurement, oxalate, 248
Chronic renal failure
  epithelial transport, 224
  hyperoxaluria type I, 189
Citrate
  inhibitors of crystallization, 32
  nephrolithiasis, animal model, 350
Clostridium, bacteria, oxalate-degrading, 135
Colon
  bacteria, oxalate-degrading, 138
  bacterial populations, bacteria, oxalate-
      degrading, 135

epithelial transport, 221, 222
Colonization incidence in humans, bacteria, oxalate-degrading, 137
Colorimetry-spectrophotometry, measurement, oxalate, 241
Concentration in red blood cells, measurement, oxalate, 240
Constant composition crystallization, crystallization *in vitro,* 13
Critical flocculation concentrations, aggregation of calcium oxalate, 44
Crohn's disease, bacteria, oxalate-degrading, 138
Crystal. See also Stone
　agglomeration, inhibitors of crystallization, 24, 25, 27, 32, 33
　aggregates, 344
　　crystal, cell, interaction of, 327
　　nephrolithiasis, animal model, 344
　cell, interaction of, 235, 323–342, 330, 343
　　aggregates, 327
　　attachment, 326
　　calcific deposits, 324
　　calcium, 324, 329
　　calculous attachment, 324
　　cellular factors in retention, 326
　　epithelial ovine kidney, 329
　　fate of crystal-cell attachment, 327
　　glycosaminoglycans, 329
　　human urinary stones, 343
　　hyperoxaluric rats, 326
　　inner medullary late collecting duct (IMCD), 326
　　kidney, 323, 324
　　lipid peroxidation, 330
　　loops of Henle, 324
　　Madin-Darby Canine kidney, 326
　　modulation of, 328
　　papillary calcium plaques, 324
　　papillary injury, 324
　　papillary stones, 235
　　porcine kidney, 326
　　renal epithelial cells, 326
　　renal tubular damage, 324
　　role of oxalate in crystal-cell attachment, 329
　　scanning electron microscopic examinations, 325
　stone
　　disease, 236, 323
　　formation, 329
　　retention, 325

supersaturation, 330
Tamm-Horsfall glycoprotein, 329
uptake of calcium, 330
urolithiasis, 325, 327, 330, 343
deposit, 348, 354
　crystal, cell, interaction of, in kidney, 343
　nephrolithiasis, animal model, 348, 354
distribution, fungi, calcium oxalate in, 100
fixation, 352
　nephrolithiasis, animal model, 352
formation
　in interstitium, 352
　nephrolithiasis, animal model, in interstitium, 352
　plants, higher, oxalate formation, 64
ghosts, lipid matrix, 293
growth, inhibitors of crystallization, 24, 25, 27
idioblasts, plants, higher, oxalate formation, 60, 65, 68
lipid matrix, in urine, 293
nucleation, 349
　nephrolithiasis, animal model, 349
plastids, plants, higher, oxalate formation, 60
retentin, 351
retention, nephrolithiasis, animal model, 353
sand, plants, higher, oxalate formation, 56
urate excretion, in urine, 309
Crystal-face, inhibitors of crystallization, 26
Crystallization, modulators, 350
　nephrolithiasis, animal model, 350
Crystallization *in vitro,* 1–21, 2, 3, 4, 12, 13, 14
calcium oxalate
　crystallization, 1, 3
　dihydrate, 8
　monohydrate, 8
　trihydrate, 12
constant composition crystallization, 13
desaturation, 9
effectors of, 23
gel crystallization model, 13
metastable limit, 8, 9, 10
mixed suspension mixed product removal, 14
nephrolithiasis, animal model, 1–21
particle size distribution, 5
regulation of crystallization, 23
seeded systems, 5

supersaturation, 2, 3, 4, 11, 12
unseeded batch crystallizers, 8
urolithiasis, 1
Crystal-phase-specificity, inhibitors of
crystallization, 26
Cunninghamella echinulata, fungi, calcium
oxalate in, 104
Cytoplasmic edema, nephrolithiasis, animal
model, 347

D

DeBary, Anton, fungi, calcium oxalate in,
73
Defense mechanisms, biosynthesis of
calcium oxalate, 120
Degrading enzymes in plants, biosynthesis
of calcium oxalate, 122
Dental plaque, lipid matrix, 291
Dentin, lipid matrix, 291
D-erythroascorbic acid, biosynthesis of
calcium oxalate, 116
Desaturation, crystallization *in vitro,* 9
Diabetes, urate excretion, 313
Diagnosis, hyperoxaluria type I, 197
Diet
hyperoxaluia, measurement, oxalate, 249
hyperoxaluria, epithelial transport, 224
oxalate, bacteria, oxalate-degrading, 156
urate excretion, 306
urolithiasis, 176
Drinking water, 180
Druse, plants, higher, oxalate formation, 56
Druse-like crystals, fungi, calcium oxalate
in, 89, 90, 96

E

Earthworm intestine, bacteria, oxalate-
degrading, 147
Effector
definition, inhibitors of crystallization, 23
inhibitors of crystallization, 24
Enamel, lipid matrix, 291
Endocrystallosis, urate excretion, 310
Endoperidium, fungi, calcium oxalate in,
100, 102, 103
Endoplasmic reticulum, nephrolithiasis,
animal model, dilatation of, 347
End-stage renal failure, measurement,
oxalate, 250
Enteric hyperoxaluria, bacteria, oxalate-
degrading, 138

Environmental factors, urolithiasis, 176
Enzymatic procedures, measurement,
oxalate, 242
Enzymatic techniques, measurement,
oxalate, 246
Enzyme replacement therapy, hyperoxaluria
type I, 200
Enzymic phenotypes, hyperoxaluria type I,
191
Enzymuria, nephrolithiasis, animal model,
347
Epinephrine, epithelial transport, 221
Epitaxy, urate excretion, 308
Epithelial basement membrane,
nephrolithiasis, animal model,
352
Epithelial brush border membrane,
nephrolithiasis, animal model,
347
Epithelial cell surface, nephrolithiasis,
animal model, 352
Epithelial ovine kidney, crystal, cell,
interaction of, 329
Epithelial transport
calcium, 224
calcium oxalate
absorption, 218
flux, 218, 221
transport, 222, 228
enteric hyperoxaluria, 223
large intestine, 221
luminal membrane, 229
rabbit, ileal brush border, 219
Epstein Barr virus, hyperoxaluria type I,
193
Estrogen, nephrolithiasis, animal model,
348
Ethylene glycol, nephrolithiasis, animal
model, 344
Eubacterium lentum, bacteria, oxalate-
degrading, 137
Excretion
with age, measurement, oxalate, 252
of calcium, nephrolithiasis, animal model,
344
Extracellular fluids, measurement, oxalate,
240

F

Face- and phase-specificity, inhibitors of
crystallization, 29
Familial predisposition, urolithiasis, 175

Family history
  nephrolithiasis, cellular abnormalities, 208
  urolithiasis, 175
Floc formation, aggregation of calcium
    oxalate, 44
Flocculation, aggregation of calcium oxalate,
    42, 46
Fluids, measurement, oxalate, and tissues,
    255
Formigenes, biosynthesis of calcium oxalate,
    120
Function
  biosynthesis of calcium oxalate, 117
  oxalic acid in fungi, biosynthesis of
    calcium oxalate, 121
Fungi, 53, 131
  calcium oxalate in, 73–111, 78, 80, 82,
      84, 86, 90, 92, 96, 107
  Agaricus bisporus, 73
  Agaricus campestris, 73
  Agaricus carminescens, 78
  Anton deBary, 73
  Aspergillus, 92
  bacterial spores, 92
  Bulbillomyces farinosus, 76
  crystal distribution, 100
  crystallinity, 82
  Cunninghamella echinulata, 104
  druse-like crystals, 89, 90, 96
  endoperidium, 100, 102, 103
  Geastrum, 74, 89, 96, 98
  Gilbertella persicaria, 106
  hyphal cell wall, 88
  Hysterangium crassa, 76
  lichens, 73, 106
  litter, 84, 88
  microorganisms, 73
  morphology, 100, 107
  Mucor hiemalis, 106
  mycorrhizal fungus, 76
  needle-like crystal, 73, 86, 88
  needle-shaped crystals, 107
  Neurospora, 92
  Penicillium, 92
  Phallus caninus, 74
  Pinus ponderosa, 90
  plant diseases, 80
  Psalliota campestris, 73
  pyramidal, 82
  raphide-like crystal, 96
  Rhizopus oryzae, 106
  rhombohedrals, 82
  rodlet, 92, 96

  on spores, 92
  rod-shaped structures, 92
  Schizophyllum commune, 92
  slime molds, 73
  sporangia, 106
  systematics, 74
  Tilletia indica, 92
  Trichophyton, 92
  Tubulicium clematidis, 76
  twins, 88, 106
  measurement, oxalate, 242

G

Gas chromatography, measurement, oxalate,
    243
Gastrointestinal disease
  bacteria, oxalate-degrading, 138
  epithelial transport, 223
Gastrointestinal habitats, bacteria, oxalate-
    degrading, 147
Gastrointestinal microbes, bacteria, oxalate-
    degrading, 132
Gastrointestinal populations, bacteria,
    oxalate-degrading, 156
Geastrum, fungi, calcium oxalate in, 74, 89,
    96, 98
Gel crystallization model, crystallization *in
    vitro*, 13
Gene therapy, hyperoxaluria type I, 201
Genetically determined, nephrolithiasis,
    cellular abnormalities, 208
Genotypes, hyperoxaluria type I, 193
Geriatric populations, urolithiasis, 174
Ginkgo biloba, plants, higher, oxalate
    formation, 58
Glycosaminoglycan
  crystal, cell, interaction of, 329
  nephrolithiasis
    animal model, 350
    cellular abnormalities, 207
  urate excretion, 311
Gout, urate excretion, 305
Gymnosperms, plants, higher, oxalate
    formation, 53, 56

H

Heparin, inhibitors of crystallization, 32
Hereditary syndromes, urolithiasis, 176
Heredity, urolithiasis, 175
Heterocoagulation, aggregation of calcium
    oxalate, 41

Heterogeneous nucleation, nephrolithiasis,
    animal model, 349
High grade chronic hyperoxaluria,
    nephrolithiasis, animal model,
    344
High oxalate diets, bacteria, oxalate-
    degrading, 132
Hind gut, bacteria, oxalate-degrading, 133,
    156
Horses, bacteria, oxalate-degrading, 133
Hydrogen peroxide, measurement, oxalate,
    246
Hydroxyapatite, aggregation of calcium
    oxalate, particle, 41
Hyperabsorption of calcium, 217
    epithelial transport, 217
Hyperglycolic aciduria, hyperoxaluria type I,
    189
Hyperoxaluria, 179, 254, 299, 343, 344, 353
    crystal, cell, interaction of, 343
    lipid matrix, 299
    nephrolithiasis, animal model, 343, 344,
        352, 353
    type I, 189–205, 193, 197, 200
        biochemical prenatal diagnosis,
            198
        chronic renal failure, 189
        diagnosis, 197
        enzyme
            diagnosis, 197
            phenotypes, 191
            prenatal diagnosis, 198
        enzyme replacement therapy, 200
        Epstein Barr virus, 193
        gene therapy, 201, 202
        genotypes, 193, 196
        human kidney stones, 207
        hyperglycolic aciduria, 189, 197
        hyperoxaluria, 189, 197
        kidney, transplantations, 200
        liver
            parenchymal cells, 192
            transplantation, 200
        mitochondria, 192
        molecular genetic diagnosis, 198
        mutation, 195, 196
            polymorphisms, in AGT Gene,
                195
        prenatal diagnosis, 198
        restriction-fragment-length
            polymorphism (RFLP), 193
        symptomatic treatment, 200
        transplantations, kidney, 200

Hyperuricaemia, 306
    urate excretion, 306
Hyperuricosuria, 305, 306
    and calcium stones, 306
    urate excretion, 305
Hyphal cell wall, fungi, calcium oxalate in,
    88
Hysterangium crassum, bacteria, oxalate-
    degrading, 144

I

Ileal resection, epithelial transport, 223
Immunohistological techniques,
    nephrolithiasis, animal model,
    351
Impaired renal function, measurement,
    oxalate, 253
Indirect radionuclide-dilution procedures,
    measurement, oxalate, 244
Infants, measurement, oxalate, 252
Inflammatory bowel disease, bacteria,
    oxalate-degrading, 138
Inhibitors of crystallization, 23–36, 24, 25,
    26, 29
    adsorption characteristics, 25
    agglomeration, 27
    bio-regulation of crystallization, 31
    citrate, 32, 33
    crystal
        agglomeration, 24, 25, 27, 32, 33
        growth, 27
        -phase-specificity, 26
    crystal-face, 26
    crystallization process, 33
    effector, 24
    definition, 23
    face- and phase-specificity, 29
    heparin, 32
    nucleation, 24
    phase transition, 24
    supersaturation, 24
    urine, 31, 32
Inner medullary late collecting duct (IMCD)
    crystal, cell, interaction of, 326
    lipid matrix, 300
Insufficient polymer, aggregation of calcium
    oxalate, 42
Interchangeable, urate excretion, 305
Intestinal microbes, bacteria, oxalate-
    degrading, 132
Intestinal oxalate-degrading bacteria,
    bacteria, oxalate-degrading, 155

Intestine
  excretion in chronic renal failure,
      epithelial transport, 224
  handling of oxalate, 217
Intracellular fluids, measurement, oxalate,
      240
Ion balance in plants, biosynthesis of
      calcium oxalate, 117
Ion transport, nephrolithiasis, cellular
      abnormalities, 209
Isolations
  bacteria, oxalate-degrading, 137
  intestinal anaerobes, bacteria, oxalate-
      degrading, 133

J

Jejuno-ileal bypass, epithelial transport, 223
Jejunoileal bypass surgery, bacteria, oxalate-
      degrading, 138

K

Kidney
  crystal, cell, interaction of, 323
  lipid matrix, 297
  measurement, oxalate, 240, 253
  nephrolithiasis, animal model, 345, 354
Kidney stone
  aggregation of calcium oxalate, 37, 38
  lipid matrix, 300
  measurement, oxalate, 248
  nephrolithiasis, animal model, 348, 351,
      353
  urate excretion, 313
Kinase, nephrolithiasis, cellular
      abnormalities, 210, 211

L

Laboratory rats, bacteria, oxalate-degrading,
      139
Large bowel, bacteria, oxalate-degrading,
      156
Large intestine, epithelial transport, 221, 222
L-ascorbic acid, biosynthesis of calcium
      oxalate, 114
Lathyrus sativus, biosynthesis of calcium
      oxalate, 123
Leafhoppers, biosynthesis of calcium
      oxalate, 120
Lemna minor, biosynthesis of calcium
      oxalate, 114

Lichens, 53
  fungi, calcium oxalate in, 73, 106
Lipid
  peroxidation, crystal, cell, interaction of,
      330
  profile, lipid matrix, 292
Lipid matrix, 291–304, 293
  animal model, 300
  biochemical extraction of lipids, 295
  bone, 291
  brush border, 299
  calcification sites, 291
  calcium oxalate, 300
      crystallization *in vitro*, 297
      crystallization *in vivo*, 299
  calculus formation, 291
  cellular degradation products, 292, 299
  crystal
      ghosts, 293
      matrix, 293
      in urine, 293
  dental plaque, 291
  dentin, 291
  enamel, 291
  hyperoxaluria, 299
  inner medullary late collecting duct
      (IMCD), 300
  kidney, 297, 300
      stone, 300
  lipid profile, 292
  Mandel's cell culture studies, 300
  membranes and lipids in stone matrix,
      293
  molecular interaction between membrane
      lipids
  phosphatidylserine, 300
  transmission electron microscopy, 295
  urinary lipids, 292
  urinary stone matrix, 297
  urolithiasis, 291
  uterine fluid, 291
Liquid chromatography, measurement,
      oxalate, 243
Lithogen
  crystal, cell, interaction of, 343
  nephrolithiasis, animal model, 343
Litter
  crystals, fungi, calcium oxalate in, 84
  fungi, fungi, calcium oxalate in, 88
  samples, fungi, calcium oxalate in, 88
Liver
  parenchymal cells, hyperoxaluria type I,
      192

transplantation, hyperoxaluria type I, 200
Loops of Henle
   crystal, cell, interaction of, 324
   nephrolithiasis, animal model, 351
Lumen
   epithelial transport, 218
   membrane, epithelial transport, 227
   nephrolithiasis, animal model, 347
   oxalate exchange, epithelial transport, 228
   plasma membrane, nephrolithiasis, animal
         model, 352

M

Macromolecules, 265–290, 266, 267, 268,
         269, 272, 273, 276, 277, 278,
         279, 281, 283, 294
   abundance in urine, 275
   amino acid sequence analysis, 279
   aspartic acids, 281
   bacteria, 267
      infection, 274
   biomineralised tissues, 265
   blood coagulation, 280
   bone, 265
      mineralization, 278
   breast, 278
   cartilage, 265
   chemical composition of matrix, 270
   chondroitin sulphate, 269
   components of matrix, 268
   crystal matrix, 271, 273
      protein, 280, 293
   crystal nucleation, 268
   crystalline particles, 266, 268
   dentin, 265
   D-glucuronic, 281
   erythrocyte, 266
   fluid, drinking greater quantities of, 268
   gallbladder, 278
   gastrointestinal tract, 278
   glutamic, 281
   glycine, 281
   glycosaminoglycans, 270, 271
   high-pressure liquid chromatography
         (HPLC), 270
   human prothrombin, 280
   human serum albumin, 269
   human urinary protein, 281
   inorganic ash, 270
   insolubility, 266
   kidney, 278, 283
      stone, 265

large serum proteins, 267
lipids, 267, 270
loop of Henle, 279, 280
lung, 278
macromolecular products, 267
malbumin, 269
matrix, 266
   components, 268, 2670
mitochondrial ghosts, 267
molecular biology, 270
mucoprotein, 272
myxomavirus, 274
nephrocalcin, 273, 276
nephrocalcinosis, 275
osteocalcin, 276
osteopontin, 274, 278
Ouchterlony method, 282
pancreas, 278
physical structure of matrix, 266
physicochemistry, 270, 273
protein, 272
   implicated in stone formation,
         274
   inhibitor, 277
prothrombin, 279
red blood cell, 266
renal cell carcinoma, 276
reproductive tracts, 278
salivary glands and sweat glands, 278
shell, 265
stone, 268, 273, 276
   formation, 269, 271, 279
   matrix, 270, 271, 274, 275, 282,
         283
   pathogenesis, 270, 283
   ultrastructure, 266
supersaturation, 269
Tamm-Horsfall glycoprotein, 272, 274
uric acid, 270
urinary excretion, 275
urinary macromolecule, 274, 275, 276
urinary proteins, 279
urinary saturation, 269
urinary tract, 274, 278
urine, 266, 268, 270, 271, 272, 276, 282
urolithiasis, 274, 280
uroliths, 265
uromucoid, 272
uropontin, 278
valine, 281
Western blotting, 279
Madin-Darby Canine kidney, crystal, cell,
         interaction of, 326

Magnesium, 179
  biosynthesis of calcium oxalate, 113
  nephrolithiasis, animal model, 350
Malabsorption, epithelial transport, of bile
      salts, and fatty acids, 223
Mammalian urine, nephrolithiasis, animal
      model, 350
Mandel's cell culture studies, lipid matrix,
      300
Measurement, oxalate, 242, 246, 248, 249,
      252
  abnormal oxalate excretion, 249
  amniotic fluid, 255
  anion exchange chromatography,
      246
  atomic absorption spectrometry, 241
  in biological fluids, 239–263
  bone specimens, 255
  calcium oxalate, 239, 243
    in blood, 244
    in body fluids, 240
    in clinical practice, 248
    concentration, 245
    decarboxylase, 242, 247
    excretion, 252
    oxidase, 242, 246
    in plasma and urine, 239
    related hyperacidurias, 254
    retention, 250
    in urine, 241
  charcoal
    adsorption, 242
    treatment, 247
  chemical techniques, 241
  chromatography, 243, 248
  colorimetry-spectrophotometry, 241
  concentration in red blood cells, 240
  dietary hyperoxaluia, 249
  end-stage renal failure, 250
  excretion with age, 252
  extracellular fluids, 240, 248
  fluids, and tissues, 255
  fungi, 242
  gas chromatography, 243
  gas-liquid chromatography (GLC), 243
  hydrogen peroxide, 246
  hyperoxaluia, 249, 250
  impaired renal function, 253
  indirect radionuclide-dilution procedures,
      244
  infants and children, 252
  intracellular fluids, 240
  kidney, 240, 248, 253

  stone, 248
  liquid chromatography, 243
  plasma
    oxalate, 252
    samples, 246
  renal failure, 253
  sample collection, 245
  seasonal variations, 252
  separated plasma, 245
  sex profiles, 250, 252
  tissue accumulation, oxalate, 240
  ultrafiltration, 245, 246, 247
  urinary oxalate assay, 243
  urine, 248, 251
    oxalate
      in adults, 250
      in infants and children, 251
  vitamin D, 253
Membrane
  in stone matrix, lipid matrix, 293
  vesicle studies, epithelial transport,
      227
Metastable limit, crystallization *in vitro*,
      8, 9, 10
Microbes, 131
  bacteria, oxalate-degrading, in human
      feces, 135
Microorganisms, 53
  fungi, calcium oxalate in, 73
  plants, higher, oxalate formation, 67
Microscopic examination, urate excretion,
      310
Mineral elements, biosynthesis of calcium
      oxalate, 118
Ministones, nephrolithiasis, animal model,
      344
Mitochondria
  hyperoxaluria type I, 192
  plants, higher, oxalate formation, 54
Modulation, crystal, cell, interaction of,
      328
Molecular genetic diagnosis, hyperoxaluria
      type I, 198
Monocotyledonous plants, plants, higher,
      oxalate formation, 62
Monosodium urate, urate excretion, 308
Morphology, fungi, calcium oxalate in, 100,
      107
Mucor hiemalis, fungi, calcium oxalate in,
      106
Mutation, hyperoxaluria type I, 195, 196
Mycorrhizal fungus, fungi, calcium oxalate
      in, 76

N

Needle-like crystal, fungi, calcium oxalate
    in, 73, 88
Nephrocalcin, 273, 276, 350
Nephrocalcinosis, 275
Nephrolithiasis, 351, 352
    animal model, 313, 343–359, 347, 348,
        349, 350, 351, 352, 353, 354
        aggregate size, 351
        ammonium chloride, 344
        calcium, 349
            oxalate nephrolithiasis, 348
            phosphate as nucleator, 349
            phosphate nephrolithiasis, 348
        cellular membranes as nucleator,
            350
        chronic hyperoxaluria, 344, 347, 352
        citrate, 350
        crystal
            aggregate, 344, 351, 352
            deposition, 348 in kidney, 343
            fixation, 352
            formation, in interstitium, 352
            nucleation, 349
            retentin, 351
            retention, 353
        crystallization
            modulators, 350
            in vitro, 1–21
        cytoplasmic edema, 347
        endoplasmic reticulum, dilatation of,
            347
        enzymuria, 347
        epithelial basement membrane, 352
        epithelial brush border membrane,
            347
        epithelial cell surface, 352
        estrogen, 348
        ethylene glycol, 344
        excretion of calcium, 344
        glycosaminoglycans, 350
        heterogeneous nucleation, 349
        high grade chronic hyperoxaluria,
            344
        hyperoxaluria, 343, 344, 345, 353
        immunohistological techniques, 351
        kidney, 345, 349, 354
            stone, 348, 351, 353
        lithogen, 343
        loops of Henle, 351, 352
        low grade chronic hyperoxaluria, 347
        lumen, 347

        luminal plasma membrane, 352
        magnesium, 350
        mammalian urine, 350
        ministones, 344
        morphological changes, 345
        nephrocalcin (NC), 350
        osteopontin (OP), 350
        ovariectomy, 348
        phosphocitrate, 350
        proximal tubular brush border, 347
        pyrophosphate, 350
        renal epithelial injury, 345
        renal papillary interstitium, 352
        renal tubular epithelium, 345
        sex hormone and nephrolithiasis,
            348
        Sprague-Dawley rats, 344
        swelling of mitochondria, 347
        tissue culture studies have, 353
        urinary acidification, 348
        urinary excretion, 344, 347
        urinary macromolecules, 350
        urinary stones, development of, 353
        urine, 347, 354
        urolithiasis, 343, 348
        uronic acid rich protein (UAP), 350
        vacuolation, 347
    cellular abnormalities, 207–216, 210
        anion, 209
        calcium urolithiasis, 207
        caseinkinase, 211
        cation transport, 209
        family history, 208
        genetically determined, 208
        glycosaminoglycans, 207
        idiopathic calcium nephrolithiasis
            (ICN), 207
        ion transport, 209
        kinase, 210
        kinase activities, 211
        proteinkinases, 211
        red blood cell, 208
        renal enzymes, 207
        renal stone disease, 208, 209
        tyrosinekinase, 211
Nephron, aggregation of calcium oxalate,
    37, 39
Neurospora, fungi, calcium oxalate in, 92
Nucleation
    inhibitors of crystallization, 24
    potential of cellular membranes, 297
Nymphaea, plants, higher, oxalate formation,
    67, 69

O

Oath of Hippocrites, urolithiasis, 169
Opportunistic disease, aggregation of
        calcium oxalate, 37
Osteopontin (OP), nephrolithiasis, animal
        model, 350
Ovariectomy, nephrolithiasis, animal model,
        348
Oxalobacter
    bacteria, oxalate-degrading, 135
    oxalate-degrading, vibrioformis,
        141

P

Papillary calcium plaques, crystal, cell,
        interaction of, 324
Particle aggregation in simple electrolytes,
        aggregation of calcium oxalate,
        40
Particle size distribution, crystallization *in
        vitro*, 5
Pelargonium crispum, biosynthesis of
        calcium oxalate, 115
Penicillium, fungi, calcium oxalate in, 92
Phallus caninus, fungi, calcium oxalate in,
        74
Phase transition, inhibitors of crystallization,
        24
Phosphatidylserine, lipid matrix, 300
Phosphocitrate, nephrolithiasis, animal
        model, 350
Phosphorous cycling, bacteria, oxalate-
        degrading, 144
Photosynthesis, biosynthesis of calcium
        oxalate, 119
Phytopathogenic fungi, biosynthesis of
        calcium oxalate, 121
Pilots, urolithiasis, 180
Pinus ponderosa, fungi, calcium oxalate in,
        90
Plants, higher, oxalate formation, 53–72, 54,
        55, 56, 60, 62, 66
    angiosperms, 53, 54, 58
    Anton von Leewenhoek, 54
    ascorbic acid, 54
    Begonia, 59
    biomineralization, 67
    callose, 57
    Canavalia leaves, 59
    cell walls, 56
    cellulose, 57

crystal
    formation, 64
    idioblasts, 60, 62, 65, 67, 68
    plastids, 60
    sand, 56
druses, 56
endoplasmic, 54
Ginkgo biloba, 58
gymnosperms, 53, 54, 56
microorganisms, 67
mitochondria, 54
monocotyledonous plants, 62
Nymphaea, 67, 69
plant cell cytoplasm, 55
prisms, 56
psychotria, 61
reticulum, 54
stylods, 56
sugar beet, 61
sweet pepper, 61, 63, 67
vacuoles, 54, 60
Plasma oxalate, measurement, oxalate, 252
Plasma samples, measurement, oxalate,
        246
Polyethyleneimine, aggregation of calcium
        oxalate, 44
Polygalacturonase, biosynthesis of calcium
        oxalate, 122
Population surveys, urolithiasis, 171
Porcine kidney, crystal, cell, interaction of,
        326
Potassium
    biosynthesis of calcium oxalate, 113
    intake, 179
Premature infants, urolithiasis, 174
Prenatal diagnosis, hyperoxaluria type I, 198
Prickly-pear cactus, biosynthesis of calcium
        oxalate, 121
Prisms, plants, higher, oxalate formation, 56
Proteinkinases, nephrolithiasis, cellular
        abnormalities, 211
Proximal tubular brush border,
        nephrolithiasis, animal model,
        347
Psalliota campestris, fungi, calcium oxalate
        in, 73
Pseudomonas, bacteria, oxalate-degrading,
        149
Psychotria, plants, higher, oxalate formation,
        61
Pyramidal, fungi, calcium oxalate in, 82
Pyrophosphate, nephrolithiasis, animal
        model, 350

R

Rabbit, epithelial transport
distal colon, 222
renal brush border membranes, 228
Raphide-like crystal surfaces, fungi, calcium
oxalate in, 96
Red beets, biosynthesis of calcium oxalate,
114
Red blood cell, nephrolithiasis, cellular
abnormalities, 208
Relationship between affluence, urolithiasis,
176
Renal enzymes, nephrolithiasis, cellular
abnormalities, 207
Renal epithelial injury, nephrolithiasis,
animal model, 345
Renal failure, measurement, oxalate,
253
Renal handling of oxalate, epithelial
transport, 225
Renal papillary interstitium, nephrolithiasis,
animal model, 352
Renal stone disease, nephrolithiasis, cellular
abnormalities, 208
Renal tubular damage, crystal, cell,
interaction of, 324
Restriction-fragment-length polymorphism
(RFLP), hyperoxaluria type I,
193
Reticulum, plants, higher, oxalate formation,
54
Rhizomorphs
bacteria, oxalate-degrading, 144
fungi, calcium oxalate in, 78, 84, 96
Rhizosphere soils, bacteria, oxalate-
degrading, 143
Rhombohedrals, fungi, calcium oxalate in,
82
Rodlet pattern, fungi, calcium oxalate in,
96
Rod-shaped structures, fungi, calcium
oxalate in, 92
Rumen, bacteria, oxalate-degrading, 132,
155

S

Salt, urate excretion, 308
Salting-out, 315
urate excretion, 313
Sample collection, measurement, oxalate,
245

Scanning electron
microscopic examinations, crystal, cell,
interaction of, 325
urate excretion, 310
Schizophyllum commune, fungi, calcium
oxalate in, 92
Sclerotium rolfsii, biosynthesis of calcium
oxalate, 116
Seasonal variations, measurement, oxalate,
252
Sedentary occupations, urolithiasis, 180
Sediments, bacteria, oxalate-degrading,
140
Seeded systems, crystallization *in vitro,* 5
Separated plasma, measurement, oxalate,
245
Sex profile, measurement, oxalate, 250
Significance of colonization by *O.
formigenes*, bacteria, oxalate-
degrading, 138
Slime molds, fungi, calcium oxalate in, 73
Small intestine
bacteria, oxalate-degrading, 156
epithelial transport, 218
Socioeconomic status and geographical
location, urate excretion, 306
Sodium
biosynthesis of calcium oxalate, 113
intake, 179
urate, 315
excretion, 314
urate excretion, 314
Spinach, biosynthesis of calcium oxalate,
114
Sporangia, fungi, calcium oxalate in, 106
Spores, bacteria, fungi, calcium oxalate in,
92
Sprague-Dawley rats, nephrolithiasis, animal
model, 344
Steatorrhea, bacteria, oxalate-degrading,
138
Stern potential, aggregation of calcium
oxalate, 40
Stomach, epithelial transport, 218
Stone. See also Crystal
aggregation of calcium oxalate, 37
crystal, cell, interaction of, 323, 329
matrix, lipid matrix, 295
retention, crystal, cell, interaction of,
325
urate excretion, 305
Streptomyces, bacteria, oxalate-degrading,
146

Stylods, plants, higher, oxalate formation,
    56
Sugar beet, plants, higher, oxalate formation,
    61
Supersaturation
    aggregation of calcium oxalate, 37
    crystal, cell, interaction of, 330
    crystallization *in vitro*, 2, 3, 4, 11, 12
    inhibitors of crystallization, 24
Sweet pepper, plants, higher, oxalate
    formation, 61, 63, 67
Swelling of mitochondria, nephrolithiasis,
    animal model, 347
Swine, bacteria, oxalate-degrading, 133
Symptomatic treatment, hyperoxaluria type
    I, 200
Systematics, fungi, calcium oxalate in, 74

T

Tamm-Horsfall glycoprotein, crystal, cell,
    interaction of, 329
Terrestrial ecosystems, bacteria, oxalate-
    degrading, 143
Thermophilic habitats, bacteria, oxalate-
    degrading, 141
Thiobacillus novellus, bacteria, oxalate-
    degrading, 147
Tilletia indica, fungi, calcium oxalate in, 92
Tissue
    accumulation of oxalate, measurement,
        oxalate, 240
    culture studies have, nephrolithiasis,
        animal model, 353
Transmission electron microscopy, lipid
    matrix, 295
Transport, epithelial, 217–238, 221, 225,
    229
    absorptive or dietary hyperoxaluria, 224
    calcium, 223, 224
        oxalate, 229
            absorption, 218
            bioavailability, 224
            flux, 218, 219, 221
            and oxalate homeostasis, 222
            transport, 222, 227, 228
            in urine, 217
        restriction, 217
    chronic renal failure, 224
    colon, 221, 222
    enteric hyperoxaluria, 223
    epinephrine, 221
    gastrointestinal diseases, 223

hyperabsorption of calcium, 217
ileal resection, 223
intestinal excretion in chronic renal
    failure, 224
jejuno-ileal bypass, 223
large intestine, 221, 222
lumen, 218
    membrane, 227, 229
    oxalate exchange, 228
malabsorption, of bile salts, and fatty
    acids, 223
membrane vesicle studies, 227
rabbit
    distal colon, 222
    ileal brush border, 219
    renal brush border membranes, 228
renal handling of oxalate, 225
small intestine, 218
stomach, 218
transport, epithelial, calcium oxalate
    excretion, 218
    uptake, 219
    in urine, 224
Trichophyton, fungi, calcium oxalate in,
    92
Tubulicium clematidis, fungi, calcium
    oxalate in, 76
Twins, fungi, calcium oxalate in, 88
Tyrosinekinase, nephrolithiasis, cellular
    abnormalities, 211

U

Ultrafiltration, measurement, oxalate, 245
Ultraviolet absorption spectroscopy, 314
Unseeded batch crystallizers, crystallization
    in vitro, 8
Upper tract
    calculi, urolithiasis, 170
    urolithiasis in adults, urolithiasis, 170
Uptake of calcium, crystal, cell, interaction
    of, 330
Urate excretion, 305–321, 306, 307, 308,
    314
    allopurinol and calcium stones, 307
    body weight, 306
    calcium, 308
        stone, 307
            in gouty patients, 305
    crystal, in urine, 309
    diabetes, 313
    diet, 306
    endocrystallosis, 310

epitaxy, 308
glycosaminoglycan, 311
gout, 305
human urine, 312, 313
hyperuricaemia, 306
hyperuricosuria, 305, 306
interchangeable, 305
kidney stone, 313
microscopic examination, 310
monosodium urate, 308
salt, 308
salting-out, 313
scanning electron, 310
socioeconomic status and geographical
        location, 306
sodium, 314
sodium urate, 314
stone formation, 305
uric acid, 306, 309, 313
urinary inhibitors, 308
urinary urate excretion, 305
urine, 308
urolithiasis, 305
Uric acid, 315
    urate excretion, 306, 313
Urinary tract calculi, urolithiasis, 171
Urine
    aggregation of calcium oxalate, 40
    crystal, cell, interaction of, 343
    inhibitors of crystallization, 31, 32
    measurement, oxalate, 248
    nephrolithiasis, animal model, 347, 354
    oxalate, measurement, oxalate
        in adults, 250
        in infants and children, 251
    urate excretion, 308
Urolithiasis, 1, 169–188, 170, 171, 172, 174,
        176, 274, 280, 291, 305, 325,
        327, 330, 343
    age, 173
    bacteria, oxalate-degrading, 156
    bladder
        calculi, 169
        stone, 169
            in children, 169
    calcium, 178, 179
    crystal, cell, interaction of, 325, 330
    diet, 176
    endemic bladder calculi, 169

    endemic stones, 169
    environmental factors, 176
    expenditure, 176
    exposure to cadmium, 180
    familial predisposition, 175
    family history, 175
    geriatric populations, 174
    heredity, 175
    incidence and prevalence of urolithiasis,
        171
    lipid matrix, 291
    low birth weight, 174
    nephrolithiasis, animal model, 343, 348
    oath of Hippocrites, 169
    pilots, 180
    population surveys, 171, 172
    premature infants, 174, 175
    relationship between affluence, 176
    sedentary occupations, 180
    upper tract
        calculi, 170
        urolithiasis in adults, 170
    urinary tract calculi, 171
    urolithiasis, 171, 172
        population surveys, 172
Uronic acid rich protein (UAP),
        nephrolithiasis, animal model,
        350
Uterine fluid, lipid matrix, 291

V

Vacuolation, nephrolithiasis, animal model,
        347
Vacuoles, plants, higher, oxalate formation,
        54, 60
Vitamin D, measurement, oxalate, 253
Von Leewenhoek, Anton, 54

W

Wood-rot, biosynthesis of calcium oxalate,
        122

X

Xanthobacter, bacteria, oxalate-degrading,
        147
Xylem, biosynthesis of calcium oxalate, 118

9 780367 448868